NASA Reference Publication 1398

1997

Total Solar Eclipse of 1999 August 11

Fred Espenak
Goddard Space Flight Center
Greenbelt, Maryland
USA

Jay Anderson
Environment Canada
Winnipeg, Manitoba
CANADA

National Aeronautics and
Space Administration

Goddard Space Flight Center
Greenbelt, Maryland

PREFACE

This work is the latest in a series of NASA publications containing detailed predictions, maps and meteorological data for future central solar eclipses of interest. Published as part of NASA's Reference Publication (RP) series, the eclipse bulletins are prepared in cooperation with the Working Group on Eclipses of the International Astronomical Union and are provided as a public service to both the professional and lay communities, including educators and the media. In order to allow a reasonable lead time for planning purposes, eclipse bulletins are published 24 to 36 months before each event.

Single copies of the bulletins are available at no cost and may be ordered by sending a 9 x 12 inch self addressed stamped envelope (SASE) with sufficient postage (12 oz. or 340 g.). Use stamps only; cash or checks cannot be accepted. Requests within the U. S. may use the Postal Service's Priority Mail for $3.00. Please print the eclipse date (year & month) or the NASA RP number in the lower left corner of the return SASE. Requests from outside the U. S. and Canada may send nine international postal coupons to cover postage. Exceptions to the postage requirements will be made to professional researchers and scientists, or for international requests where political or economic restraints prevent the transfer of funds to other countries. All requests should be accompanied by a copy of the request form on the last page.

The 1999 bulletin uses two detailed mapping data bases for the path figures. World Vector Shoreline (WVS) and World Data Bank II (WDBII), were developed by the Defense Mapping Agency (U. S. Department of Defense) and the U. S. Central Intelligence Agency, respectively. The WDBII outline files were digitized from navigational charts to a working scale of approximately 1:3,000,000, and represent the "state of the art" in the mid 1970s. The WVS data sets are given at several resolutions, including 1:1,000,000, 1:250,000 and 1:100,000. For maximum efficiency and speed, these data have been compressed and reformatted into direct access files by Jan C. Depner (U. S. Naval Oceanographic Office) and James A. Hammak (NORDA). WDBII and WVS are available through the *Global Relief Data CD-ROM* from the National Geophysical Data Center. These vector data have made it possible to generate eclipse path figures at resolutions greater than 1:10,000,000. The more detailed path figures include curves of constant duration of totality within the umbral path. This permits the user to quickly estimate the duration of totality at various locations shown in the figures.

The geographic coordinates data base includes over 90,000 cities and locations. This permits the identification of many more cities within the umbral path and their subsequent inclusion in the local circumstances tables. These same coordinates are plotted in the path figures and are labeled when the scale allows. The source of these coordinates is Rand McNally's *The New International Atlas*. A subset of these coordinates is available in a digital form which we've augmented with population data.

The bulletins have undergone a great deal of change since their inception in 1993. The expansion of the mapping and geographic coordinates data bases have significantly improved the coverage and level of detail demanded by eclipse planning. Some of these changes are the direct result of suggestions from our readers. We strongly encourage you to share your comments, suggestions and criticisms on how to improve the content and layout in subsequent editions. Although every effort is made to ensure that the bulletins are as accurate as possible, an error occasionally slips by. We would appreciate your assistance in reporting all errors, regardless of their magnitude.

A special thanks goes to Dr. B. Ralph Chou for a new and expanded discussion on solar eclipse eye safety. Dr. Chou is Professor of Optometry at the University of Waterloo and he has over twenty-five years of eclipse observing experience. With so much fear and misinformation about watching eclipses, we thought it appropriate to invite a leading authority on the subject to provide some definitive answers.

Dr. Joe Gurman (GSFC/Solar Physics Branch) has made this and previous eclipse bulletins available over the Internet. They can be read or downloaded via the World-Wide Web using a browser from Goddard's Solar Data Analysis Center eclipse information page: *http://umbra.nascom.nasa.gov/eclipse/*. Most of the files are also available via anonymous ftp. Umbral path data for all central eclipses through the year 2005 are also available. Complete details may be found elsewhere in this publication.

During 1996, Espenak developed a new web site which provides general information on both solar and lunar eclipses occurring during the next two decades. Hints on eclipse photography and eye safety may be found there as well as links to other eclipse related web sites. The URL for the site is: *http://planets.gsfc.nasa.gov/eclipse/eclipse.html*.

In addition to the general information web site above, a special web site devoted to the 1999 total solar eclipse has been set up: *http://planets.gsfc.nasa.gov/eclipse/TSE1999/TSE1999.html*. It includes supplemental predictions, figures and maps which could not be included in the present publication. NASA RP1398 is the already the largest eclipse bulletin yet published and represents an upper limit on the size of all future eclipse bulletins. For specialized or more detailed eclipse predictions useful to a smaller audience, this information will be served via the Web.

Since the eclipse bulletins are of a limited and finite size, they cannot include everything needed by every scientific investigation. Some investigators may require exact contact times which include lunar limb effects or for a specific observing site not listed in the bulletin. Other investigations may need customized predictions for an aerial rendezvous or from the path limits for grazing eclipse experiments. We would like to assist such investigations by offering to calculate additional predictions for any professionals or large groups of amateurs. Please contact Espenak with complete details and eclipse prediction requirements.

We would like to acknowledge the valued contributions of a number of individuals who were essential to the success of this publication. The format and content of the NASA eclipse bulletins has drawn heavily upon over 40 years of eclipse *Circulars* published by the U. S. Naval Observatory. We owe a debt of gratitude to past and present staff of that institution who have performed this service for so many years. The many publications and algorithms of Dr. Jean Meeus have served to inspire a life-long interest in eclipse prediction. We thank Francis Reddy, who helped develop the original data base of geographic coordinates and to Rique Pottenger (Astro Communications Service) for his assistance in expanding the data base to over 90,000 cities. Dr. Wayne Warren provided a draft copy of the *IOTA Observer's Manual* for use in describing contact timings near the path limits.

Prof. Jay M. Pasachoff reviewed the manuscript and offered many helpful suggestions. The availability of the eclipse bulletins via the Internet is due to the efforts of Dr. Joseph B. Gurman. The support of Environment Canada is acknowledged in the acquisition and arrangement of the weather data.

The names and spellings of countries, cities and other geopolitical regions are not authoritative, nor do they imply any official recognition in status. Corrections to names, geographic coordinates and elevations are actively solicited in order to update the data base for future eclipses. All calculations, diagrams and opinions are those of the authors and they assume full responsibility for their accuracy.

Fred Espenak
NASA/Goddard Space Flight Center
Planetary Systems Branch, Code 693
Greenbelt, MD 20771
USA

email: espenak@lepvax.gsfc.nasa.gov
FAX: (301) 286-0212

Jay Anderson
Environment Canada
123 Main Street, Suite 150
Winnipeg, MB,
CANADA R3C 4W2

email: jander@cc.umanitoba.ca
FAX: (204) 983-0109

Current and Future NASA Solar Eclipse Bulletins

NASA Eclipse Bulletin	RP #	Publication Date
Annular Solar Eclipse of 1994 May 10	*1301*	*April 1993*
Total Solar Eclipse of 1994 November 3	*1318*	*October 1993*
Total Solar Eclipse of 1995 October 24	*1344*	*July 1994*
Total Solar Eclipse of 1997 March 9	*1369*	*July 1995*
Total Solar Eclipse of 1998 February 26	*1383*	*April 1996*
Total Solar Eclipse of 1999 August 11	*1398*	*March 1997*
- - - - - - - - - - - future - - - - - - - - - - -		
Total Solar Eclipse of 2001 June 21	—	*Spring 1998*
Total Solar Eclipse of 2002 December 4	—	*Spring 1999*
Annular and Total Solar Eclipses of 2003	—	*Spring 2000*

Total Solar Eclipse of 1999 August 11

Table of Contents

TOTAL SOLAR ECLIPSE OF 1999 AUGUST 11

Figures, Tables and Maps

Eclipse Predictions

Introduction

On Wednesday, 1999 August 11, a total eclipse of the Sun will be visible from within a narrow corridor which traverses the Eastern Hemisphere. The path of the Moon's umbral shadow begins in the Atlantic and crosses central Europe, the Middle East, and India (Figures 3, 4 and 5) where it ends at sunset in the Bay of Bengal. A partial eclipse will be seen within the much broader path of the Moon's penumbral shadow, which includes northeastern North America, all of Europe, northern Africa and the western half of Asia (Figures 1 and 2).

Umbral Path And Visibility

The last total solar eclipse of the 20th century begins in the North Atlantic about 300 kilometers south of Nova Scotia where the Moon's umbral shadow first touches down on Earth at 09:30:57 UT. Along the sunrise terminator, the maximum duration is a mere 47 seconds as seen from the center of the narrow 49 kilometers wide path. No major landfall occurs for the first forty minutes as the shadow sweeps across the North Atlantic. The umbra finally reaches the Isles of Scilly off the southwestern coast of England at 10:10 UT (Figure 6). At this locale, it is already mid-morning with the Sun 45° above the eastern horizon. The center line duration is 2 minutes and the path width has expanded to 103 kilometers as the shadow pursues its eastern track with a ground velocity of 0.91 km/s.

One minute later (10:11 UT), the umbra arrives along the shores of the Cornwall Peninsula. In the following four minutes, the shadow skirts the southern coast giving eager observers a brief taste of totality. Plymouth, the largest English city in the path, is north of center line and witnesses a total phase lasting 1 minute 39 seconds. London misses the total phase but experiences partiality with a maximum magnitude of 0.968. By 10:16 UT, the umbra leaves England as it quickly traverses the English Channel. The Channel Islands of Guernsey and Jersey lie just south of the path and witness a partial eclipse of magnitude >0.995. To the north, Alderney is deep in the path and enjoys over one and a half minutes of totality.

Not since 1961 has the Moon cast its dark shadow upon central Europe. The southern edge of the umbra first reaches the Normandy coast just as the northern edge leaves England (10:16 UT). But another four minutes elapse before the center line makes landfall in northern France (Figure 7). As the shadow sweeps through the French countryside, its southern edge passes 30 kilometers north of Paris. The City of Lights will bear witness to a partial event of magnitude 0.992 at 10:23 UT. Continuing on its eastward track, the path's northern limit crosses into southern Belgium, Luxembourg and Germany. Meanwhile, the center line cuts through Champagne where the citizens of Metz witness a total eclipse lasting 2 minutes 13 seconds (10:29 UT). Four minutes later, the entire umbra crosses into southern Germany (Figure 8) and the picturesque Rhine Valley. North of the path, Frankfurt witnesses a 0.979 magnitude partial eclipse, while Stuttgart lies near path center for 2 minutes 17 seconds of totality. At 10:35 UT, the Sun's altitude stands at 55°, the path width is 109 kilometers and the ground velocity is 0.74 km/s. Although München (Munich) lies 20 kilometers south of the center line, the city's two million citizens will still witness more than two minutes of totality, provided the winds of good fortune bring clear skies on eclipse day.

At 10:41 UT, the umbra leaves Germany and crosses into Austria where it encounters the Eastern Alps. Wien (Vienna) is almost 40 kilometers north of the path and experiences a 0.990 magnitude partial eclipse. The southern edge of the path grazes northeastern Slovenia as the shadow enters Hungary at 10:47 UT (Figure 9). Lake Balaton lies wholly within the path where the central duration lasts 2 minutes 22 seconds (10:50 UT). Like Wein, Budapest is also located about 40 kilometers north of the path where a 0.991 magnitude partial eclipse will occur. As the shadow leaves Hungary, the southern third briefly sweeps through northern Yugoslavia before continuing on into Romania.

The instant of greatest eclipse[1] occurs at 11:03:04 UT when the axis of the Moon's shadow passes closest to the center of Earth (gamma[2] =0.506). At that moment, the shadow's epicenter is located among the rolling hills of south-central Romania very near Rîmnicu-Vîlcea (Figure 10). The length of totality

---

[1] The instant of greatest eclipse occurs when the distance between the Moon's shadow axis and Earth's geocenter reaches a minimum. Although greatest eclipse differs slightly from the instants of greatest magnitude and greatest duration (for total eclipses), the differences are usually quite small.

[2] Minimum distance of the Moon's shadow axis from Earth's center in units of equatorial Earth radii.

reaches its maximum duration of 2 minutes 23 seconds, the Sun's altitude is 59°, the path width is 112 kilometers and the umbra's velocity is 0.680 km/s. Four minutes later (11:07 UT), Romania's capital city Bucuresti (Bucharest) is engulfed by the shadow. Since Bucharest lies on the center line near the instant of greatest eclipse, it enjoys a duration nearly as long at 2 minutes 22 seconds. Traveling south-southeast, the path encompasses the Romania-Bulgaria border before leaving land and heading out across the Black Sea.

The next landfall occurs along the Black Sea coast of northern Turkey at 11:21 UT (Figure 11). Ankara lies 150 kilometers south of the path and witnesses a 0.969 magnitude partial eclipse. The track diagonally bisects Turkey as it moves inland while the center line duration begins a gradual but steady decrease. At 11:29 UT, Turhal falls deep within the shadow for 2 minutes 15 seconds. The umbra reaches Turkey's southeastern border at 11:45 UT and briefly enters northwestern Syria as it crosses into Iraq (Figure 12). The center line duration is now 2 minutes 5 seconds with the Sun's altitude at 50°. Baghdad lies 220 kilometers south of the path and experiences a 0.940 magnitude partial eclipse (Figure 13). Arriving at Iran's western boundary at 11:52 UT, the shadow spends the next half hour crossing sparsely populated mountain ranges and deserts (Figures 13, 14 and 15). Tehran lies north of the path where its eight million inhabitants witness a 0.943 magnitude partial eclipse. At 12:22 UT, the shadow enters Pakistan and skirts the shores of the Arabian Sea (Figure 16). Karachi is near the center line and experiences 1 minute 13 seconds of total eclipse with the Sun 22° above the western horizon. The path width has shrunk to 85 kilometers while the shadow's speed has increased to 2 km/s.

The umbra arrives in India, the last nation in its path, at 12:28 UT (Figure 17). As the shadow sweeps across the sub-continent, its velocity rapidly increases while the center line duration drops below one minute and the Sun's altitude decreases to 7° (Figures 18 and 19). The eleven million inhabitants of Calcutta will witness a 0.879 magnitude partial eclipse with the Sun a scant 2° above the western horizon. Leaving India just north of Vishakhapatnam at 12:36 UT, the shadow sweeps into the Bay of Bengal where it departs Earth and races back into space (12:36:23 UT), not to return until the next millennium. Over the course of 3 hours and 7 minutes, the Moon's umbra travels along a path approximately 14000 kilometers long and covering 0.2% of Earth's surface area.

General Maps of the Eclipse Path

Orthographic Projection Map of the Eclipse Path

Figure 1 is an orthographic projection map of Earth [adapted from Espenak, 1987] showing the path of penumbral (partial) and umbral (total) eclipse. The daylight terminator is plotted for the instant of greatest eclipse with north at the top. The sub-Earth point is centered over the point of greatest eclipse and is indicated with an asterisk-like symbol. The sub-solar point at that instant is also shown.

The limits of the Moon's penumbral shadow define the region of visibility of the partial eclipse. This saddle shaped region often covers more than half of Earth's daylight hemisphere and consists of several distinct zones or limits. At the northern and/or southern boundaries lie the limits of the penumbra's path. Partial eclipses have only one of these limits, as do central eclipses when the shadow axis falls no closer than about 0.45 radii from Earth's center. Great loops at the western and eastern extremes of the penumbra's path identify the areas where the eclipse begins/ends at sunrise and sunset, respectively. If the penumbra has both a northern and southern limit, the rising and setting curves form two separate, closed loops. Otherwise, the curves are connected in a distorted figure eight. Bisecting the 'eclipse begins/ends at sunrise and sunset' loops is the curve of maximum eclipse at sunrise (western loop) and sunset (eastern loop). The exterior tangency points **P1** and **P4** mark the coordinates where the penumbral shadow first contacts (partial eclipse begins) and last contacts (partial eclipse ends) Earth's surface. The path of the umbral shadow bisects the penumbral path from west to east and is shaded dark gray.

A curve of maximum eclipse is the locus of all points where the eclipse is at maximum at a given time. They are plotted at each half hour Universal Time (UT), and generally run from northern to southern penumbral limits, or from the maximum eclipse at sunrise or sunset curves to one of the limits. The outline of the umbral shadow is plotted every ten minutes in UT. Curves of constant eclipse magnitude[3] delineate the locus of all points where the magnitude at maximum eclipse is constant. These curves run

[3] Eclipse magnitude is defined as the fraction of the Sun's diameter occulted by the Moon. It is strictly a ratio of *diameters* and should not be confused with eclipse obscuration which is a measure of the Sun's surface *area* occulted by the Moon. Eclipse magnitude may be expressed as either a percentage or a decimal fraction (e.g.: 50% or 0.50).

exclusively between the curves of maximum eclipse at sunrise and sunset. Furthermore, they are quasi-parallel to the northern/southern penumbral limits and the umbral paths of central eclipses. Northern and southern limits of the penumbra may be thought of as curves of constant magnitude of 0%, while adjacent curves are for magnitudes of 20%, 40%, 60% and 80%. The northern and southern limits of the path of total eclipse are curves of constant magnitude of 100%.

At the top of Figure 1, the Universal Time of geocentric conjunction between the Moon and Sun is given followed by the instant of greatest eclipse. The eclipse magnitude is given for greatest eclipse. For central eclipses (both total and annular), it is equivalent to the geocentric ratio of diameters of the Moon and Sun. Gamma is the minimum distance of the Moon's shadow axis from Earth's center in units of equatorial Earth radii. The shadow axis passes south of Earth's geocenter for negative values of Gamma. Finally, the Saros series number of the eclipse is given along with its relative sequence in the series.

Stereographic Projection Map of the Eclipse Path

The stereographic projection of Earth in Figure 2 depicts the path of penumbral and umbral eclipse in greater detail. The map is oriented with the north up with the point of greatest eclipse near the center. International political borders are shown and circles of latitude and longitude are plotted at 20° increments. The region of penumbral or partial eclipse is identified by its northern and southern limits, curves of eclipse begins or ends at sunrise and sunset, and curves of maximum eclipse at sunrise and sunset. Curves of constant eclipse magnitude are plotted for 20%, 40%, 60% and 80%, as are the limits of the path of total eclipse. Also included are curves of greatest eclipse at every half hour Universal Time.

Figures 1 and 2 may be used to determine quickly the approximate time and magnitude of maximum eclipse at any location within the eclipse path.

Equidistant Conic Projection Maps of the Eclipse Path

Figures 3, 4 and 5 are equidistant conic projection maps chosen to minimize distortion, and which isolate specific regions of the umbral path. Once again, curves of maximum eclipse and constant eclipse magnitude are plotted and labeled. A linear scale is included for estimating approximate distances (kilometers). Within the northern and southern limits of the path of totality, the outline of the umbral shadow is plotted at ten minute intervals. The duration of totality (minutes and seconds) and the Sun's altitude correspond to the local circumstances on the center line at each shadow position.

The scale used in these figures is ~1:15,904,000. The positions of larger cities and metropolitan areas in and near the umbral path are depicted as black dots. The size of each city is logarithmically proportional to its population using 1990 census data (Rand McNally, 1991).

Equidistant Cylindrical Projection Maps of the Eclipse Path

Figures 6 through 19 all use a simple equidistant cylindrical projection scaled for the central latitude of each map. They all use high resolution coastline data from the World Data Base II (WDB) and World Vector Shoreline (WVS) data bases and have a scale of 1:2,784,000. These maps were chosen to isolate small regions along the entire land portion of the eclipse path. Once again, curves of maximum eclipse and constant eclipse magnitude are included as well as the outline of the umbral shadow. A special feature of these maps are the curves of constant umbral eclipse duration (i.e., totality) which are plotted within the path. These curves permit fast determination of approximate durations without consulting any tables. Furthermore, city data from a recently enlarged geographic data base of over 90,000 positions are plotted to give as many locations as possible in the path of totality. Local circumstances have been calculated for these positions and can be found in Tables 9 through 35.

ELEMENTS, SHADOW CONTACTS AND ECLIPSE PATH TABLES

The geocentric ephemeris for the Sun and Moon, various parameters, constants, and the Besselian elements (polynomial form) are given in Table 1. The eclipse elements and predictions were derived from the DE200 and LE200 ephemerides (solar and lunar, respectively) developed jointly by the Jet Propulsion Laboratory and the U. S. Naval Observatory for use in the *Astronomical Almanac* for 1984 and thereafter. Unless otherwise stated, all predictions are based on center of mass positions for the Moon and Sun with no corrections made for center of figure, lunar limb profile or atmospheric refraction. The predictions depart from normal IAU convention through the use of a smaller constant for the mean lunar radius k for all umbral contacts (see: LUNAR LIMB PROFILE). Times are expressed in either Terrestrial Dynamical Time (TDT) or in Universal Time (UT), where the best value of ΔT[4] available at the time of preparation is used.

From the polynomial form of the Besselian elements, any element can be evaluated for any time t_1 (in decimal hours) via the equation:

$$\mathbf{a} = a_0 + a_1{*}t + a_2{*}t^2 + a_3{*}t^3 \quad (\text{or } \mathbf{a} = \Sigma\, [a_n{*}t^n];\ n = 0 \text{ to } 3)$$

where: $\mathbf{a} = x, y, d, l_1, l_2, \text{ or } \mu$
$t = t_1 - t_0$ (decimal hours) and $t_0 = 11.000$ TDT

The polynomial Besselian elements were derived from a least-squares fit to elements rigorously calculated at five separate times over a six hour period centered at t_0. Thus, the equation and elements are valid over the period $8.00 \le t_0 \le 14.00$ TDT.

Table 2 lists all external and internal contacts of penumbral and umbral shadows with Earth. They include TDT times and geodetic coordinates with and without corrections for ΔT. The contacts are defined:

P1 - Instant of first external tangency of penumbral shadow cone with Earth's limb. (partial eclipse begins)
P4 - Instant of last external tangency of penumbral shadow cone with Earth's limb. (partial eclipse ends)

U1 - Instant of first external tangency of umbral shadow cone with Earth's limb. (umbral eclipse begins)
U2 - Instant of first internal tangency of umbral shadow cone with Earth's limb.
U3 - Instant of last internal tangency of umbral shadow cone with Earth's limb.
U4 - Instant of last external tangency of umbral shadow cone with Earth's limb. (umbral eclipse ends)

Similarly, the northern and southern extremes of the penumbral and umbral paths, and extreme limits of the umbral center line are given. The IAU longitude convention is used throughout this publication (i.e., for longitude, east is positive and west is negative; for latitude, north is positive and south is negative).

The path of the umbral shadow is delineated at five minute intervals in Universal Time in Table 3. Coordinates of the northern limit, the southern limit and the center line are listed to the nearest tenth of an arc-minute (~185 m at the Equator). The Sun's altitude, path width and umbral duration are calculated for the center line position. Table 4 presents a physical ephemeris for the umbral shadow at five minute intervals in UT. The center line coordinates are followed by the topocentric ratio of the apparent diameters of the Moon and Sun, the eclipse obscuration[5], and the Sun's altitude and azimuth at that instant. The central path width, the umbral shadow's major and minor axes and its instantaneous velocity with respect to Earth's surface are included. Finally, the center line duration of the umbral phase is given.

Local circumstances for each center line position listed in Tables 3 and 4 are presented in Table 5. The first three columns give the Universal Time of maximum eclipse, the center line duration of totality and the altitude of the Sun at that instant. The following columns list each of the four eclipse contact times followed by their related contact position angles and the corresponding altitude of the Sun. The four contacts identify significant stages in the progress of the eclipse. They are defined as follows:

---

[4] ΔT is the difference between Terrestrial Dynamical Time and Universal Time.

[5] Eclipse obscuration is defined as the fraction of the Sun's surface area occulted by the Moon.

First Contact	–	Instant of first external tangency between the Moon and Sun. (partial eclipse begins)
Second Contact	–	Instant of first internal tangency between the Moon and Sun. (central or umbral eclipse begins; total or annular eclipse begins)
Third Contact	–	Instant of last internal tangency between the Moon and Sun. (central or umbral eclipse ends; total or annular eclipse ends)
Fourth Contact	–	Instant of last external tangency between the Moon and Sun. (partial eclipse ends)

The position angles **P** and **V** identify the point along the Sun's disk where each contact occurs[6]. Second and third contact altitudes are omitted since they are always within 1° of the altitude at maximum eclipse.

Table 6 presents topocentric values from the central path at maximum eclipse for the Moon's horizontal parallax, semi-diameter, relative angular velocity with respect to the Sun, and libration in longitude. The altitude and azimuth of the Sun are given along with the azimuth of the umbral path. The northern limit position angle identifies the point on the lunar disk defining the umbral path's northern limit. It is measured counter-clockwise from the north point of the Moon. In addition, corrections to the path limits due to the lunar limb profile are listed. The irregular profile of the Moon results in a zone of 'grazing eclipse' at each limit that is delineated by interior and exterior contacts of lunar features with the Sun's limb. This geometry is described in greater detail in the section LIMB CORRECTIONS TO THE PATH LIMITS: GRAZE ZONES. Corrections to center line durations due to the lunar limb profile are also included. When added to the durations in Tables 3, 4, 5 and 7, a slightly longer central total phase is predicted along most of the path.

To aid and assist in the plotting of the umbral path on large scale maps, the path coordinates are also tabulated at 1° intervals in longitude in Table 7. The latitude of the northern limit, southern limit and center line for each longitude is tabulated to the nearest hundredth of an arc-minute (~18.5 m at the Equator) along with the Universal Time of maximum eclipse at each position. Finally, local circumstances on the center line at maximum eclipse are listed and include the Sun's altitude and azimuth, the umbral path width and the central duration of totality.

In applications where the zones of grazing eclipse are needed in greater detail, Table 8 lists these coordinates over land based portions of the path at 30' intervals in longitude. The time of maximum eclipse is given at both northern and southern limits as well as the path's azimuth. The elevation and scale factors are also given (See: LIMB CORRECTIONS TO THE PATH LIMITS: GRAZE ZONES).

---

[6] P is defined as the contact angle measured counter-clockwise from the *north* point of the Sun's disk.
V is defined as the contact angle measured counter-clockwise from the *zenith* point of the Sun's disk.

Local Circumstances Tables

Local circumstances for approximately 1460 cities, metropolitan areas and places in North America, Europe, Africa and Asia are presented in Tables 9 through 36. These tables give the local circumstances at each contact and at maximum eclipse[7] for every location. The coordinates are listed along with the location's elevation (meters) above sea-level, if known. If the elevation is unknown (i.e., not in the data base), then the local circumstances for that location are calculated at sea-level. In any case, the elevation does not play a significant role in the predictions unless the location is near the umbral path limits and the Sun's altitude is relatively small (<10°). The Universal Time of each contact is given to a tenth of a second, along with position angles **P** and **V** and the altitude of the Sun. The position angles identify the point along the Sun's disk where each contact occurs and are measured counter-clockwise (i.e., eastward) from the north and zenith points, respectively. Locations outside the umbral path miss the umbral eclipse and only witness first and fourth contacts. The Universal Time of maximum eclipse (either partial or total) is listed to a tenth of a second. Next, the position angles **P** and **V** of the Moon's disk with respect to the Sun are given, followed by the altitude and azimuth of the Sun at maximum eclipse. Finally, the corresponding eclipse magnitude and obscuration are listed. For umbral eclipses (both annular and total), the eclipse magnitude is identical to the topocentric ratio of the Moon's and Sun's apparent diameters. The eclipse magnitude is always less than 1 for annular eclipses and equal to or greater than 1 for total eclipses. The final column gives the duration of totality if this location lies in the path of the Moon's umbral shadow. The effects of refraction have not been included in these calculations, nor have there been any corrections for center of figure or the lunar limb profile.

Locations were chosen based on general geographic distribution, population, and proximity to the path. The primary source for geographic coordinates is *The New International Atlas* (Rand McNally, 1991). Elevations for major cities were taken from *Climates of the World* (U. S. Dept. of Commerce, 1972). In this rapidly changing political world, it is often difficult to ascertain the correct name or spelling for a given location. Therefore, the information presented here is for location purposes only and is not meant to be authoritative. Furthermore, it does not imply recognition of status of any location by the United States Government. Corrections to names, spellings, coordinates and elevations is solicited in order to update the geographic data base for future eclipse predictions.

For countries in the path of totality, expanded versions of the local circumstances tables listing many more locations are available via a special web site of supplemental material for the total solar eclipse of 1999 *(http://planets.gsfc.nasa.gov/eclipse/TSE1999/TSE1999.html).* This site also has tables of local circumstances for all astronomical observatories listed in the *Astronomical Almanac for 1996.*

Detailed Maps of the Umbral Path

The path of totality has been plotted by hand on a set of ten detailed maps appearing in the last section of this publication. The maps are Global Navigation and Planning Charts or GNC's from the Defense Mapping Agency, which use a Lambert conformal conic projection. More specifically, GNC-4 covers Europe and western Asia while GNC-12 covers the Middle East and south Asia. GNC's have a scale of 1:5,000,000 (1 inch ~ 69 nautical miles), which is adequate for showing major cities, highways, airports, rivers, bodies of water and basic topography required for eclipse expedition planning including site selection, transportation logistics and weather contingency strategies.

Northern and southern limits, as well as the center line of the path, are plotted using data from Table 7. Although no corrections have been made for center of figure or lunar limb profile, they have little or no effect at this scale. Atmospheric refraction has not been included, as its effects play a significant role only at very low solar altitudes. In any case, refraction corrections to the path are uncertain since they depend on the atmospheric temperature-pressure profile, which cannot be predicted in advance. If observations from the graze zones are planned, then the zones of grazing eclipse must be plotted on higher scale maps using coordinates in Table 8. See PLOTTING THE PATH ON MAPS for sources and more information. The GNC paths also depict the curve of maximum eclipse at five and/or ten minute increments in UT from Table 3.

Selected sections of the path are plotted on higher resolution maps (i.e., ONC maps *with a scale of* 1:1,000,000) which are available via a special web site of supplemental material for the total solar eclipse of 1999 *(http://planets.gsfc.nasa.gov/eclipse/TSE1999/TSE1999.html).*

---

[7] For partial eclipses, maximum eclipse is the instant when the greatest fraction of the Sun's diameter is occulted. For total eclipses, maximum eclipse is the instant of mid-totality.

Estimating Times of Second And Third Contacts

The times of second and third contact for any location not listed in this publication can be estimated using the detailed maps found in the final section. Alternatively, the contact times can be estimated from maps on which the umbral path has been plotted. Table 7 lists the path coordinates conveniently arranged in 1° increments of longitude to assist plotting by hand. The path coordinates in Table 3 define a line of maximum eclipse at five minute increments in time. These lines of maximum eclipse each represent the projection diameter of the umbral shadow at the given time. Thus, any point on one of these lines will witness maximum eclipse (i.e., mid-totality) at the same instant. The coordinates in Table 3 should be added to the map in order to construct lines of maximum eclipse.

The estimation of contact times for any one point begins with an interpolation for the time of maximum eclipse at that location. The time of maximum eclipse is proportional to a point's distance between two adjacent lines of maximum eclipse, measured along a line parallel to the center line. This relationship is valid along most of the path with the exception of the extreme ends, where the shadow experiences its largest acceleration. The center line duration of totality **D** and the path width **W** are similarly interpolated from the values of the adjacent lines of maximum eclipse as listed in Table 3. Since the location of interest probably does not lie on the center line, it is useful to have an expression for calculating the duration of totality **d** as a function of its perpendicular distance **a** from the center line:

$$\mathbf{d} = \mathbf{D}\,(1 - (2\,\mathbf{a}/\mathbf{W})^2)^{1/2} \text{ seconds} \qquad [1]$$

where:
- **d** = duration of totality at desired location (seconds)
- **D** = duration of totality on the center line (seconds)
- **a** = perpendicular distance from the center line (kilometers)
- **W** = width of the path (kilometers)

If $\mathbf{t_m}$ is the interpolated time of maximum eclipse for the location, then the approximate times of second and third contacts ($\mathbf{t_2}$ and $\mathbf{t_3}$, respectively) are:

Second Contact:	$\mathbf{t_2 = t_m - d/2}$	[2]
Third Contact:	$\mathbf{t_3 = t_m + d/2}$	[3]

The position angles of second and third contact (either **P** or **V**) for any location off the center line are also useful in some applications. First, linearly interpolate the center line position angles of second and third contacts from the values of the adjacent lines of maximum eclipse as listed in Table 5. If $\mathbf{X_2}$ and $\mathbf{X_3}$ are the interpolated center line position angles of second and third contacts, then the position angles $\mathbf{x_2}$ and $\mathbf{x_3}$ of those contacts for an observer located **a** kilometers from the center line are:

Second Contact:	$\mathbf{x_2 = X_2} - \arcsin\,(\mathbf{2\,a/W})$	[4]
Third Contact:	$\mathbf{x_3 = X_3} + \arcsin\,(\mathbf{2\,a/W})$	[5]

where:
- $\mathbf{x_n}$ = interpolated position angle (either **P** or **V**) of contact **n** at location
- $\mathbf{X_n}$ = interpolated position angle (either **P** or **V**) of contact **n** on center line
- **a** = perpendicular distance from the center line (kilometers) (use negative values for locations south of the center line)
- **W** = width of the path (kilometers)

Mean Lunar Radius

A fundamental parameter used in eclipse predictions is the Moon's radius k, expressed in units of Earth's equatorial radius. The Moon's actual radius varies as a function of position angle and libration due to the irregularity in the limb profile. From 1968 through 1980, the Nautical Almanac Office used two separate values for k in their predictions. The larger value (k=0.2724880), representing a mean over topographic features, was used for all penumbral (exterior) contacts and for annular eclipses. A smaller value (k=0.272281), representing a mean minimum radius, was reserved exclusively for umbral (interior) contact calculations of total eclipses [*Explanatory Supplement*, 1974]. Unfortunately, the use of two different values of k for umbral eclipses introduces a discontinuity in the case of hybrid or annular-total eclipses.

In August 1982, the International Astronomical Union (IAU) General Assembly adopted a value of k=0.2725076 for the mean lunar radius. This value is now used by the Nautical Almanac Office for all solar eclipse predictions [Fiala and Lukac, 1983] and is currently the best mean radius, averaging mountain peaks and low valleys along the Moon's rugged limb. The adoption of one single value for k eliminates the discontinuity in the case of annular-total eclipses and ends confusion arising from the use of two different values. However, the use of even the best 'mean' value for the Moon's radius introduces a problem in predicting the true character and duration of umbral eclipses, particularly total eclipses. A total eclipse can be defined as an eclipse in which the Sun's disk is completely occulted by the Moon. This cannot occur so long as any photospheric rays are visible through deep valleys along the Moon's limb [Meeus, Grosjean and Vanderleen, 1966]. But the use of the IAU's mean k guarantees that some annular or annular-total eclipses will be misidentified as total. A case in point is the eclipse of 3 October 1986. Using the IAU value for k, the *Astronomical Almanac* identified this event as a total eclipse of 3 seconds duration when it was, in fact, a beaded annular eclipse. Since a smaller value of k is more representative of the deeper lunar valleys and hence the minimum solid disk radius, it helps ensure the correct identification of an eclipse's true nature.

Of primary interest to most observers are the times when umbral eclipse begins and ends (second and third contacts, respectively) and the duration of the umbral phase. When the IAU's value for k is used to calculate these times, they must be corrected to accommodate low valleys (total) or high mountains (annular) along the Moon's limb. The calculation of these corrections is not trivial but must be performed, especially if one plans to observe near the path limits [Herald, 1983]. For observers near the center line of a total eclipse, the limb corrections can be more closely approximated by using a smaller value of k which accounts for the valleys along the profile.

This publication uses the IAU's accepted value of k=0.2725076 for all penumbral (exterior) contacts. In order to avoid eclipse type misidentification and to predict central durations which are closer to the actual durations at total eclipses, we depart from standard convention by adopting the smaller value of k=0.272281 for all umbral (interior) contacts. This is consistent with predictions in *Fifty Year Canon of Solar Eclipses: 1986 - 2035* [Espenak, 1987]. Consequently, the smaller k produces shorter umbral durations and narrower paths for total eclipses when compared with calculations using the IAU value for k. Similarly, predictions using a smaller k result in longer umbral durations and wider paths for annular eclipses than do predictions using the IAU's k.

Lunar Limb Profile

Eclipse contact times, magnitude and duration of totality all depend on the angular diameters and relative velocities of the Moon and Sun. Unfortunately, these calculations are limited in accuracy by the departure of the Moon's limb from a perfectly circular figure. The Moon's surface exhibits a rather dramatic topography, which manifests itself as an irregular limb when seen in profile. Most eclipse calculations assume some mean radius that averages high mountain peaks and low valleys along the Moon's rugged limb. Such an approximation is acceptable for many applications, but if higher accuracy is needed, the Moon's actual limb profile must be considered. Fortunately, an extensive body of knowledge exists on this subject in the form of Watts' limb charts [Watts, 1963]. These data are the product of a photographic survey of the marginal zone of the Moon and give limb profile heights with respect to an adopted smooth reference surface (or datum). Analyses of lunar occultations of stars by Van Flandern [1970] and Morrison [1979] have shown that the average cross-section of Watts' datum is slightly elliptical rather than circular. Furthermore, the implicit center of the datum (i.e., the center of figure) is displaced from the Moon's center of mass. In a follow-up analysis of 66,000 occultations, Morrison and Appleby [1981] have found that the radius of the datum appears to vary with libration. These variations produce systematic errors in Watts' original limb profile heights that attain 0.4 arc-seconds at some position angles. Thus, corrections to Watts' limb data are necessary to ensure that the reference datum is a sphere with its center at the center of mass.

The Watts charts have been digitized by Her Majesty's Nautical Almanac Office in Herstmonceux, England, and transformed to grid-profile format at the U. S. Naval Observatory. In this computer readable form, the Watts limb charts lend themselves to the generation of limb profiles for any lunar libration. Ellipticity and libration corrections may be applied to refer the profile to the Moon's center of mass. Such a profile can then be used to correct eclipse predictions which have been generated using a mean lunar limb.

Along the path, the Moon's topocentric libration (physical + optical) in longitude ranges from l=+5.5° to l=+4.0°. Thus, a limb profile with the appropriate libration is required in any detailed analysis of contact times, central durations, etc.. But a profile with an intermediate value is useful for planning purposes and may even be adequate for most applications. The lunar limb profile presented in Figure 20 includes corrections for center of mass and ellipticity [Morrison and Appleby, 1981]. It is generated for 11:00 UT, which corresponds to western Romania just before greatest eclipse. The Moon's topocentric libration is l=+4.81°, and the topocentric semi-diameters of the Sun and Moon are 946.8 and 973.9 arc-seconds, respectively. The Moon's angular velocity with respect to the Sun is 0.379 arc-seconds per second.

The radial scale of the limb profile in Figure 20 (at bottom) is greatly exaggerated so that the true limb's departure from the mean lunar limb is readily apparent. The mean limb with respect to the center of figure of Watts' original data is shown (dashed) along with the mean limb with respect to the center of mass (solid). Note that all the predictions presented in this publication are calculated with respect to the latter limb unless otherwise noted. Position angles of various lunar features can be read using the protractor marks along the Moon's mean limb (center of mass). The position angles of second and third contact are clearly marked along with the north pole of the Moon's axis of rotation and the observer's zenith at mid-totality. The dashed line with arrows at either end identifies the contact points on the limb corresponding to the northern and southern limits of the path. To the upper left of the profile are the Sun's topocentric coordinates at maximum eclipse. They include the right ascension R.A., declination *Dec.*, semi-diameter *S.D.* and horizontal parallax *H.P.*. The corresponding topocentric coordinates for the Moon are to the upper right. Below and left of the profile are the geographic coordinates of the center line at 11:00 UT while the times of the four eclipse contacts at that location appear to the lower right. Directly below the profile are the local circumstances at maximum eclipse. They include the Sun's altitude and azimuth, the path width, and central duration. The position angle of the path's northern/southern limit axis is *PA(N.Limit)* and the angular velocity of the Moon with respect to the Sun is *A.Vel.(M:S)*. At the bottom left are a number of parameters used in the predictions, and the topocentric lunar librations appear at the lower right.

In investigations where accurate contact times are needed, the lunar limb profile can be used to correct the nominal or mean limb predictions. For any given position angle, there will be a high mountain (annular eclipses) or a low valley (total eclipses) in the vicinity that ultimately determines the true instant of contact. The difference, in time, between the Sun's position when tangent to the contact point on the mean limb and tangent to the highest mountain (annular) or lowest valley (total) at actual contact is the desired correction to the predicted contact time. On the exaggerated radial scale of Figure 20, the Sun's limb can be represented as an epicyclic curve that is tangent to the mean lunar limb at the point of contact and departs from the limb by **h** through:

$$\mathbf{h} = \mathbf{S}\,(\mathbf{m}-1)\,(1-\cos[\mathbf{C}]) \qquad [6]$$

where: **h** = departure of Sun's limb from mean lunar limb
S = Sun's semi-diameter
m = eclipse magnitude
C = angle from the point of contact

Herald [1983] has taken advantage of this geometry to develop a graphical procedure for estimating correction times over a range of position angles. Briefly, a displacement curve of the Sun's limb is constructed on a transparent overlay by way of equation [6]. For a given position angle, the solar limb overlay is moved radially from the mean lunar limb contact point until it is tangent to the lowest lunar profile feature in the vicinity. The solar limb's distance **d** (arc-seconds) from the mean lunar limb is then converted to a time correction Δ by:

$$\Delta = \mathbf{d}\,\mathbf{v}\cos[\mathbf{X} - \mathbf{C}] \qquad [7]$$

where: Δ = correction to contact time (seconds)
d = distance of Solar limb from Moon's mean limb (arc-sec)
v = angular velocity of the Moon with respect to the Sun (arc-sec/sec)
X = center line position angle of the contact
C = angle from the point of contact

This operation may be used for predicting the formation and location of Baily's beads. When calculations are performed over a large range of position angles, a contact time correction curve can then be constructed.

Since the limb profile data are available in digital form, an analytical solution to the problem is possible that is quite straightforward and robust. Curves of corrections to the times of second and third contact for most position angles have been computer generated and are plotted in Figure 20. The circular protractor scale at the center represents the nominal contact time using a mean lunar limb. The departure of the contact correction curves from this scale graphically illustrates the time correction to the mean predictions for any position angle as a result of the Moon's true limb profile. Time corrections external to the circular scale are added to the mean contact time; time corrections internal to the protractor are subtracted from the mean contact time. The magnitude of the time correction at a given position angle is measured using any of the four radial scales plotted at each cardinal point.

For example, Table 17 gives the following data for München, Germany:

Second Contact = 10:37:12.3 UT $P2=130°$
Third Contact = 10:39:20.1 UT $P_3=263°$

Using Figure 20, the measured time corrections and the resulting contact times are:

$C2=-3.1$ seconds; Second Contact = 10:37:12.3 –3.1s = 10:37:09.2 UT
$C_3=-1.7$ seconds; Third Contact = 10:39:20.1 –1.7s = 10:39:18.4 UT

The above corrected values are within 0.1 seconds of a rigorous calculation using the actual limb profile.

Lunar limb profile diagrams for a number of other positions/times along the path of totality are available via a special web site of supplemental material for the total solar eclipse of 1999 (*http://planets.gsfc.nasa.gov/eclipse/TSE1999/TSE1999.html*).

Limb Profile Effects on the Duration of Totality

As was previously discussed, the Moon's center of figure (i.e., the geometric center of the Watts' datum) is displaced from the Moon's center of mass. A case in point is the lunar limb geometry at 11:00 UT (Figure 20) where the center of figure is displaced –0.28 and +0.38 arc-seconds in ecliptic latitude and longitude, respectively. This shift is fairly characteristic along much of the 1999 umbral path but varies considerably between eclipses due to different libration geometries. Since most predictions appearing in this publication are calculated with respect to the Moon's center of mass, the center of figure offset has a small but significant consequence on the duration of totality. When compounded with the irregularities of the lunar limb profile, the overall result is to shift the maximum duration of totality south of the center line by 2-6 kilometers along the European path, and 3-10 kilometers along the Middle Eastern path.

Furthermore, the true limb profile of the 1999 eclipse actually produces a longer duration of totality than the one calculated using a mean limb. This results in maximum durations lasting 1 to 3 seconds longer than the nominal center line durations along much of the path.

Figure 21 shows a series of calculations for the duration of totality within ±30 kilometers of the center line and spaced at ten minute intervals along the path through Europe (a) and the Middle East (b). For a given time, the duration of totality is calculated at 1 kilometer intervals perpendicular to the path within a 60 kilometer zone centered on the center line. Predictions using the Moon's center of mass and mean limb are represented by the dotted curves. Predictions using the actual limb profile to calculate corrected contact times and the resulting duration of totality are plotted as solid curves. What becomes immediately apparent upon inspection of Figure 21, is the asymmetry of the true limb duration curves and is a consequence of the complex Sun/Moon limb geometry which changes quickly with path position.

Observers wishing to witness the maximum possible duration of totality from a given section of the path can use Figure 21 to optimize their location with respect to the center line.

Limb Corrections To The Path Limits: Graze Zones

The northern and southern umbral limits provided in this publication were derived using the Moon's center of mass and a mean lunar radius. They have not been corrected for the Moon's center of figure or the effects of the lunar limb profile. In applications where precise limits are required, Watts' limb data must be used to correct the nominal or mean path. Unfortunately, a single correction at each limit is not possible since the Moon's libration in longitude and the contact points of the limits along the Moon's limb each vary as a function of time and position along the umbral path. This makes it necessary to calculate a unique correction to the limits at each point along the path. Furthermore, the northern and southern limits of the umbral path are actually paralleled by a relatively narrow zone where the eclipse is neither penumbral nor umbral. An observer positioned here will witness a slender solar crescent that is fragmented into a series of bright beads and short segments whose morphology changes quickly with the rapidly varying geometry between the limbs of the Moon and the Sun. These beading phenomena are caused by the appearance of photospheric rays that alternately pass through deep lunar valleys and hide behind high mountain peaks as the Moon's irregular limb grazes the edge of the Sun's disk. The geometry is directly analogous to the case of grazing occultations of stars by the Moon. The graze zone is typically five to ten kilometers wide and its interior and exterior boundaries can be predicted using the lunar limb profile. The interior boundaries define the actual limits of the umbral eclipse (both total and annular) while the exterior boundaries set the outer limits of the grazing eclipse zone.

Table 6 provides topocentric data and corrections to the path limits due to the true lunar limb profile. At five minute intervals, the table lists the Moon's topocentric horizontal parallax, semi-diameter, relative angular velocity of the Moon with respect to the Sun and lunar libration in longitude. The Sun's center line altitude and azimuth is given, followed by the azimuth of the umbral path. The position angle of the point on the Moon's limb which defines the northern limit of the path is measured counter-clockwise (i.e., eastward) from the north point on the limb. The path corrections to the northern and southern limits are listed as interior and exterior components in order to define the graze zone. Positive corrections are in the northern sense while negative shifts are in the southern sense. These corrections (minutes of arc in latitude) may be added directly to the path coordinates listed in Table 3. Corrections to the center line umbral durations due to the lunar limb profile are also included and they are mostly positive. Thus, when added to the central durations given in Tables 3, 4, 5 and 7, a slightly longer central total phase is predicted.

Detailed coordinates for the zones of grazing eclipse at each limit for all land based sections of the path are presented in Table 8. Given the uncertainties in the Watts data, these predictions should be accurate to ±0.3 arc-seconds. The interior graze coordinates take into account the deepest valleys along the Moon's limb which produce the simultaneous second and third contacts at the path limits. Thus, the interior coordinates define the true edge of the path of totality. They are calculated from an algorithm which searches the path limits for the extreme positions where no photospheric beads are visible along a ±30° segment of the Moon's limb, symmetric about the extreme contact points at the instant of maximum eclipse. The exterior graze coordinates are somewhat arbitrarily defined and calculated for the geodetic positions where an unbroken photospheric crescent of 60° in angular extent is visible at maximum eclipse.

In Table 8, the graze zone latitudes are listed every 1° in longitude (at sea level) and include the time of maximum eclipse at the northern and southern limits as well as the path's azimuth. To correct the path for locations above sea level, *Elev Fact*[8] is a multiplicative factor by which the path must be shifted north perpendicular to itself (i.e., perpendicular to path azimuth) for each unit of elevation (height) above sea level. To calculate the shift, a location's elevation is multiplied by the *Elev Fact* value. Negative values (usually the case for eclipses in the Northern Hemisphere) indicate that the path must be shifted south. For instance, if one's elevation is 1000 meters above sea level and the *Elev Fact* value is –0.20, then the shift is –200m (= 1000m x –0.20). Thus, the observer must shift the path coordinates 200 meters in a direction perpendicular to the path and in a negative or southerly sense.

The final column of Table 8 lists the *Scale Fact* (km/arc-second). This scaling factor provides an indication of the width of the zone of grazing phenomena, due to the topocentric distance of the Moon and the projection geometry of the Moon's shadow on Earth's surface. Since the solar chromosphere has an apparent thickness of about 3 arc-seconds, and assuming a *Scale Fact* value of 2 km/arc-seconds, then the chromosphere should be visible continuously during totality for any observer in the path who is within 6 kilometers (=2x3) of each interior limit. However, the most dynamic beading phenomena occurs within 1.5 arc-seconds of the Moon's limb. Using the above Scale Factor, this translates into the first 3 kilometers inside the interior limits. But observers should position themselves at least 1 kilometer inside the interior limits (south of the northern interior limit or north of the southern interior limit) in order to ensure that they are inside the path due of to small uncertainties in Watts' data and the actual path limits.

For applications where the zones of grazing eclipse are needed at a higher frequency in longitude interval, tables of coordinates every 7.5' in longitude are available via a special web site for the total solar eclipse of 1999 *(http://planets.gsfc.nasa.gov/eclipse/TSE1999/TSE1999.html)*.

---

[8] is the product, tan(90-A) * sin(D), where A is the altitude of the Sun and D is the difference between the azimuth of the Sun and the azimuth of the limit line, with the sign selected to be positive if the path should be shifted north with positive elevations above sea level.

Saros History

The total eclipse of 1999 August 11 is the twenty-first member of Saros series 145 (Table 37), as defined by van den Bergh [1955]. All eclipses in the series occur at the Moon's ascending node and gamma[9] decreases with each member in the family. The series is a young one which began with a minuscule partial eclipse at high northern hemisphere latitudes on 1639 Jan 04. After fourteen partial eclipses each of increasing magnitude, the first central eclipse occurred on 1891 Jun 06. The event was a six second annular eclipse with a path sweeping through eastern Siberia and the Arctic Ocean. Although the vertex of the umbral shadow fell just short of Earth's surface, the Moon's distance was gradually decreasing with each subsequent eclipse in the series. In fact, the very next eclipse was a hybrid or annular/total eclipse on 1909 Jun 17. Greatest eclipse occurred in the Arctic Ocean and lasted 24 seconds.

The third central eclipse of Saros 145 occurred on 1927 Jun 29. It was the first total eclipse of the family and coincidentally passed through England in addition to Scandinavia and Siberia. On 1945 Jul 09, the path of totality began in Idaho and quickly swept northeast through Montana, Saskatchewan and Manitoba. After crossing Hudson Bay, Greenland and the North Atlantic, the umbra returned to Scandinavia and Siberia. The fifth central eclipse occurred on 1963 Jul 20 and is well known to many eclipse observers. Its path crossed Alaska, central and eastern Canada and Maine. The event drew a great deal of media attention and a beautiful article about the eclipse appeared months later in the pages of National Geographic [November 1963]. In fact, one of the authors (Espenak) has fond memories of watching the partial phases of this eclipse as a boy from his grandmother's home in Long Island.

The most recent eclipse of the series took place on 1981 Jul 31 and its path crossed central Siberia, Sakhalin Island and the Pacific Ocean where it ended north of Hawaii. After 1999, the following member occurs on 2017 Aug 21. *This* is the first total solar eclipse visible from the continental United States since 1979 Feb 26. The path of totality stretches from Oregon through Idaho, Wyoming, Nebraska, Missouri, Illinois, Kentucky, Tennessee and the Carolinas and has a greatest duration of 2m 40s (*http://planets.gsfc.nasa.gov/eclipse/*eclipse*/SEmap/TSENorAm2001.GIF*).

During the 21st through 24th centuries, Saros 145 continues to produce total solar eclipses of increasing duration as the path of each event shifts southward. By the time the midpoint of the series is reached (2324 Feb 25), the duration of totality exceeds four minutes. The duration continues to increase into the 25th and 26th centuries. The maximum duration of totality peaks at 7m 12s on 2522 Jun 25. In the remaining six umbral eclipses, the duration rapidly drops but still lasts almost three minutes during the final total eclipse on 2648 Sep 09.

For the next three and a half centuries, twenty partial eclipses of progressively decreasing magnitude occur. The final event takes place on 3009 Apr 17 from the polar regions of the Southern Hemisphere. A detailed list of eclipses in Saros series 145 appears in Table 37.

In summary, Saros series 145 includes 77 eclipses with the following distribution:

Saros 145	*Partial*	*Annular*	*Ann/Total*	*Total*
Non-Central	34	0	0	0
Central	—	1	1	41

[9] Minimum distance of the Moon's shadow axis from Earth's center in units of equatorial Earth radii. Gamma defines the instant of greatest eclipse and takes on negative values south of the Earth's center.

WEATHER PROSPECTS FOR THE ECLIPSE

INTRODUCTION

The lunar shadow begins its three hour trek off the coast of Nova Scotia in the cloudy skies of the North Atlantic, just beyond Canada's Sable Island. A third of the globe later it reaches land in southwest England. While this is the best location in Great Britain, better weather can be found farther along the track where the frequency of sunny weather improves steadily into the Middle East. Beyond Iran, the Indian monsoon rapidly diminishes the chances of a view of this spectacle and sunset finds the shadow over the very cloudy skies of the Bay of Bengal.

THE BRITISH ISLES

Of all of the countries of Europe, the British Isles are most exposed to the varying weather from the mobile westerly air flow off of the North Atlantic. While blessed with mild winters, the dampness, cool summers and meteorological exposure to the westerlies has lead to the pronouncement that the Isles have "no climate, only weather." Historically, Britain has appealed more to invading northern peoples rather than those from the south. The former found the warm moist climate an improvement on their homelands, while to the latter, especially the French, it appeared illogical to covet such a cloudy, damp and windy land (Manley, 1970). While good for the "English lawn," the damp gray skies will not find favor at eclipse time.

But fairness dictates that we note the eclipse comes at the height of summer over the sunniest parts of the English countryside. Land's End and the south coast of England offer the best prospects for sunshine anywhere in the Isles, and so by good fortune the eclipse track is located to its best advantage.

The upper air currents which carry the weather systems onto the islands are predominantly westerly. The main track of invading low pressure centers is north of Scotland, but trailing cold fronts frequently drag across England as these disturbances pass, bringing persistent cloudiness and precipitation. The cooler, drier high pressure systems following these fronts do not always improve sky conditions, as the colder airmass is often unstable and speckles the afternoon landscape with showers or thundershowers.

Periodic blocking patterns occasionally interrupt the alternating stream of highs and lows, bringing episodes of stable unchanging weather. This interruption in the variable weather pattern may or may not favor the eclipse observer depending on where the block develops. If the circulation over the islands is anticyclonic, as when a high builds northward from the Azores anticyclone or takes up residence over western France, then sunny dry weather will dominate, likely for several days. However, when cyclonic circulations settle in, especially if the low center lingers over England, dull gray skies will dominate.

Long term studies suggest that a variable westerly flow can be found about 28% of the time, cyclonic blocking for 24% and anticyclonic patterns for 23%. Suitable eclipse-viewing weather will be found during anticyclonic days and possibly in the wake of cold fronts during the changeable westerly weather. In the wake of a cold frontal passage, coastal areas are more likely to have sunshine than interior locations as the cooling effect of the nearby ocean suppresses the development of convective cloud. This distinction is often enhanced by the formation of sea breeze circulations which can bring a zone of clear skies along the waterfront to an otherwise cloudy day. The zone of clearing will extend inland at best for only a few kilometers depending on the precise nature of the onshore winds.

Realistically, however, climate statistics suggest that England is the least suitable land location from which to view the eclipse except for parts of India and Pakistan where monsoon cloudiness dominates. The satellite-derived climatology drawn in Figure 22 shows a mean cloud cover of 55 to 65% across southern England. Surface observations show an average of 5.8 to 6.5 days of the month with scattered cloud and good visibility (Table 38), slightly lower than much of the rest of western Europe, but only a third of the cloud-free days available in Bulgaria and Romania, and a quarter to a fifth the number of days in the best locations in Turkey and Iran. This rather heavy cloudiness culminates in a meager 43% of the possible sunshine in England.

Land's End and The Lizard are among the sunniest spots in the England thanks to the suppression of convective cloud by sea breezes and a slightly closer position to the semi-permanent anticyclone over the Azores. The effect of these two factors is minor but detectable and can be seen by comparing statistics for Plymouth and The Lizard in Table 38. Data collected at the former show a frequency of scattered cloud and good visibilities of 5.8 days per month; at The Lizard this statistic improves slightly to 6.5 days per month. Most climatological studies concede a five to ten percent increase in sunshine to the coastal sites over those inland, but much of this advantage is only realized during unstable convective days, such as those found behind a cold front. Figure 23 promises a stationary observer only a 45% chance of seeing the eclipse in southern England.

Whatever the cause, low level stratiform clouds are the most likely type of cloud along the south coast (Warren et. al, 1986). These clouds are formed as warm maritime air approaches from the south and moves over the colder waters of the nearby English Channel. Such clouds typically cover three-quarters of the sky when present, making a clear patch difficult to find. The best location to find a gap in this kind of cloud cover is to seek shelter against the southerly flow behind the low range of hills which define the spine of Cornwall so that low level winds are flowing downhill and drying the air. The hills near the center line between Penzance and The Lizard do not accord much protection, in part because they are not quite oriented to lie across the likely direction of the wind. In the event of a heavy layer of low cloud, the eclipse watcher is best advised to sacrifice some of the duration of the eclipse and head northeast along the highway toward Bodmin, leaving the center line behind. It's a small chance, but the best that can be offered under the circumstances.

Western Europe

During the summer months, the "European monsoon" sets in across the western half of the continent – a period of cloudy and showery weather which begins with the onset of westerly winds in mid June and continues into September. At other times of the year the prevailing wind direction aloft is more variable, bringing alternating spells of warm and cold weather. This does not imply that cold air does not spread southward in summer, for cold spells with unstable weather come at intervals of a week or two, just as in North America. In fact, the weather along the eclipse track is similar in character to that found in the northern plains States and Canadian prairie provinces, with frequent weak disturbances bringing showers and thundershowers. Intervals between disturbances often come with fine and dry weather, though these are less frequent in Europe than in North America. Prolonged dry weather in Europe comes when a high pressure system builds northward from the Azores anticyclone.

The varied topography through Germany and Austria creates a wealth of climatic sub-regions, but these are defined more by temperature and precipitation differences than in cloudiness. Cloud patterns are strongly modified by the higher mountain ranges, particularly the Alps and Jura Mountains to the south of the eclipse track, but the main effect of these barriers is to create cloud rather than dissipate it. The prevailing westerly and northwesterly winds are forced to rise up these slopes, cooling and causing clouds to form. Southerly winds, flowing over the mountains toward the eclipse track, would tend to dissipate the cloud and bring dry sunny weather on the downhill side (a process known as a chinook in North America and a foehn wind in Europe) but southerlies are relatively rare during August and the benefits of the mountain barrier are not likely to be realized.

In spite of the influence of the Alps, there is a steady trend to sunnier skies as the eclipse track proceeds across Europe, and the distance from the Atlantic moisture source increases. The percent frequency of clear skies is about 18% over Normandy on the Channel coast and barely rises above 20% throughout the length of the eclipse track over the rest of France. However, beyond this point the maritime character of the westerly winds becomes more continental, and sunshine frequency begins a slow climb through Germany and Austria. This improving trend is interrupted briefly through central Austria where a branch of the Alps reaches down to the eclipse track and brings an increase in cloudiness. The effect is quite dramatic – the number of hours of sunshine in summer drops nearly 25% across middle of Austria in comparison with its eastern and western borders.

Figure 22 and Table 38 show that the mean August cloudiness decreases from 60% near Cornwall to 50% near Paris. The mean number of hours of sunshine grows from 6.5 over France to just over 8 hours in Austria, with a matching increase in the frequency of scattered cloud conditions. Figure 23 shows little difference in the prospects between Land's End and Paris, but then a slowly growing probability of seeing the eclipse through Germany and Austria.

There is not much to recommend one part of the track over another other than to suggest that eclipse viewers head eastward to take advantage of the slow climatological improvement. A much more reliable option is to simply await the weather forecasts in the days ahead of the eclipse and pick a site which is forecast to be sunny. While long range forecasts are available out to ten days, they do not become particularly reliable until about five days in advance of the event. More will be said about this later.

Summing up then, and extracting the smallest details from the statistical record, the best sites along the center line in France will be found from Compiègne, past Reims, to Metz, though the advantage gained is very small. In Germany the most suitable climatology is found from Ulm past Munich to the Austrian border. In Austria, climatology favors a location near the Hungarian border south of Vienna, though a location near the German border comes in as a close second.

Eastern Europe to the Black Sea

As the track leaves Austria, it also draws away from the influence of the westerly winds which have controlled the meteorology up to this point. In summer, the Danubian plains over southern Hungary are affected more by the Mediterranean climate advancing northward from the Adriatic Sea than from the march of Atlantic disturbances. The pronounced effect on cloud cover and the amount of sunshine leaves little doubt that the best European eclipse conditions will be found in Hungary, Romania and Bulgaria.

For the most part, the path follows the lowlands along the Danube River and is protected from stronger weather systems by the Carpathian Mountains in the north and the Balkan Mountains in the south. Prevailing winds blow lightly from the north or northwest, being drawn into a large low pressure system which forms over Iran in the summer. These etesian winds bring dry invigorating air which is constant in direction and speed. Precipitation is mostly in the form of showers and thundershowers, and tends to be greatest where the winds blow upslope – generally on the northern slopes of the Balkan Mountains.

Figure 22 shows that the eclipse track begins to cross the cloud isopleths at a sharper angle eastward from Hungary, with the result that the mean August cloud amount has dropped to about 45% at the shores of the Black Sea. The amount of sunshine climbs above ten hours per day, more than 70% of the maximum possible. The number of days with scattered cloud or less at eclipse time rises from about half the month near the Austrian border to nearly two-thirds over Bulgaria. The probability of seeing the eclipse reaches 63% at the Black Sea ports of Varna and Constanta, popular summer destinations with beaches and fine Roman ruins to attract visitors and eclipse-seekers.

Thunderstorms have a relatively high frequency of occurrence in eastern Europe and bring considerable cloudiness when present. But this convective cloud relies on the heating of the ground for much of its development, unless pushed by cold fronts or other weather disturbances. Because of this, cloud cover statistics tend to underestimate the chances of seeing an eclipse which takes place in the morning hours or in the early afternoon. The maximum time of thunderstorm development occurs at 6 PM local time, well after the noon hour date for the eclipse. Indeed, the cooling associated with the eclipse may delay the onset of convection for another hour or two.

The path crosses Hungary's Lake Balaton, a popular resort area of warm water and sandy beaches. At one time the border between the Ottoman and Hapsburg empires ran down the center of the lake, and a number of ruined castles dot the northern hills above the lake. The location is easily reached from Budapest.

According to Dr. Szécsényi-Nagy of Loránd Eötvös University, the weather in Hungary is normally stable in summer with occasional long periods (3-4 weeks) of high pressure, cloudless skies and a dry atmosphere. Dr. Szécsényi-Nagy studied 20 years of weather records for the week on either side of the eclipse date and noted that only three days of the 300 were without sunshine at stations along the track. According to his results, sunshine is the overwhelming character of the day, with nearly two thirds of the records showing more than ten hours each day.

His conclusions are mirrored by the statistics for Keszthely ("cast-eye"), a resort and service town on the western shore of Lake Balaton. The 14 days of the month in which scattered cloud is reported at eclipse time at this location is the best in Hungary and comparable with sites in Romania and Bulgaria. Lake Balaton seems to have more than its share of sunny weather, likely a result of the protection afforded by the surrounding heights. The eclipse viewing probabilities in Figure 23 do not show this advantage, but this is probably caused by the smoothing of the climatological record of cloud cover by the computer model. The real differences are best seen in the actual record of observations.

Just after its point of maximum eclipse, the Moon's shadow crosses Bucharest, the capital of Romania. This city of two million promises to be a prime eclipse-viewing site, in part because of the comfort and ease of access, and partly because of the excellent weather prospects and the long eclipse duration. Since the center line neatly bisects the city, eclipse-viewing can be done from the wide boulevards or one of the many city parks.

The sum of all of the climatological measures points to Hungary, Bulgaria and Romania as the choicest locations in Europe for viewing this event. The shores of the Black Sea offer the greatest prospects of success, with a generous sunny climate enhanced by sea breeze circulations. Other suitable spots can be found all along the track from Lake Balaton eastward. To get any better weather prospects, the eclipse location will have to move to the other side of the Black Sea.

Turkey, Iraq and Iran

Once across the Black Sea, the lunar shadow's eastward track moves into the best weather conditions. Weather in the region is dominated by an extension of the large monsoon low over India and Pakistan.

While low pressure systems are normally associated with cloud and rain, this particular low lies beneath an upper level high which suppresses the formation of cloud. A weak frontal system extends from northern Iran to the Mediterranean coast, separating moderate temperatures and moisture on the north side from the semi-arid airmasses to the south. Upper level winds tend to flow from the west in northern Turkey and from the east in more southerly parts along the track.

The narrow coastal plain along the Black Sea coast of Turkey has a Mediterranean climate and represents the airmasses north of the front. This area has sufficient moisture to grow figs, olives, tea and tobacco. Most of the rain falls in winter, but steady northwest winds in the summer season bring occasional convective clouds with showers and thundershowers.

The eclipse comes ashore at a remote part of the Turkish coastline near the town of Cide. To the east of the shadow path is Sinop, an ancient city of Greek and Byzantine origins. Its most famous native son was Diogenes the Cynic, who is reputed to have replied to Alexander the Great when asked what Alexander could do for him, "Yes, stand aside, you're blocking my light," – a perfect quotation (though not sentiment) to go with a solar eclipse [Ayliffe et al. 1994]. The scenic road to the center line winds closely along the coast, forced to the edge of the sea by a range of 2000 meter mountains which line the Black Sea coast. Cide, a few kilometers inland from the sea, has a 10 km long pebbly beach nearby which could provide a site to watch the eclipse. Travel time from Sinop to Cide is about two hours. West of the center line is the regional capital Zonguldak, slightly closer than Sinop, and with good transportation connections to Ankara.

The steady onshore winds along the coast promote the development of convective cloudiness as they rise up the mountain slopes which line the sea, but cloud statistics for Zonguldak and Sinop promise weather at least as favorable as on the Bulgarian coast. Farther south, the track reaches Sivas on the eastern edge of the Anatolian Plateau. This city of magnificent Selçuk monuments has a history extending from Hittite times to the formation of modern Turkey. The frequency of days with scattered cloud rises rapidly here as downslope winds from the surrounding mountains dry the air. Much of the meager cloudiness is due to occasional summer thundershowers, though the incidence of rainy days is dwindling rapidly. Nevertheless, some of these thunderstorms can bring violent weather with widespread cloudiness. As in eastern Europe, they are primarily an afternoon event, but because the eclipse is later in Turkey and the heating of the ground more pronounced, thunderstorms are more likely here than in Europe at eclipse time. Beyond Sivas the shadow bounces across the eastern limb of the Taurus Mountains, with gradually improving prospects for good eclipse weather.

When the shadow path finally departs the Taurus Mountains it descends into the Tigris Valley and reaches Dyarbakir, a sprawling city with a large Kurdish population. The political turmoil of the area has converted Dyarbakir into a large military outpost, enhanced by the presence of a large NATO air base with shrieking jets and thumping helicopter rotors. Weather prospects are excellent (in spite of the anomalous "% of Possible Sunshine" in Table 38) as August is the hottest and driest month of the year. Statistics for the area show that only scattered clouds can be found at eclipse time on 24 to 28 days of the month, and nearly 30 days at the border with Syria. The viewing probability in Figure 23 rises above 80% as thunderstorms almost cease to be a threat, and the eclipse track moves into its most promising arena.

Mountainous terrain intervenes again as the eclipse path moves across Kurdistan into Iraq. It is a volatile area and best left alone, in spite of the clear skies. Travelers here must contend with a region in turmoil and will have to have a special quest for adventure. Even better weather and a more stable political climate comes in Iran where the track leaves the Zagros Mountains and moves onto the desert plateau. This is an arid area with temperatures approaching 50° C. The low humidity makes the heat somewhat more bearable (as will the cooling with an eclipse), but by the time the track reaches the Pakistan border, sultry humidities from the Gulf of Oman promise less pleasant observing conditions.

Skies are nearly cloudless over Iran for the most part, except for the occasional patches of scattered convective clouds and light showers. These are usually weak enough that they will succumb readily to the cooling which comes with the eclipse. Figure 23 shows that the apex of the eclipse viewing prospects is reached at Esfahan in southern Iran. Sites here have a 96% chance of seeing the Sun on August 11.

Pakistan and India

In August the southwest monsoon season over India is just past its peak, though still in full sway. This humid, wet and cloudy season does not retreat from eastern Pakistan until early September, too late to affect the eclipse. Cloud conditions over western Pakistan are heavier than over Iran because of the abundant moisture available from the Arabian Sea and the convection-promoting influence of the monsoon. Statistics for Karachi show a dramatic decline in the frequency of days with scattered cloud, though not in the number of days with rain.

As the track moves into India, the low Sun angle, late hour, and the extensive cloudiness of the Indian monsoon bring the poorest conditions of anywhere along the track. Satellite pictures reveal a land cloaked in cloud day after day and the probability of seeing the eclipse declines rapidly. By the time the shadow reaches the sunset terminator, the chances of seeing the eclipse have dropped almost to zero.

The Atlantic Ocean and the Black Sea

The proximity of the sunrise eclipse to Nova Scotia will undoubtedly tempt some observers to try a ship-board expedition from the eastern seaboard of North America. Though skies have a high frequency of cloud cover, the mobility offered by a ship should be able to overcome this deficiency, to some extent, provided good weather advice is available. The very low Sun angle at the start of the eclipse will seriously impede the search for a hole in any cloud cover which might be there, but provided the excursion is not just a day trip with little time for exploration, the effort has a good chance of being rewarded.

Black Sea prospects are comparable to nearby lands enjoying favorable climate statistics and the diminished influence of the variable westerlies. Mobility offers advantages similar to those described above.

Mean wave heights in the western Atlantic range between 1 and 1.5 meters off of the Nova Scotia coast, making time-exposure photography a little challenging, particularly through a telescope. Waves on the much smaller Black Sea tend to keep under 0.5 meters except near the Turkish coast where the prevailing winds have the entire length of the sea to build wave heights.

Coping with the Weather on Eclipse Day

From England to Romania and Bulgaria, eclipse day will be subject to the vagaries of the westerly winds which carry highs and lows across the European continent. While climatology offers some advice for long range planning, mobile eclipse observers will almost certainly be able to find a view of the Sun on August 11. Particularly in western Europe, where the collection of climate statistics over the decades does not favor one location over another, the ability to respond to short and long-range weather forecasts will be very helpful. Modern meteorology is capable of producing detailed computer forecasts which stretch for ten days into the future. These forecasts can be assessed early enough to make general travel plans which can then be refined as eclipse day approaches.

Long range (5-10 day) computer forecasts are notoriously inaccurate and must be used with care. They should be used only for long range planning, to pick a general area for travel, as they are likely to show many changes before August 11 arrives. The best sites will lie beneath an upper level ridge ahead of, or in, a surface high pressure region, if these structures are forecast. This does not guarantee good weather, but it does greatly enhance one's prospects. If the numerical models don't predict these conditions, then more sophisticated decisions must be made which will probably require the services of a meteorologist.

Computer forecasts become much more reliable within five days, and still more accurate within 48 hours. The amount of detail they reveal also increases, with fields of relative humidity, precipitation, cloud cover, winds and temperature becoming available as the critical date approaches. A logical approach would be to select two or three specific sites within a convenient travel distance at this time, and then make a decision between them at the last possible moment.

Long range forecasts for Europe from the National Weather Service (NWS) in the United States and the European Centre for Medium Range Weather Forecasting (ECMWF) are readily available on the World Wide Web from Purdue University and a number of other locations. At Purdue the U.S. medium range forecast (MRF) model is usually available each morning after 12:30 UT. The ECMWF forecast out to six days is also available at the same site.

MRF charts show the 500 millibar flow (about 5000 m above the surface) and the surface pressure pattern. Look for an upper level ridge on the 500 mb chart, and the position of highs and lows on the other. These two charts will allow an early evaluation of the weather over Europe (most models are global, though not all parts of the globe are shown in the Web sites). The really hard part is whether or not the models are trustworthy, especially eight or ten days into the future. What clues are available to indicate that the numerical weather patterns will match the real ones on eclipse day?

Consistency is one clue – is the same general pattern forecast from one day to the next? Another is whether or not two different models forecast the same general patterns. Most likely there will be small differences between the two, but if highs and lows are hundreds of kilometers apart, or upper features don't line up, then use the charts with caution. Try to find a spot which looks good on both charts, and wait another day to see if the agreement improves. Don't finalize long range plans at this stage.

Once the eclipse is within five days, the greater reliability and detail in the models allows serious planning, for now we know not only the location of weather systems, but also how the cloud is draped around them. But be careful! Predictions are not gospel, and thunderstorms especially are difficult to predict with accuracy when more than a day or two away. Low level cloud is often unforecast, or indicated only in a subtle variation in the humidity pattern. Keep looking for the best spot within your travel range according to the position of the low and the upper ridge. By now of course, normal meteorological forecasts will be available and you may simply have to watch the television.

Long range forecasts over Turkey and the Middle East are not likely to be as informative as those over Europe, where weather systems are more active and changeable. Individuals and groups who are not mobile may wish to consider sites where climate statistics work in their favor, from Hungary eastward through Turkey, rather than take their chances with the westerlies in western Europe or England.

The Probability of Seeing the Eclipse

When the Sun is high, the probability of seeing an eclipse depends only on the proportion of the sky which is covered by cloud, a factor which is represented by the mean cloudiness of a site. As the altitude of the Sun declines however, the depth of the cloud becomes more important as a restriction to visibility. This is easily understood by considering the effects of a towering thunderstorm – if overhead, the blocking capacity depends on the area of the base of the storm. If the thunderstorm is on the horizon, the blocking effect depends on the width of the storm perpendicular to the line of sight and its height.

If the average height of clouds in an area and the mean cloud amount are known, then the probability of seeing the eclipse can be calculated for a given solar altitude. Such a calculation is shown in Figure 23, using climatological values for these parameters. Adjustments are made for the time of day, since a large part of the cloud cover along the track is convective in nature, and thus dependent on the hour.

As with any calculation, the results of Figure 23 should be used cautiously. The climatological data used in the modeling are smoothed out from their true values, and small scale variations are lost or muted. Actual forecasts on and ahead of eclipse day will provide much more information than this figure, though of course they do not permit planning years ahead of time.

Summary

Weather prospects for the eclipse begin in the dismal cloudiness of the North Atlantic and improve steadily along the eclipse track as far as the Middle East. Beyond Iran, cloud cover thickens again and the eclipse ends as it began with diminished promises. The best prospects in Europe are found along the shores of the Black Sea. Iran offers the most favorable weather along the entire track, though cautious observers may prefer the skies of central Turkey.

Weather Web Sites

1. http://www.tvweather.com
A good starting point. This site has many links to current and past weather around the world.

2. http://shark1.esrin.esa.it:80/q_interface.html
A source of past satellite pictures around the world from 1992 onward (with gaps). These color pictures will give a feel for the weather over Europe and the Middle East in mid August.

3. http://wxp.atms.purdue.edu/
Purdue University, where the various model forecasts can be obtained. The MRF model is available over Europe for 10 days into the future. The ECMWF forecast can be examined for 6 days ahead. The MRF is available as a 9 panel display of 500 mb and surface features. Both models go as far east as Iraq. Another model, the AVN is available for 72 hours into the future and includes a chart of relative humidity (where 70% RH or higher implies cloud).

4. http://www.meteo.fr/tpsreel/images/satt0.jpg
A current European satellite image is available at Meteo France, as well as other sites.

OBSERVING THE ECLIPSE

EYE SAFETY AND SOLAR ECLIPSES

B. Ralph Chou, MSc, OD
Associate Professor, School of Optometry, University of Waterloo
Waterloo, Ontario, Canada N2L 3G1

A total solar eclipse is probably the most spectacular astronomical event that most people will experience in their lives. There is a great deal of interest in watching eclipses, and thousands of astronomers (both amateur and professional) travel around the world to observe and photograph them.

A solar eclipse offers students a unique opportunity to see a natural phenomenon that illustrates the basic principles of mathematics and science that are taught through elementary and secondary school. Indeed, many scientists (including astronomers!) have been inspired to study science as a result of seeing a total solar eclipse. Teachers can use eclipses to show how the laws of motion and the mathematics of orbital motion can predict the occurrence of eclipses. The use of pinhole cameras and telescopes or binoculars to observe an eclipse leads to an understanding of the optics of these devices. The rise and fall of environmental light levels during an eclipse illustrate the principles of radiometry and photometry, while biology classes can observe the associated behavior of plants and animals. It is also an opportunity for children of school age to contribute actively to scientific research - observations of contact timings at different locations along the eclipse path are useful in refining our knowledge of the orbital motions of the Moon and earth, and sketches and photographs of the solar corona can be used to build a three-dimensional picture of the Sun's extended atmosphere during the eclipse.

However, observing the Sun can be dangerous if you do not take the proper precautions. The solar radiation that reaches the surface of Earth ranges from ultraviolet (UV) radiation at wavelengths longer than 290 nm to radio waves in the meter range. The tissues in the eye transmit a substantial part of the radiation between 380 and 1400 nm to the light-sensitive retina at the back of the eye. While environmental exposure to UV radiation is known to contribute to the accelerated aging of the outer layers of the eye and the development of cataracts, the concern over improper viewing of the Sun during an eclipse is for the development of "eclipse blindness" or retinal burns.

Exposure of the retina to intense visible light causes damage to its light-sensitive rod and cone cells. The light triggers a series of complex chemical reactions within the cells which damages their ability to respond to a visual stimulus, and in extreme cases, can destroy them. The result is a loss of visual function which may be either temporary or permanent, depending on the severity of the damage. When a person looks repeatedly or for a long time at the Sun without proper protection for the eyes, this photochemical retinal damage may be accompanied by a thermal injury - the high level of visible and near-infrared radiation causes heating that literally cooks the exposed tissue. This thermal injury or photocoagulation destroys the rods and cones, creating a small blind area. The danger to vision is significant because photic retinal injuries occur without any feeling of pain (there are no pain receptors in the retina), and the visual effects do not occur for at least several hours after the damage is done [Pitts, 1993].

The only time that the Sun can be viewed safely with the naked eye is during a total eclipse, when the Moon completely covers the disk of the Sun. ***It is never safe to look at a partial or annular eclipse, or the partial phases of a total solar eclipse, without the proper equipment and techniques.*** Even when 99% of the Sun's surface (the photosphere) is obscured during the partial phases of a solar eclipse, the remaining crescent Sun is still intense enough to cause a retinal burn, even though illumination levels are comparable to twilight [Chou, 1981, 1996; Marsh, 1982]. Failure to use proper observing methods may result in permanent eye damage or severe visual loss. This can have important adverse effects on career choices and earning potential, since it has been shown that most individuals who sustain eclipse-related eye injuries are children and young adults [Penner and McNair, 1966; Chou and Krailo, 1981].

The same techniques for observing the Sun outside of eclipses are used to view and photograph annular solar eclipses and the partly eclipsed Sun [Sherrod, 1981; Pasachoff & Menzel 1992; Pasachoff & Covington, 1993; Reynolds & Sweetsir, 1995]. The safest and most inexpensive method is by projection. A pinhole or small opening is used to form an image of the Sun on a screen placed about a meter behind the opening. Multiple openings in perfboard, in a loosely woven straw hat, or even between interlaced fingers can be used to cast a pattern of solar images on a screen. A similar effect is seen on the ground below a broad-leafed tree: the many "pinholes" formed by overlapping leaves creates hundreds of crescent-shaped images.

Binoculars or a small telescope mounted on a tripod can also be used to project a magnified image of the Sun onto a white card. All of these methods can be used to provide a safe view of the partial phases of an eclipse to a group of observers, but care must be taken to ensure that no one looks through the device. The main advantage of the projection methods is that nobody is looking directly at the Sun. The disadvantage of the pinhole method is that the screen must be placed at least a meter behind the opening to get a solar image that is large enough to see easily.

The Sun can only be viewed directly when filters specially designed to protect the eyes are used. Most such filters have a thin layer of chromium alloy or aluminum deposited on their surfaces that attenuates both visible and near-infrared radiation. A safe solar filter should transmit less than 0.003% (density~4.5)[10] of visible light (380 to 780 nm) and no more than 0.5% (density~2.3) of the near-infrared radiation (780 to 1400 nm). Figure 24 shows the spectral response for a selection of safe solar filters.

One of the most widely available filters for safe solar viewing is shade number 14 welder's glass, which can be obtained from welding supply outlets. A popular inexpensive alternative is aluminized mylar manufactured specifically for solar observation. ("Space blankets" and aluminized mylar used in gardening are *not* suitable for this purpose!) Unlike the welding glass, mylar can be cut to fit any viewing device, and doesn't break when dropped. Many experienced solar observers use one or two layers of black-and-white film that has been fully exposed to light and developed to maximum density. The metallic silver contained in the film emulsion is the protective filter. Some of the newer black and white films use dyes instead of silver and these are *unsafe*. Black-and-white negatives with images on it (e.g., medical x-rays) are also *not* suitable. More recently, solar observers have used floppy disks and compact disks (both CDs and CD-ROMs) as protective filters by covering the central openings and looking through the disk media. However, the optical quality of the solar image formed by a floppy disk or CD is relatively poor compared to mylar or welder's glass. Some CDs are made with very thin aluminum coatings which are not safe - if you can see through the CD in normal room lighting, don't use it!! No filter should be used with an optical device (e.g. binoculars, telescope, camera) unless it has been specifically designed for that purpose and is mounted at the front end (i.e., end towards the Sun). Some sources of solar filters are listed in the following section.

Unsafe filters include all color film, black-and-white film that contains no silver, photographic negatives with images on them (x-rays and snapshots), smoked glass, sunglasses (single or multiple pairs), photographic neutral density filters and polarizing filters. Most of these transmit high levels of invisible infrared radiation which can cause a thermal retinal burn (see Figure 24). The fact that the Sun appears dim, or that you feel no discomfort when looking at the Sun through the filter, is no guarantee that your eyes are safe. Solar filters designed to thread into eyepieces that are often provided with inexpensive telescopes are also unsafe. These glass filters can crack unexpectedly from overheating when the telescope is pointed at the Sun, and retinal damage can occur faster than the observer can move the eye from the eyepiece. Avoid unnecessary risks. Your local planetarium, science center, or amateur astronomy club can provide additional information on how to observe the eclipse safely.

There has been concern expressed about the possibility that UVA radiation (wavelengths between 315 and 380 nm) in sunlight may also adversely affect the retina [Del Priore, 1991]. While there is some experimental evidence for this, it only applies to the special case of aphakia, where the natural lens of the eye has been removed because of cataract or injury, and no UV-blocking spectacle, contact or intraocular lens has been fitted. In an intact normal human eye, UVA radiation does not reach the retina because it is absorbed by the crystalline lens. In aphakia, normal environmental exposure to solar UV radiation may indeed cause chronic retinal damage. However, the solar filter materials discussed in this article attenuate solar UV radiation to a level well below the minimum permissible occupational exposure for UVA (ACGIH, 1994), so an aphakic observer is at no additional risk of retinal damage when looking at the Sun through a proper solar filter.

In the days and weeks preceding a solar eclipse, there are often news stories and announcements in the media, warning about the dangers of looking at the eclipse. Unfortunately, despite the good intentions behind these messages, they frequently contain misinformation, and may be designed to scare people from seeing the eclipse at all. However, this tactic may backfire, particularly when the messages are intended for students. A student who heeds warnings from teachers and other authorities not to view the eclipse because of the danger to vision, and learns later that other students did see it safely, may feel cheated out of the experience. Having now learned that the authority figure was wrong on one occasion, how is this student

---

[10] In addition to the term transmittance (in percent), the energy transmission of a filter can also be described by the term density (unitless) where density 'd' is the common logarithm of the reciprocal of transmittance 't' or $d = \log_{10}[1/t]$. A density of '0' corresponds to a transmittance of 100%; a density of '1' corresponds to a transmittance of 10%; a density of '2' corresponds to a transmittance of 1%, etc....

going to react when other health-related advice about drugs, alcohol, AIDS, or smoking is given [Pasachoff, 1997]? Misinformation may be just as bad, if not worse than no information at all.

In spite of these precautions, the *total* phase of an eclipse can and should be viewed without any filters whatsoever. The naked eye view of totality is not only completely safe, it is truly and overwhelmingly awe-inspiring!

Sources for Solar Filters

The following is a brief list of sources for mylar and/or glass filters specifically designed for safe solar viewing with or without a telescope. The list is not meant to be exhaustive, but is simply a representative sample of sources for solar filters currently available in North America and Europe. For additional sources, see advertisements in *Astronomy* and/or *Sky & Telescope* magazines. The inclusion of any source on this list does not imply an endorsement of that source by the authors or NASA.

- ABELexpress - Astronomy Division, 230-Y E. Main St., Carnegie, PA 15106. (412) 279-0672
- Celestron International, 2835 Columbia St., Torrance, CA 90503. (310) 328-9560
- Edwin Hirsch, 29 Lakeview Dr., Tomkins Cove, NY 10986. (914) 786-3738
- Meade Instruments Corporation, 16542 Millikan Ave., Irvine, CA 92714. (714) 756-2291
- Orion Telescope Center, 2450 17th Ave., PO Box 1158-S, Santa Cruz, CA 95061. (408) 464-0446
- Pocono Mountain Optics, 104 NP 502 Plaza, Moscow, PA 18444. (717) 842-1500
- Rainbow Symphony, Inc., 6860 Canby Ave., #120, Reseda, CA 91335 (800) 821-5122
- Roger W. Tuthill, Inc., 11 Tanglewood Lane, Mountainside, NJ 07092. (908) 232-1786
- Telescope and Binocular Center, P.O. Box 1815, Santa Cruz, CA 95061-1815. (408) 763-7030
- Thousand Oaks Optical, Box 5044-289, Thousand Oaks, CA 91359. (805) 491-3642
- Khan Scope Centre, 3243 Dufferin Street, Toronto, Ontario, Canada M6A 2T2(416) 783-4140
- Perceptor Telescopes TransCanada, Brownsville Junction Plaza, Box 38, Schomberg, Ontario, Canada L0G 1T0 (905) 939-2313
- Eclipse 99 Ltd., Belle Etoile, Rue du Hamel, Guernsey GY5 7QJ. 001 44 1481 64847

IAU Solar Eclipse Education Committee

In order to ensure that astronomers and public health authorities have access to information related to safe viewing practices, the International Astronomical Union, the international organization for professional astronomers, set up a Solar Eclipse Education Committee. Under Prof. Jay M. Pasachoff of Williams College, the Committee has assembled information on safe methods of observing the Sun and solar eclipses, eclipse-related eye injuries, and samples of educational materials on solar eclipses.

For more information, contact Prof. Jay M. Pasachoff, Hopkins Observatory, Williams College, Williamstown, MA 01267, USA (e-mail: jay.m.pasachoff@williams.edu). Information on safe solar filters can be obtained by contacting Dr. B. Ralph Chou (e-mail: bchou@sciborg.uwaterloo.ca).

Eclipse Photography

The eclipse may be safely photographed provided that the above precautions are followed. Almost any kind of camera with manual controls can be used to capture this rare event. However, a lens with a fairly long focal length is recommended to produce as large an image of the Sun as possible. A standard 50 mm lens yields a minuscule 0.5 mm image, while a 200 mm telephoto or zoom produces a 1.9 mm image. A better choice would be one of the small, compact catadioptic or mirror lenses that have become widely available in the past ten years. The focal length of 500 mm is most common among such mirror lenses and yields a solar image of 4.6 mm. With one solar radius of corona on either side, an eclipse view during totality will cover 9.2 mm. Adding a 2x tele-converter will produce a 1000 mm focal length, which doubles the Sun's size to 9.2 mm. Focal lengths in excess of 1000 mm usually fall within the realm of amateur telescopes. If full disk photography of partial phases on 35 mm format is planned, the focal length of the optics must not exceed 2600 mm. However, since most cameras don't show the full extent of the image in their viewfinders, a more practical limit is about 2000 mm. Longer focal lengths permit photography of only a magnified portion of the Sun's disk. In order to photograph the Sun's corona during totality, the focal length should be no longer than 1500 mm to 1800 mm (for 35 mm equipment). However, a focal length of 1000 mm requires less critical framing and can capture some of the longer coronal streamers. For any particular focal length, the diameter of the Sun's image is approximately equal to the focal length divided by 109 (Table 25).

Image Scale of a Total Solar Eclipse at Various Focal Lengths

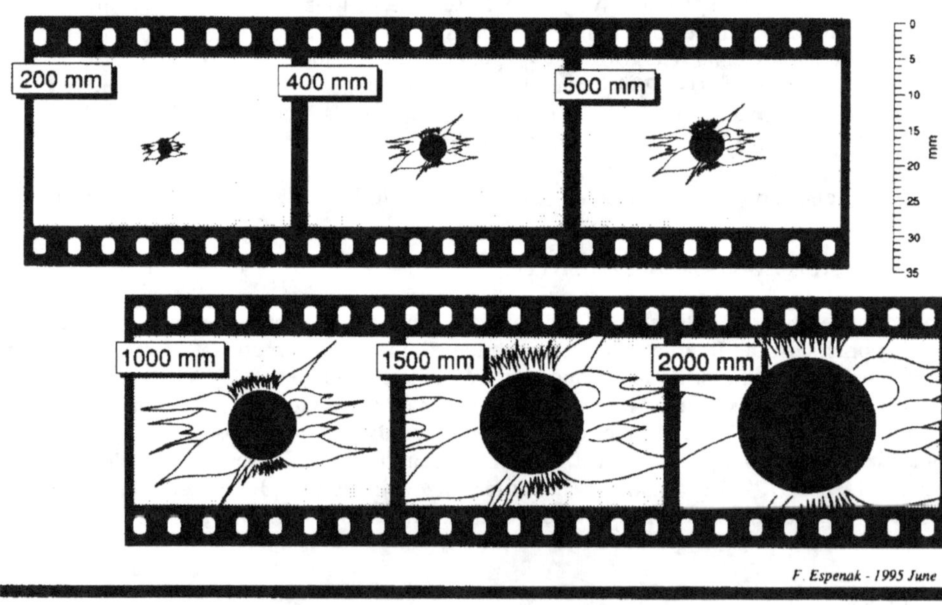

A mylar or glass solar filter must be used on the lens throughout the partial phases for both photography and safe viewing. Such filters are most easily obtained through manufacturers and dealers listed in *Sky & Telescope* and *Astronomy* magazines (see: Sources for Solar Filters). These filters typically attenuate the Sun's visible and infrared energy by a factor of 100,000. However, the actual filter factor and choice of ISO film speed will play critical roles in determining the correct photographic exposure. A low to medium speed film is recommended (ISO 50 to 100) since the Sun gives off abundant light. The easiest method for determining the correct exposure is accomplished by running a calibration test on the uneclipsed Sun. Shoot a roll of film of the mid-day Sun at a fixed aperture (f/8 to f/16) using every shutter speed between 1/1000 and 1/4 second. After the film is developed, note the best exposures and use them to photograph all the partial phases. The Sun's surface brightness remains constant throughout the eclipse, so no exposure compensation is needed except for the crescent phases which require two more stops due to solar limb darkening. Bracketing by several stops is also necessary if haze or clouds interfere on eclipse day.

Certainly the most spectacular and awe inspiring phase of the eclipse is totality. For a few brief minutes or seconds, the Sun's pearly white corona, red prominences and chromosphere are visible. The great challenge is to obtain a set of photographs which captures some aspect of these fleeting phenomena. The most important point to remember is that during the total phase, all solar filters *must be removed*! The corona has a surface brightness a million times fainter than the photosphere, so photographs of the corona are made without a filter. Furthermore, it is completely safe to view the totally eclipsed Sun directly with the naked eye. No filters are needed and they will only hinder your view. The average brightness of the corona varies inversely with the distance from the Sun's limb. The inner corona is far brighter than the outer corona. Thus, no single exposure can capture its full dynamic range. The best strategy is to choose one aperture or f/number and bracket the exposures over a range of shutter speeds (i.e., 1/1000 down to 1 second). Rehearsing this sequence is highly recommended since great excitement accompanies totality and there is little time to think.

Exposure times for various combinations of film speeds (ISO), apertures (f/number) and solar features (chromosphere, prominences, inner, middle and outer corona) are summarized in Table 26. The table was developed from eclipse photographs made by Espenak as well as from photographs published in *Sky and Telescope*. To use the table, first select the ISO film speed in the upper left column. Next, move to the right to the desired aperture or f/number for the chosen ISO. The shutter speeds in that column may be used as starting points for photographing various features and phenomena tabulated in the 'Subject' column at the far left. For example, to photograph prominences using ISO 100 at f/11, the table recommends an exposure of 1/500. Alternatively, you can calculate the recommended shutter speed using the 'Q' factors tabulated along with the exposure formula at the bottom of Table 26. Keep in mind that these exposures are based on a clear sky and a corona of average brightness. You should bracket your exposures one or more stops to take into account the actual sky conditions and the variable nature of these phenomena.

Another interesting way to photograph the eclipse is to record its phases all on one frame. This is accomplished by using a stationary camera capable of making multiple exposures (check the camera instruction manual). Since the Sun moves through the sky at the rate of 15 degrees per hour, it slowly drifts through the field of view of any camera equipped with a normal focal length lens (i.e., 35 to 50 mm). If the camera is oriented so that the Sun drifts along the frame's diagonal, it will take over three hours for the Sun to cross the field of a 50 mm lens. The proper camera orientation can be determined through trial and error several days before the eclipse. This will also insure that no trees or buildings obscure the view during the eclipse. The Sun should be positioned along the eastern (left in the northern hemisphere) edge or corner of the viewfinder shortly before the eclipse begins. Exposures are then made throughout the eclipse at ~five minute intervals. The camera must remain perfectly rigid during this period and may be clamped to a wall or post since tripods are easily bumped. If you're in the path of totality, remove the solar filter during the total phase and take a long exposure (~1 second) in order to record the corona in your sequence. The final photograph will consist of a string of Suns, each showing a different phase of the eclipse.

Finally, an eclipse effect that is easily captured with point-and-shoot or automatic cameras should not be overlooked. Use a kitchen sieve or colander and allow its shadow to fall on a piece of white card-board placed several feet away. The holes in the utensil act like pinhole cameras and each one projects its own image of the Sun. The effect can also be duplicated by forming a small aperture with one's hands and watching the ground below. The pinhole camera effect becomes more prominent with increasing eclipse magnitude. Virtually any camera can be used to photograph the phenomenon, but automatic cameras must have their flashes turned off since this would otherwise obliterate the pinhole images.

For those who choose to photograph this eclipse from one of the many cruise ships in the path, some special comments are in order. Shipboard photography puts certain limits on the focal length and shutter speeds that can be used. It's difficult to make specific recommendations since it depends on the stability of the ship as well as wave heights encountered on eclipse day. Certainly telescopes with focal lengths of 1000 mm or more can be ruled out since their small fields of view would require the ship to remain virtually motionless during totality, and this is rather unlikely even given calm seas. A 500 mm lens might be a safe upper limit in focal length. Film choice could be determined on eclipse day by viewing the Sun through the camera lens and noting the image motion due to the rolling sea. If it's a calm day, you might try an ISO 100 film. For rougher seas, ISO 400 or more might be a better choice. Shutter speeds as slow as 1/8 or 1/4 may be tried if the conditions warrant it. Otherwise, stick with a 1/15 or 1/30 and shoot a sequence through 1/1000 second. It might be good insurance to bring a wider 200 mm lens just in case the seas are rougher than expected. As worst case scenario, Espenak photographed the 1984 total eclipse aboard a 95 foot yacht in seas of 3 feet. He had to hold on with one hand and point his 350 mm lens with the other! Even at that short focal length, it was difficult to keep the Sun in the field. However, any large cruise ship will offer a far more stable platform than this.

For more information on eclipse photography, observations and eye safety, see FURTHER READING in the BIBLIOGRAPHY.

Sky At Totality

The total phase of an eclipse is accompanied by the onset of a rapidly darkening sky whose appearance resembles evening twilight about 30 to 40 minutes after sunset. The effect presents an excellent opportunity to view planets and bright stars in the daytime sky. Aside from the sheer novelty of it, such observations are useful in gauging the apparent sky brightness and transparency during totality. The Sun is in Pisces and all five naked eye planets as well as a number of bright stars will be above the horizon for observers within the umbral path. Figure 25 depicts the appearance of the sky during totality as seen from the center line at 11:00 UT. This corresponds to western Romania near the point of greatest eclipse.

Mercury (mv=+0.7) and Venus (mv=–3.5) are located 18° west and 15° east of the Sun, respectively, and both will be easily visible during totality. Venus is two months past inferior conjunction while Mercury is one month shy of superior conjunction. As the brightest planet in the sky, Venus can actually be observed in broad daylight provided that the sky is cloud free and of high transparency (i.e., no dust or particulates). Look for the planet during the partial phases by first covering the crescent Sun with an extended hand. Venus will be shining so brightly, it will be impossible to miss during totality. Mercury will prove much more challenging, but not too difficult if the sky transparency is good. Under the right circumstances, it should be possible to view all five classical planets, the Moon and the Sun (or at least its corona) as one's eyes sweep across the darkened sky during totality.

A number of the brightest winter/spring stars may also be visible during totality. Regulus (mv=+1.35) is 10° east of the Sun while Castor (mv=+1.94) and Pollux (m_V=+1.14) stand 31° and 28° to the northwest. Procyon (m_V=+0.38) and Sirius (m_V=-1.46) are located 30° and 52° to the southwest, respectively. Betelgeuse (mv=+0.5v) and Rigel (m_V=+0.12) are low in the southwest at 26° and 38°, while Aldebaran (m_V=+0.85) is 20° above the western horizon. Capella (m_V=+0.08) lies 63° to the northwest. Finally, Arcturus (m_V=+1.94) lies due east 30° above the horizon.

The following ephemeris [using Bretagnon and Simon, 1986] gives the positions of the naked eye planets during the eclipse. *Delta* is the distance of the planet from Earth (A.U.'s), *V* is the apparent visual magnitude of the planet, and *Elong* gives the solar elongation or angle between the Sun and planet.

Ephemeris: 1999 Aug 11 11:00:00 UT — Equinox = Mean Date

Planet	RA	Dec	Delta	V	Size "	Phase	Elong °
Sun	09h23m08s	+15°19'42"	1.01358	-26.7	1893.6	-	-
Mercury	08h07m35s	+18°08'56"	0.82062	0.7	8.2	0.30	18.3W
Venus	10h06m46s	+04°18'35"	0.30047	-3.5	55.5	0.04	15.4E
Mars	14h55m32s	-18°28'03"	1.06814	0.3	8.8	0.86	88.5E
Jupiter	02h11m31s	+11°47'36"	4.61411	-2.1	42.7	0.99	103.7W
Saturn	03h00m28s	+14°34'31"	9.13383	0.1	18.2	1.00	91.5W

For sky maps from other locations along the path of totality, see the special web site for the total solar eclipse of 1999: *http://planets.gsfc.nasa.gov/eclipse/TSE1999/TSE1999.html*

Contact Timings from the Path Limits

Precise timings of beading phenomena made near the northern and southern limits of the umbral path (i.e., the graze zones), are of value in determining the diameter of the Sun relative to the Moon at the time of the eclipse. Such measurements are essential to an ongoing project to monitor changes in the solar diameter. Due to the conspicuous nature of the eclipse phenomena and their strong dependence on geographical location, scientifically useful observations can be made with relatively modest equipment. A small telescope, short wave radio and portable camcorder are usually used to make such measurements. Time signals are broadcast via short wave stations WWV and CHU, and are recorded simultaneously as the eclipse is videotaped. If a video camera is not available, a tape recorder can be used to record time signals with verbal timings of each event. Inexperienced observers are cautioned to use great care in making such observations. The safest timing technique consists of observing a projection of the Sun rather than directly imaging the solar disk itself. The observer's geodetic coordinates are required and can be measured from USGS or other large scale maps. If a map is unavailable, then a detailed description of the observing site should be included which provides information such as distance and directions of the nearest towns/settlements, nearby landmarks, identifiable buildings and road intersections. The method of contact timing should be described in detail, along with an estimate of the error. The precisional requirements of these observations are ±0.5 seconds in time, 1" (~30 meters) in latitude and longitude, and ±20 meters (~60 feet) in elevation. Although GPS's (Global Positioning Satellite receivers) are commercially available (~$150 US), their positional accuracy of ±100 meters is about three times larger than the minimum accuracy required by grazing eclipse measurements. GPS receivers are also a useful source for accurate UT. The International Occultation Timing Association (IOTA) coordinates observers world-wide during each eclipse. For more information, contact:

Dr. David W. Dunham, IOTA
7006 Megan Lane
Greenbelt, MD 20770-3012, USA

E-mail: David_Dunham@jhuapl.edu
Phone: (301) 474-4722

Send reports containing graze observations, eclipse contact and Baily's bead timings, including those made anywhere near or in the path of totality or annularity to:

Dr. Alan D. Fiala
Orbital Mechanics Dept.
U. S. Naval Observatory
3450 Massachusetts Ave., NW
Washington, DC 20392-5420, USA

Plotting the Path on Maps

If high resolution maps of the umbral path are needed, the coordinates listed in Tables 7 and 8 are conveniently provided in longitude increments of 1° and 30' respectively to assist plotting by hand. The path coordinates in Table 3 define a line of maximum eclipse at five minute increments in Universal Time. If observations are to be made near the limits, then the grazing eclipse zones tabulated in Table 8 should be used. A higher resolution table of graze zone coordinates at longitude increments of 7.5' is available via a special web site for the 1999 total eclipse *(http://planets.gsfc.nasa.gov/eclipse/TSE1999/TSE1999.html).* Global Navigation Charts (1:5,000,000), Operational Navigation Charts (scale 1:1,000,000) and Tactical Pilotage Charts (1:500,000) of many parts of the world are published by the National Imagery and Mapping Agency (formerly known as Defense Mapping Agency). Sales and distribution of these maps are through the National Ocean Service (NOS). For specific information about map availability, purchase prices, and ordering instructions, contact the NOS at:

NOAA Distribution Division, N/ACC3
National Ocean Service
Riverdale, MD 20737-1199, USA

phone: 301-436-8301
FAX: 301-436-6829

It is also advisable to check the telephone directory for any map specialty stores in your city or metropolitan area. They often have large inventories of many maps available for immediate delivery.

ONC (Operational Navigation Charts) series maps have a larger scale (1:1,000,000) than GNC's appearing in this publication. However, their use here would serve to increase an already record size eclipse bulletin. Instead, we offer a list of ONC maps for plotting the path using data from tables 7 and 8. In particular, the path of totality crosses the following ONC charts:

ONC E-1	England
ONC E-2	France, Germany, Austria
ONC F-2	Austria, Hungary
ONC F-3	Romania, Turkey
ONC G-4	Turkey, Syria, Iraq
ONC G-6,H-7	Iran
ONC H-8	Pakistan, India
ONC J8, J-9	India

IAU Working Group on EclipseS

Professional scientists are asked to send descriptions of their eclipse plans to the Working Group on Eclipses of the International Astronomical Union, so that they can keep a list of observations planned. Send such descriptions, even in preliminary form, to:

International Astronomical Union/Working Group on Eclipses
Prof. Jay M. Pasachoff, Chair
Williams College–Hopkins Observatory
Williamstown, MA 01267, USA
email: jay.m.pasachoff@williams.edu
FAX: (413) 597-3200

The members of the Working Group on Eclipses of Commissions 10 and 12 of the International Astronomical Union are: Jay M. Pasachoff (USA), Chair; F. Clette (Belgium), F. Espenak (USA); Iraida Kim (Russia); V. Rusin (Slovakia); Jagdev Singh (India); M. Stavinschi (Romania); Yoshinori Suematsu (Japan); consultant: J. Anderson (Canada).

NATO Workshop Proceedings for the 1999 Total Solar Eclipse

In 1996 June 1-5, a special NATO Advanced Research Workshop for "Theoretical and Observational Problems Related to Solar Eclipses" was held in Sinaia, Romania. For the first time in the history of the eclipse observations, observers and theorists were brought together to present and discuss their projects for a future eclipse (1999). Scientific sessions during the meeting covered the following areas:

Principal scientific results from the past eclipse observation.
Small and large scale theoretical models of coronal structures.
Low temperature structures in coronal environment.
Specific problems of solar eclipse observations.
Instrumental improvement for future observations.
Tasks for Total Solar Eclipse of 11 August 1999.
Public education at eclipses and eye safety.

The proceedings from the meeting will be published in 1997 June by Kluwer Academic Publishing (Netherlands) as part of the *NATO ASI Series* (Z. Mouradian and M. Stavinschi, eds.). For ordering information, please contact:

Dr. Zadig Mouradian
Observatoire de Paris-Meudon
DASOP
92195 Meudon Principal
FRANCE
email: mouradian@obspm.fr
FAX: +33.1.4507.7959

ROMANIAN PREPARATIONS FOR 1999 ECLIPSE

Romanian astronomers have established the International Association ECLIPSA'99 for the purpose of assisting both the scientific community and the general public. In addition to carrying out a series of scientific eclipse observations, ECLIPSA'99 will also play an important role in public education so that everyone can enjoy this extraordinary astronomical event.

We plan to set up a new telescope outside the Capital, to complete the observation and data bases of the three observatories of the Astronomical Institute of the Romanian Academy, as well as to built a great Planetarium at Bucharest Observatory, in the immediate vicinity of the Park "Charles the 1st".

Through a system of scholarships and awards, ECLIPSA'99 aims to specially train the staff necessary in the eclipse observation. Naturally, we will not leave out the amateur astronomers in view of the important contribution they have brought to the development of astronomy. Throughout the preparations, ECLIPSA'99 will carry out an ample program of national and international conferences and symposia, the publication of specialized and advertising materials, as well as of its own journal "Eclipsa"; it will also conduct an ample publicity campaign both at home and abroad.

To accomplish its aims ECLIPSA'99 is carrying on collaborations with similar institutions at home and abroad, with specialists in related fields, as well as with agencies and other institutions in the fields of culture, tourism, transport, trade, etc.. The funds of ECLIPSA'99 come from subscriptions, as well as from legacies and donations from home and abroad.

ECLIPSA'99 will not stop its activity after the eclipse. At that time, it will become the International Association ASTRONOMIA 21, whose goal will be to keep alive the interest in this old and, at the same time, modern science. For more information please contact:

Dr. Magdalena Stavinschi
Astr. Inst. of the Romanian Academy
Str. Cutitul de Argint 5
RO-75212 Bucharest
ROMANIA

email: magda@roastro.astro.ro
Phone: +40.1.336 36 87
FAX: +40.1.337 33 89

JOSO WORKING GROUP FOR 1999 ECLIPSE

In October 1995, the Joint Organization for Solar Observations (JOSO), a European consortium of observing solar physicists, created a new working group dedicated to the preparation of the 1999 eclipse. This Working Group (WG7) will prompt collaborations between scientific teams coming from European countries and other parts of the world to observe the eclipse, and the local scientific and academic organizations in the path of totality, by gathering and distributing information about existing resources and requirements for the practical organisation of scientific expeditions. Another WG7 project is to disseminate to the general public basic but reliable information about the 1999 event and safe eclipse viewing .

All groups who are planning to set up a scientific eclipse program are invited to join this new community. For more information, contact:

Dr. Frederic Clette - JOSO WG 7
Observatoire Royal de Belgique
Avenue Circulaire, 3
B-1180 Bruxelles
BELGIUM

email: fred@oma.be
FAX: +32.2.373.02.24

JOSO Web site : http://joso.oat.ts.astro.it/

ECLIPSE DATA ON INTERNET

NASA ECLIPSE BULLETINS ON INTERNET

To make the NASA solar eclipse bulletins accessible to as large an audience as possible, these publications are also available via the Internet. This was made possible through the efforts and expertise of Dr. Joe Gurman (GSFC/Solar Physics Branch). All future eclipse bulletins will be available via Internet.

NASA eclipse bulletins can be read or downloaded via the World-Wide Web using a Web browser (e.g.: Netscape, Microsoft Explorer, etc.) from the GSFC SDAC (Solar Data Analysis Center) Eclipse Information home page, or from top-level URL's for the currently available eclipse bulletins themselves:

http://umbra.nascom.nasa.gov/eclipse/	(SDAC Eclipse Information)
http://umbra.nascom.nasa.gov/eclipse/941103/rp.html	(1994 Nov 3)
http://umbra.nascom.nasa.gov/eclipse/951024/rp.html	(1995 Oct 24)
http://umbra.nascom.nasa.gov/eclipse/970309/rp.html	(1997 Mar 9)
http://umbra.nascom.nasa.gov/eclipse/980226/rp.html	(1998 Feb 26)
http://umbra.nascom.nasa.gov/eclipse/990811/rp.html	(1999 Aug 11)

The original Microsoft Word text files and PICT figures (Macintosh format) are also available via anonymous ftp. They are stored as BinHex-encoded, StuffIt-compressed Mac folders with .hqx suffixes. For PC's, the text is available in a zip-compressed format in files with the .zip suffix. There are three sub directories for figures (GIF format), maps (JPEG format), and tables (html tables, easily readable as plain text). For example, NASA RP 1344 (Total Solar Eclipse of 1995 October 24 [=951024]) has a directory for these files is as follows:

file://umbra.nascom.nasa.gov/pub/eclipse/951024/RP1344text.hqx	
file://umbra.nascom.nasa.gov/pub/eclipse/951024/RP1344PICTs.hqx	
file://umbra.nascom.nasa.gov/pub/eclipse/951024/ec951024.zip	
file://umbra.nascom.nasa.gov/pub/eclipse/951024/figures	(directory with GIF's)
file://umbra.nascom.nasa.gov/pub/eclipse/951024/maps	(directory with JPEG's)
file://umbra.nascom.nasa.gov/pub/eclipse/951024/tables	(directory with html's)

Other eclipse bulletins have a similar directory format.

Current plans call for making all future NASA eclipse bulletins available over the Internet, at or before publication of each. The primary goal is to make the bulletins available to as large an audience as possible. Thus, some figures or maps may not be at their optimum resolution or format. Comments and suggestions are actively solicited to fix problems and improve on compatibility and formats.

FUTURE ECLIPSE PATHS ON INTERNET

Presently, the NASA eclipse bulletins are published 24 to 36 months before each eclipse. However, there have been a growing number of requests for eclipse path data with an even greater lead time. To accommodate the demand, predictions have been generated for all central solar eclipses from 1995 through 2005 using the JPL DE/LE 200 ephemerides. All predictions use the Moon's center of mass; no corrections have been made to adjust for center of figure. The value used for the Moon's mean radius is k=0.272281. The umbral path characteristics have been predicted at 2 minute intervals of time compared to the 6 minute interval used in *Fifty Year Canon of Solar Eclipses: 1986-2035* [Espenak, 1987]. This should provide enough detail for making preliminary plots of the path on larger scale maps. Note that positive latitudes are north and positive longitudes are west. A list of currently available eclipse paths includes:

1998 February 26	–	Total Solar Eclipse
1998 August 22	–	Annular Solar Eclipse
1999 February 16	–	Annular Solar Eclipse
1999 August 11	–	Total Solar Eclipse
2001 June 21	–	Total Solar Eclipse
2001 December 14	–	Annular Solar Eclipse

2002 June 10	–	Annular Solar Eclipse
2002 December 04	–	Total Solar Eclipse
2003 May 31	–	Annular Solar Eclipse
2003 November 23	–	Total Solar Eclipse
2005 April 08	–	Annular/Total Solar Eclipse
2005 October 03	–	Annular Solar Eclipse

URL: http://umbra.nascom.nasa.gov/eclipse/predictions/year-month-day.html

The tables can be accessed through the SDAC Eclipse Information home page, or directly from the above URL For example, the eclipse path of 1999 August 11 would use the above address with the string "year-month-day" replaced by "1999-august-11". Send comments, corrections, suggestions or requests for more detailed 'ftp' instructions, to Fred Espenak via e-mail (espenak@lepvax.gsfc.nasa.gov). For Internet related problems, please contact Joe Gurman (gurman@uvsp.nascom.nasa.gov).

Downloading Bulletins and Path Tables Via Anonymous FTP

The eclipse bulletins and path tables are also available via anonymous ftp for sites which do not have access to the World Wide Web. A user first ftp's to umbra.nascom.nasa.gov (150.144.30.134), using the username "anonymous" and password "<username>@<host>". Note that the password is your e-mail address where <username> is your name and <host> is the fully qualified Internet address of your machine (e.g.- gurman@uvsp.nascom.nasa.gov). Next, you change directory with the command "cd pub/eclipse".

There are five directories 941103, 951024, 970309, 980226, and 990811; one for each of the last five eclipse bulletins (1318, 1344, 1369, 1383, and 1398 respectively). In each, there is a flat ASCII README file and two .hqx files: RPnnnntext.hqx and RPnnnnPICTS.hqx, where "nnnn" is the Reference Publication number. All .hqx files are BinHex-encoded (ASCII), StuffIt-compressed files for the Macintosh. There's also one .zip file: ecyymmdd.zip, where "yymmdd" is the date of the eclipse. This is a zip-compressed and encoded file for PC's. There are also three subdirectories, figures, maps, and tables, with (respectively), the GIF figures, the JPEG GNC charts, and the html tables (easily readable as plain text). For example, the total solar eclipse of 970309 (= 1997 Mar 9) and published as NASA RP 1369 has a directory for these files is as follows:

file://umbra.nascom.nasa.gov/pub/eclipse/970309/README
file://umbra.nascom.nasa.gov/pub/eclipse/970309/RP1369text.hqx
file://umbra.nascom.nasa.gov/pub/eclipse/970309/RP1369PICTs.hqx
file://umbra.nascom.nasa.gov/pub/eclipse/970309/ec970309.zip
file://umbra.nascom.nasa.gov/pub/eclipse/970309/figures (directory with GIF's)
file://umbra.nascom.nasa.gov/pub/eclipse/970309/maps (directory with JPEG's)
file://umbra.nascom.nasa.gov/pub/eclipse/970309/tables (directory with html's)

Directories for analogous files for other solar eclipses are arranged similarly.

The html files should be downloaded in ASCII mode and the other files in binary (IMAGE) mode. If you are not using a Web viewer to access the ftp documents, you must first type either "ascii" or "binary" to download an ASCII or a binary file, respectively. You then download the file using the ftp protocol for your particular machine.

Special Web Site for 1999 Solar Eclipse

A special web site has been set up to supplement this bulletin with additional predictions, tables and data for the total solar eclipse of 1999. Some of the data posted there include an expanded version of Table 8 (Mapping Coordinates for the Zones of Grazing Eclipse), and local circumstance tables with many more cities as well as for astronomical observatories. Also featured will be higher resolution maps of selected sections of the path of totality and limb profile figures for a range of locatiions/times along the path. The URL of this special site is:

http://planets.gsfc.nasa.gov/eclipse/TSE1999/TSE1999.html

Total Solar Eclipse of 2001 June 21

The next total eclipse of the Sun is the first one of the twenty-first century. The path of the Moon's umbral shadow begins in the South Atlantic, off the east coast of Uruguay and continues across the Atlantic where it reaches the west coast of Africa. The shadow enters Angola in the early afternoon with a center line duration of 4 1/2 minutes (Figure 26). Traveling eastward, the path sweeps through Zambia, Zimbabwe and Mozambique. At that point, the central duration drops to three minutes with the late afternoon Sun 23° above the horizon. Swiftly crossing the Mozambique Channel, the path intercepts southern Madagascar where the central duration lasts 2 1/2 minutes with a Sun altitude of 11°. The path ends two minutes later in the Indian Ocean.

Complete details will be published in the next NASA bulletin scheduled for Fall-Winter 1997.

Predictions for Eclipse Experiments

This publication has attempted to provide comprehensive information on the 1999 total solar eclipse to both the professional and amateur/lay communities. However, certain investigations and eclipse experiments may require additional information which lies beyond the scope of this work. We invite the international professional community to contact us for assistance with any aspect of eclipse prediction including predictions for locations not included in this publication, or for more detailed predictions for a specific location (e.g.: lunar limb profile and limb corrected contact times for an observing site).

This service is offered for the 1998 eclipse as well as for previous eclipses in which analysis is still in progress. To discuss your needs and requirements, please contact Fred Espenak (espenak@lepvax.gsfc.nasa.gov).

Algorithms, Ephemerides and Parameters

Algorithms for the eclipse predictions were developed by Espenak primarily from the *Explanatory Supplement* [1974] with additional algorithms from Meeus, Grosjean and Vanderleen [1966] and Meeus [1982]. The solar and lunar ephemerides were generated from the JPL DE200 and LE200, respectively. All eclipse calculations were made using a value for the Moon's radius of k=0.2722810 for umbral contacts, and k=0.2725076 (adopted IAU value) for penumbral contacts. Center of mass coordinates were used except where noted. Extrapolating from 1996 to 1998, a value for ΔT of 64.6 seconds was used to convert the predictions from Terrestrial Dynamical Time to Universal Time. The international convention of presenting date and time in descending order has been used throughout the bulletin (i.e., *year, month, day, hour, minute, second*).

The primary source for geographic coordinates used in the local circumstances tables is *The New International Atlas* (Rand McNally, 1991). Elevations for major cities were taken from *Climates of the World* (U. S. Dept. of Commerce, 1972).

All eclipse predictions presented in this publication were generated on a Macintosh PowerPC 8500 computer. Word processing and page layout for the publication were done using Microsoft Word v5.1. Figures were annotated with Claris MacDraw Pro 1.5. Meteorological diagrams were prepared using Corel Draw 5.0 and converted to Macintosh compatible files. Finally, the bulletin was printed on a 600 dpi laser printer (Apple LaserWriter Pro).

The names and spellings of countries, cities and other geopolitical regions are not authoritative, nor do they imply any official recognition in status. Corrections to names, geographic coordinates and elevations are actively solicited in order to update the data base for future eclipses. All calculations, diagrams and opinions presented in this publication are those of the authors and they assume full responsibility for their accuracy.

Bibliography

References

Bretagnon, P., and Simon, J. L., *Planetary Programs and Tables from –4000 to +2800*, Willmann-Bell, Richmond, Virginia, 1986.

Climates of the World, U. S. Dept. of Commerce, Washington DC, 1972.

Dunham, J. B, Dunham, D. W. and Warren, W. H., *IOTA Observer's Manual*, (draft copy), 1992.

Espenak, F., *Fifty Year Canon of Solar Eclipses: 1986–2035*, NASA RP-1178, Greenbelt, MD, 1987.

Explanatory Supplement to the Astronomical Ephemeris and the American Ephemeris and Nautical Almanac, Her Majesty's Nautical Almanac Office, London, 1974.

Herald, D., "Correcting Predictions of Solar Eclipse Contact Times for the Effects of Lunar Limb Irregularities," *J. Brit. Ast. Assoc.*, 1983, **93**, 6.

Meeus, J., *Astronomical Formulae for Calculators*, Willmann-Bell, Inc., Richmond, 1982.

Meeus, J., Grosjean, C., and Vanderleen, W., *Canon of Solar Eclipses*, Pergamon Press, New York, 1966.

Morrison, L. V., "Analysis of lunar occultations in the years 1943–1974...," *Astr. J.*, 1979, **75**, 744.

Morrison, L. V., and Appleby, G. M., "Analysis of lunar occultations - III. Systematic corrections to Watts' limb-profiles for the Moon," *Mon. Not. R. Astron. Soc.*, 1981, **196**, 1013.

The New International Atlas, Rand McNally, Chicago/New York/San Francisco, 1991.

van den Bergh, G., *Periodicity and Variation of Solar (and Lunar) Eclipses*, Tjeenk Willink, Haarlem, Netherlands, 1955.

Watts, C. B., "The Marginal Zone of the Moon," *Astron. Papers Amer. Ephem.*, 1963, **17**, 1-951.

Meteorology and Travel

Ayliffe, Rosie, Marc Dubin and John Gawthrop, Turkey, The Rough Guides, London, 1994.

Chandler, T.J. and S. Gregory, eds., *The Climate of the British Isles*, Longman, London and New York, 1976.

Manley, Gordon, "The Climate of the British Isles", in Wallen, C.C. (ed.), *Climates of Northern and Western Europe*, World Survey of Climatology Volume 5, Elsevier Publishing Co., Amsterdam, London, New York, 1970.

Stanley, David, Eastern Europe, Lonely Planet Publications, Hawthorn, Australia, 1995.

Takahashi, K., and H. Arakawa, *Climates of Southern and Western Asia*, World Survey of Climatology, Volume 9. Elsevier Publishing Co., Amsterdam, London, New York, 1981.

Warren, Stephen G., Carole J. Hahn, Julius London, Robert M. Chervin and Roy L. Jenne, *Global Distribution of Total Cloud Cover and Cloud Type Amounts Over Land*, National Center for Atmospheric Research, Boulder, CO., 1986.

Eye Safety

American Conference of Governmental Industrial Hygienists, "Threshold Limit Values for Chemical Substances and Physical Agents and Biological Exposure Indices," ACGIH, Cincinnati, 1996, p.100.

Chou, B. R., "Safe Solar Filters," *Sky & Telescope*, August 1981, p. 119.

Chou, B. R., "Eye safety during solar eclipses - myths and realities," in Z. Madourian & M. Stavinschi (eds.) *Theoretical and Observational Problems Related to Solar Eclipses, Proceedings of a NATO Advanced Research Workshop*. Kluwer Academic Publishers, Dordrecht, 1996 (in press).

Chou, B. R. and Krailo M. D., "Eye injuries in Canada following the total solar eclipse of 26 February 1979," *Can. J. Optometry*, 1981, 43(1):40.

Del Priore, L. V., "Eye damage from a solar eclipse" in M. Littman and K. Willcox, *Totality: Eclipses of the Sun*, University of Hawaii Press, Honolulu, 1991, p. 130.

Marsh, J. C. D., "Observing the Sun in Safety," *J. Brit. Ast. Assoc.*, 1982, **92**, 6.

Penner, R. and McNair, J. N., "Eclipse blindness - Report of an epidemic in the military population of Hawaii," *Am. J. Ophthalmology*, 1966, 61:1452.

Pitts D. G., "Ocular effects of radiant energy," in D. G. Pitts & R. N. Kleinstein (eds.) *Environmental Vision: Interactions of the Eye, Vision and the Environment*, Butterworth-Heinemann, Toronto, 1993, p. 151.

Further Reading

Allen, D., and Allen, C., *Eclipse*, Allen & Unwin, Sydney, 1987.
Astrophotography Basics, Kodak Customer Service Pamphlet P150, Eastman Kodak, Rochester, 1988.
Brewer, B., *Eclipse*, Earth View, Seattle, 1991.
Covington, M., *Astrophotography for the Amateur*, Cambridge University Press, Cambridge, 1988.
Espenak, F., "Total Eclipse of the Sun," *Petersen's PhotoGraphic*, June 1991, p. 32.
Fiala, A. D., DeYoung, J. A., and Lukac, M. R., *Solar Eclipses, 1991–2000*, USNO Circular No. 170, U. S Naval Observatory, Washington, DC, 1986.
Golub, L., and Pasachoff, J. *M., The Solar Corona*, Cambridge University Press, Cambridge, 1997.
Harris, J., and Talcott, R., *Chasing the Shadow*, Kalmbach Pub., Waukesha, 1994.
Littmann, M., and Willcox, K., *Totality, Eclipses of the Sun*, University of Hawaii Press, Honolulu, 1991.
Lowenthal, J., *The Hidden Sun: Solar Eclipses and Astrophotography*, Avon, New York, 1984.
Mucke, H., and Meeus, J., *Canon of Solar Eclipses: –2003 to +2526*, Astronomisches Büro, Vienna, 1983.
North, G., *Advanced Amateur Astronomy*, Edinburgh University Press, 1991.
Oppolzer, T. R. von, *Canon of Eclipses*, Dover Publications, New York, 1962.
Ottewell, G., The *Under-Standing of Eclipses*, Astronomical Workshop, Greenville, NC, 1991.
Pasachoff, J. M., "Solar Eclipses and Public Education," International Astronomical Union Colloquium #162: New Trends in Teaching Astronomy, D. McNally, ed., London 1997, in press.
Willcox, *Totality: Eclipses of the Sun., University of Hawaii Press, Honolulu, 1991, p. 130.Pasachoff, J. M., and Covington, M.*, Cambridge Guide to Eclipse Photography, *Cambridge University Press, Cambridge and New York, 1993.*
Pasachoff, J. M., and Menzel, D. H., *Field Guide to the Stars and Planets*, 3rd edition, Houghton Mifflin, Boston, 1992.
Reynolds, M. D. and Sweetsir, R. A., *Observe Eclipses*, Astronomical League, Washington, DC, 1995.
Sherrod, P. C., *A Complete Manual of Amateur Astronomy*, Prentice-Hall, 1981.
Zirker, J. B., *Total Eclipses of the Sun*, Princeton University Press, Princeton, 1995.

TOTAL SOLAR ECLIPSE OF 1999 AUGUST 11

FIGURES

Total Solar Eclipse of 1999 August 11

FIGURE 1: ORTHOGRAPHIC PROJECTION MAP OF THE ECLIPSE PATH

Geocentric Conjunction = 10:51:12.1 UT J.D. = 2451401.952223
Greatest Eclipse = 11:03:04.4 UT J.D. = 2451401.960468

Eclipse Magnitude = 1.02859 Gamma = 0.50623

Saros Series = 145 Member = 21 of 77

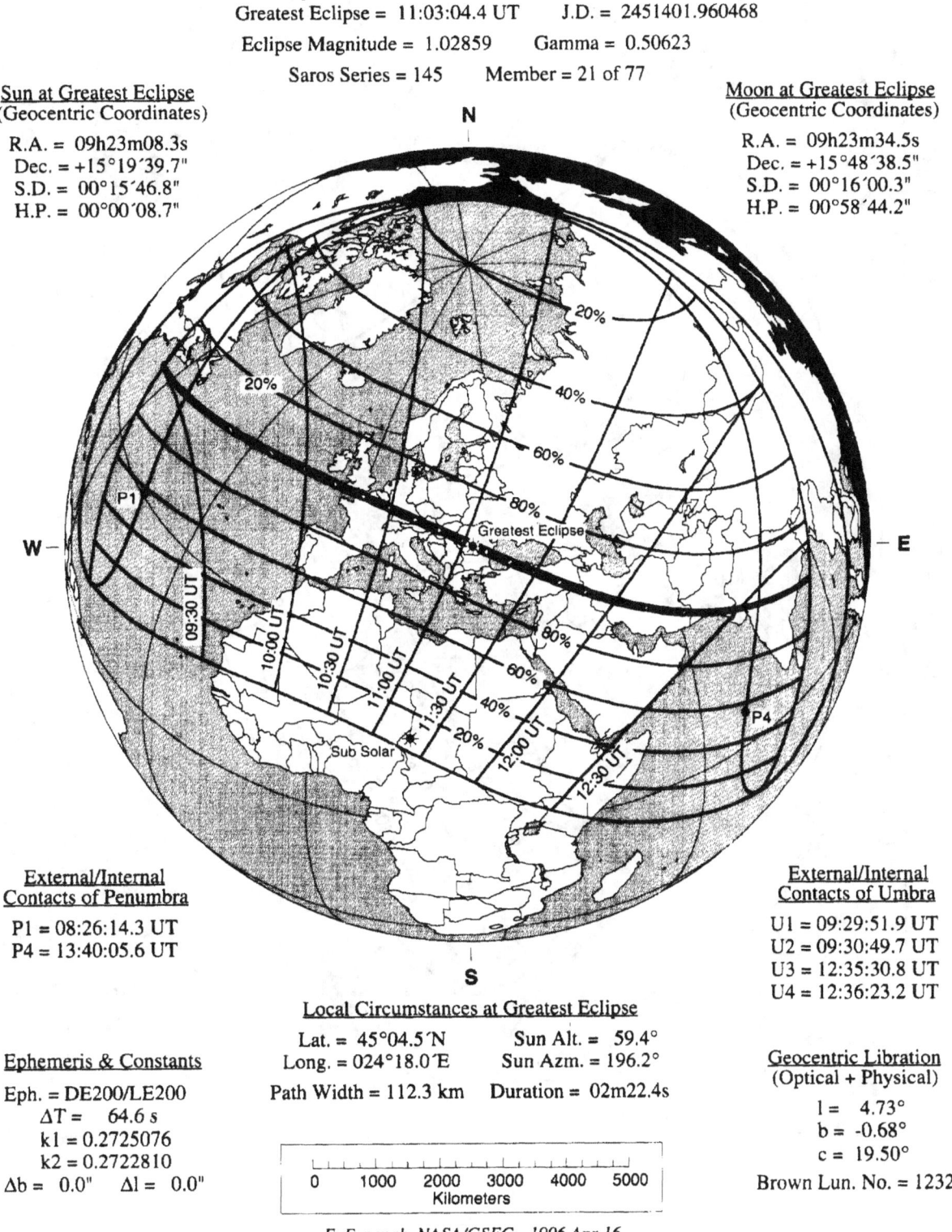

Sun at Greatest Eclipse
(Geocentric Coordinates)

R.A. = 09h23m08.3s
Dec. = +15°19′39.7″
S.D. = 00°15′46.8″
H.P. = 00°00′08.7″

Moon at Greatest Eclipse
(Geocentric Coordinates)

R.A. = 09h23m34.5s
Dec. = +15°48′38.5″
S.D. = 00°16′00.3″
H.P. = 00°58′44.2″

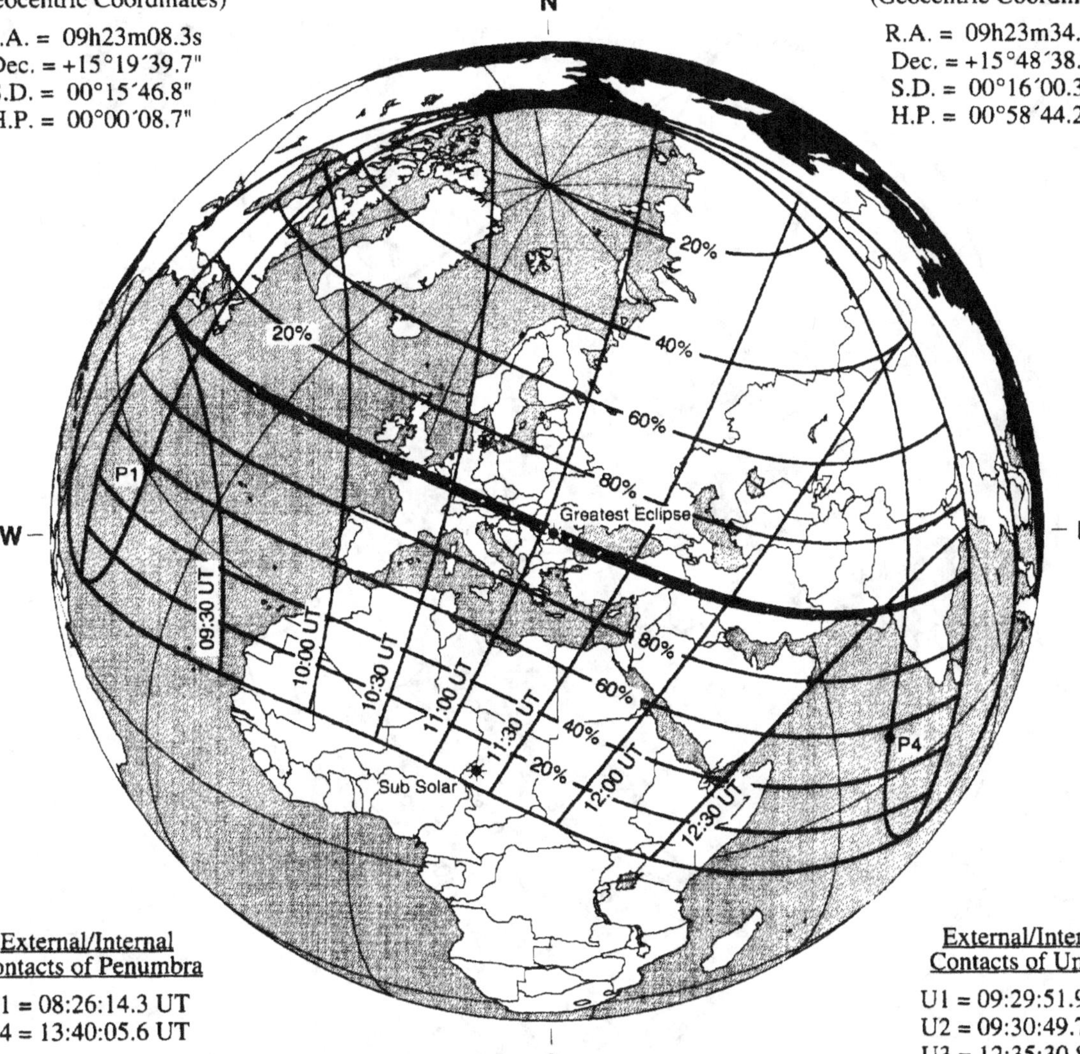

External/Internal Contacts of Penumbra

P1 = 08:26:14.3 UT
P4 = 13:40:05.6 UT

External/Internal Contacts of Umbra

U1 = 09:29:51.9 UT
U2 = 09:30:49.7 UT
U3 = 12:35:30.8 UT
U4 = 12:36:23.2 UT

Local Circumstances at Greatest Eclipse

Lat. = 45°04.5′N Sun Alt. = 59.4°
Long. = 024°18.0′E Sun Azm. = 196.2°
Path Width = 112.3 km Duration = 02m22.4s

Ephemeris & Constants

Eph. = DE200/LE200
ΔT = 64.6 s
k1 = 0.2725076
k2 = 0.2722810
Δb = 0.0″ Δl = 0.0″

Geocentric Libration
(Optical + Physical)

l = 4.73°
b = -0.68°
c = 19.50°

Brown Lun. No. = 1232

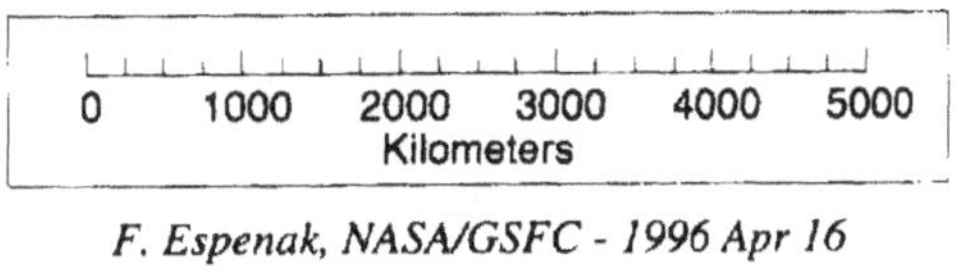

F. Espenak, NASA/GSFC - 1996 Apr 16

Total Solar Eclipse of 1999 August 11

FIGURE 2: STEREOGRAPHIC PROJECTION MAP OF THE ECLIPSE PATH

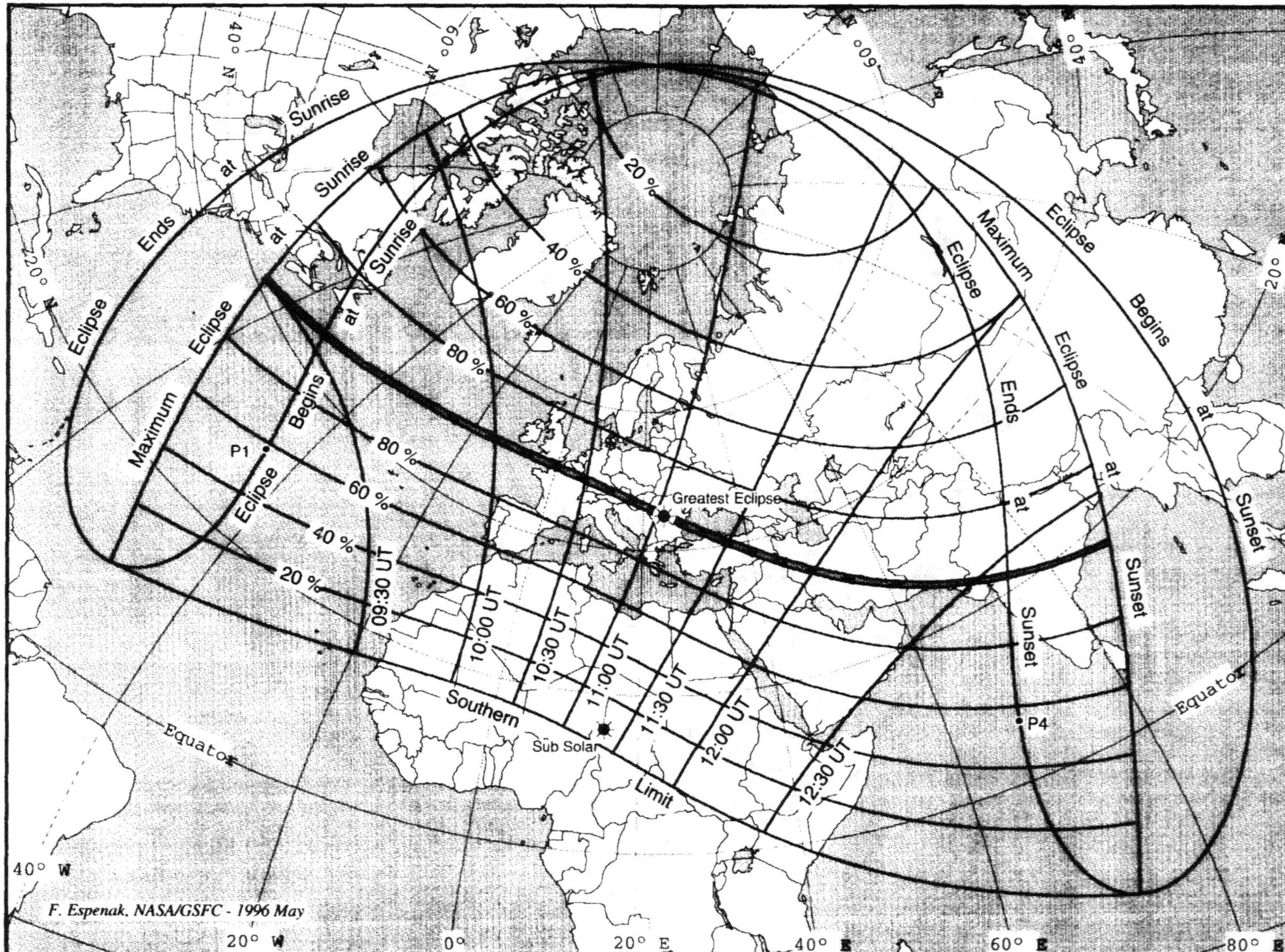

Total Solar Eclipse of 1999 August 11

FIGURE 3: THE ECLIPSE PATH THROUGH EUROPE

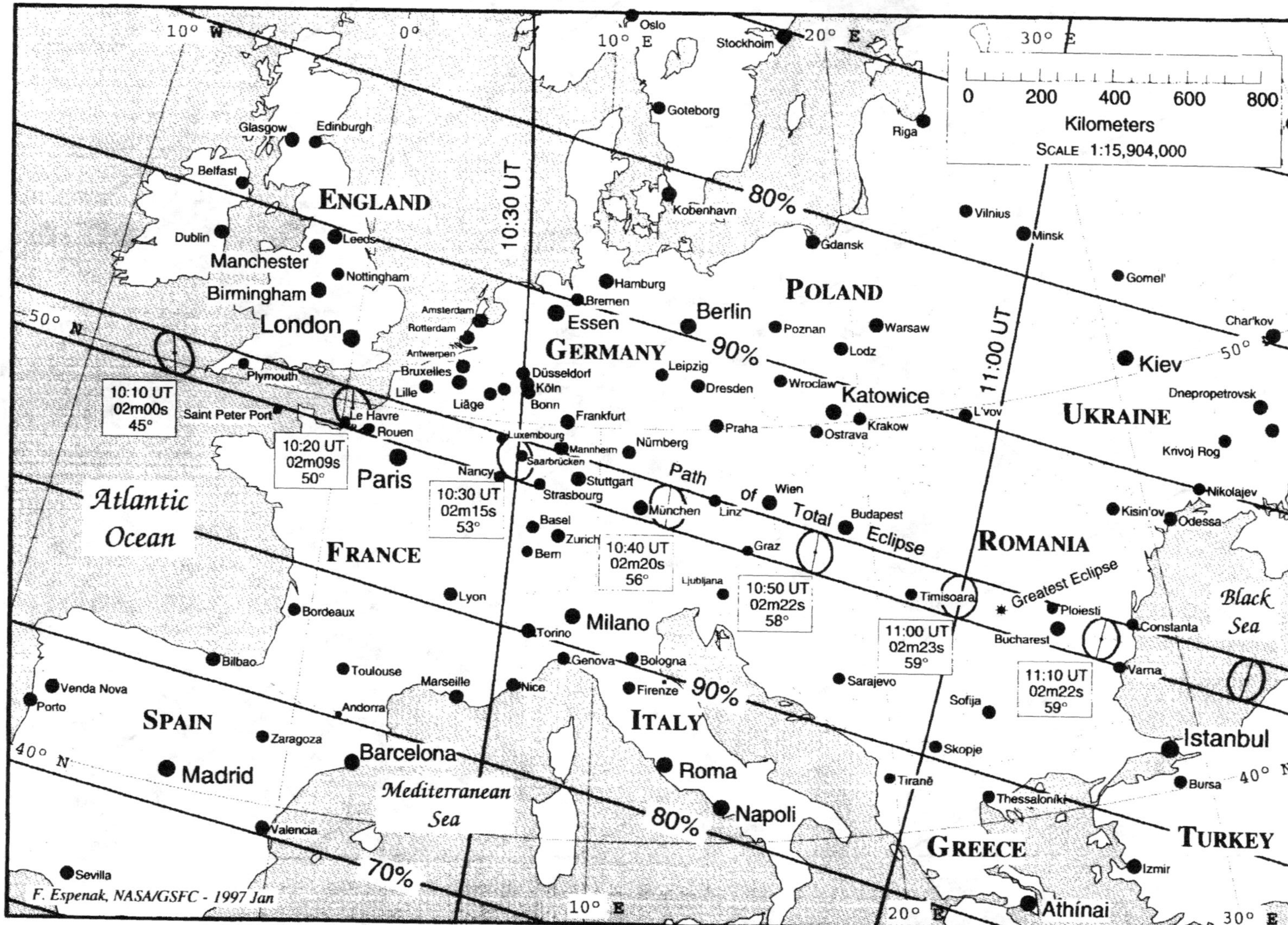

Total Solar Eclipse of 1999 August 11

Figure 4: The Eclipse Path Through the Middle East

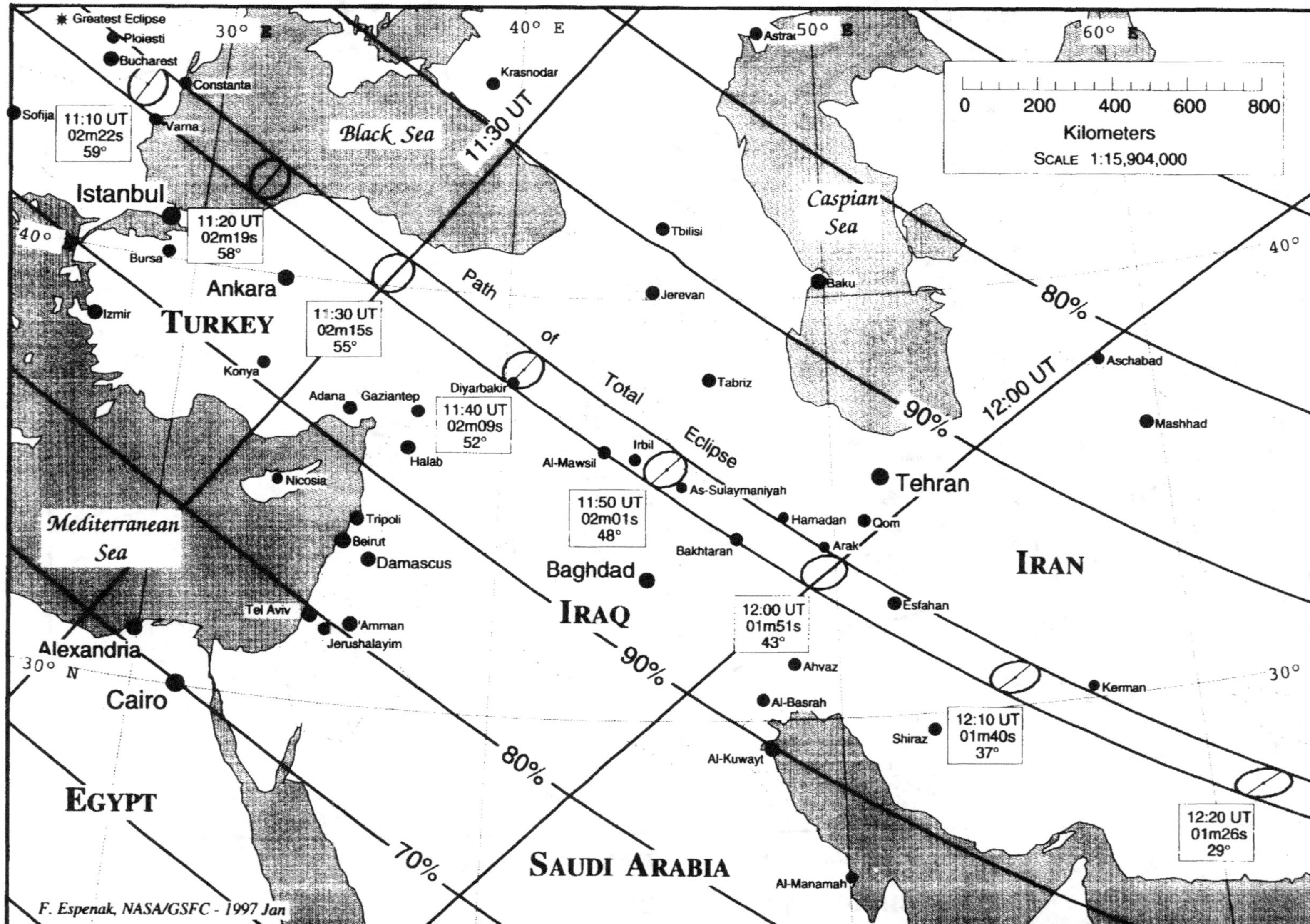

Total Solar Eclipse of 1999 August 11

Figure 5: The Eclipse Path Through South Asia

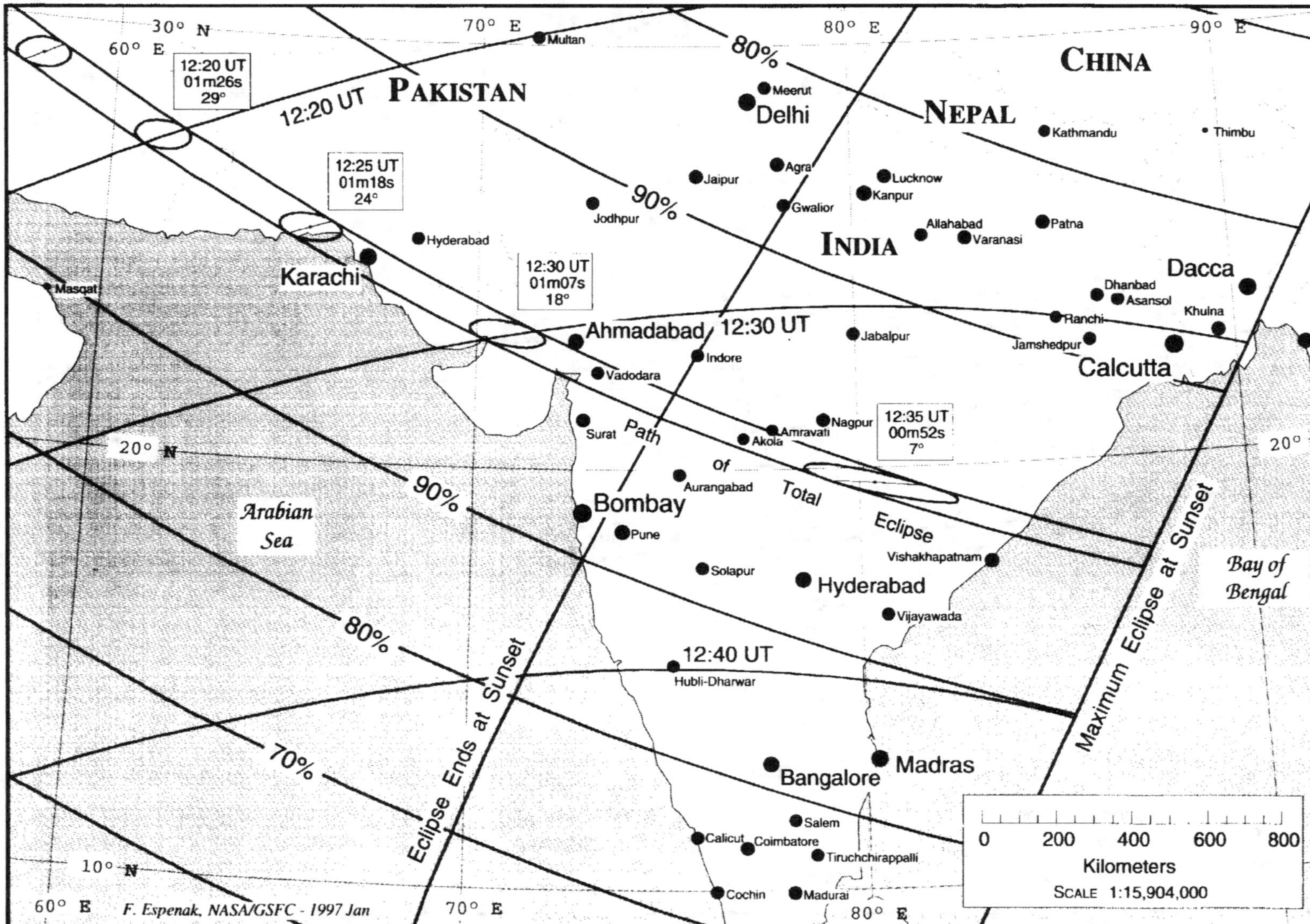

Total Solar Eclipse of 1999 August 11

FIGURE 6: THE ECLIPSE PATH THROUGH ENGLAND AND FRANCE

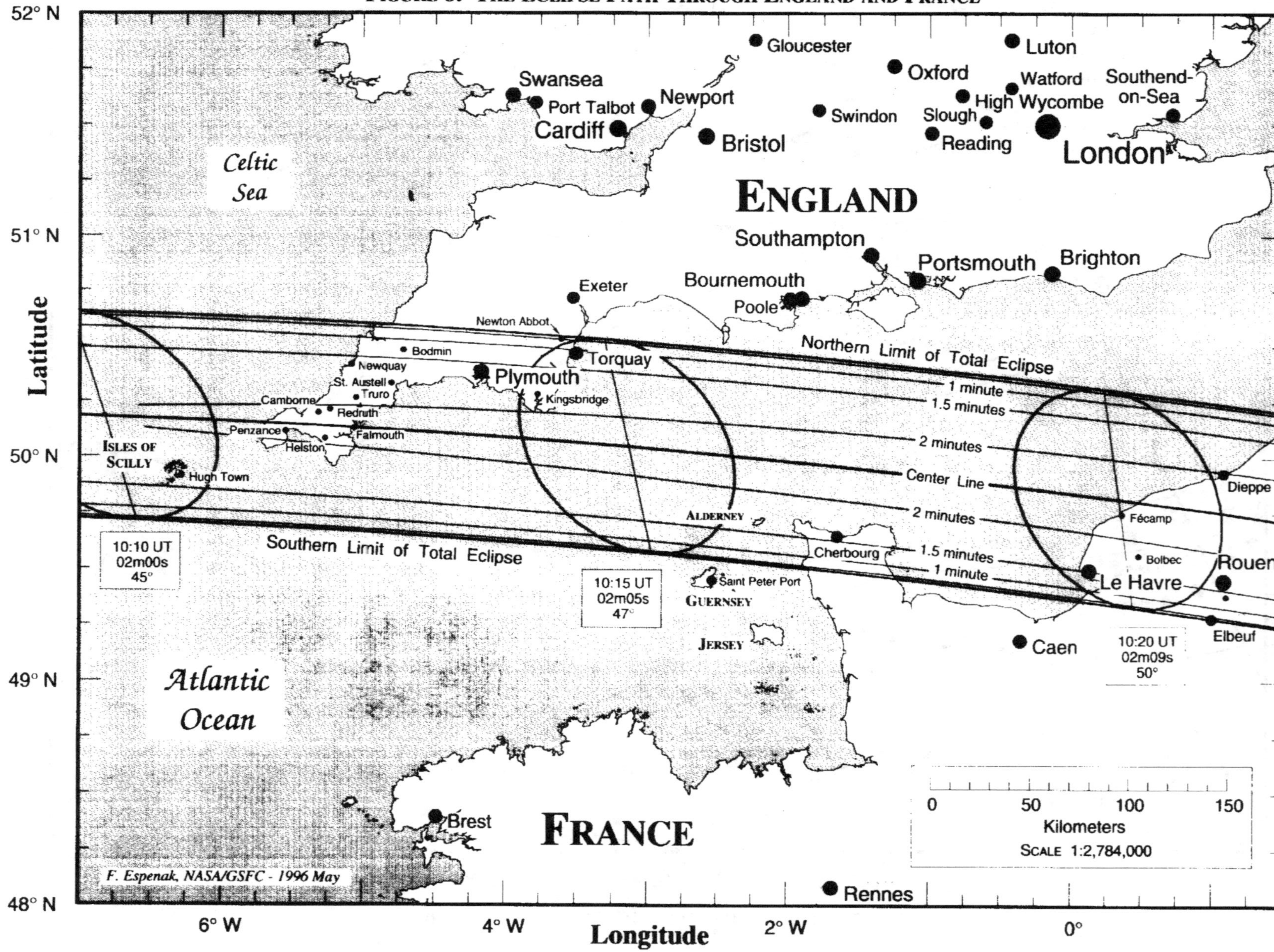

Total Solar Eclipse of 1999 August 11

FIGURE 7: THE ECLIPSE PATH THROUGH FRANCE, BELGIUM, LUXEMBOURG AND GERMANY

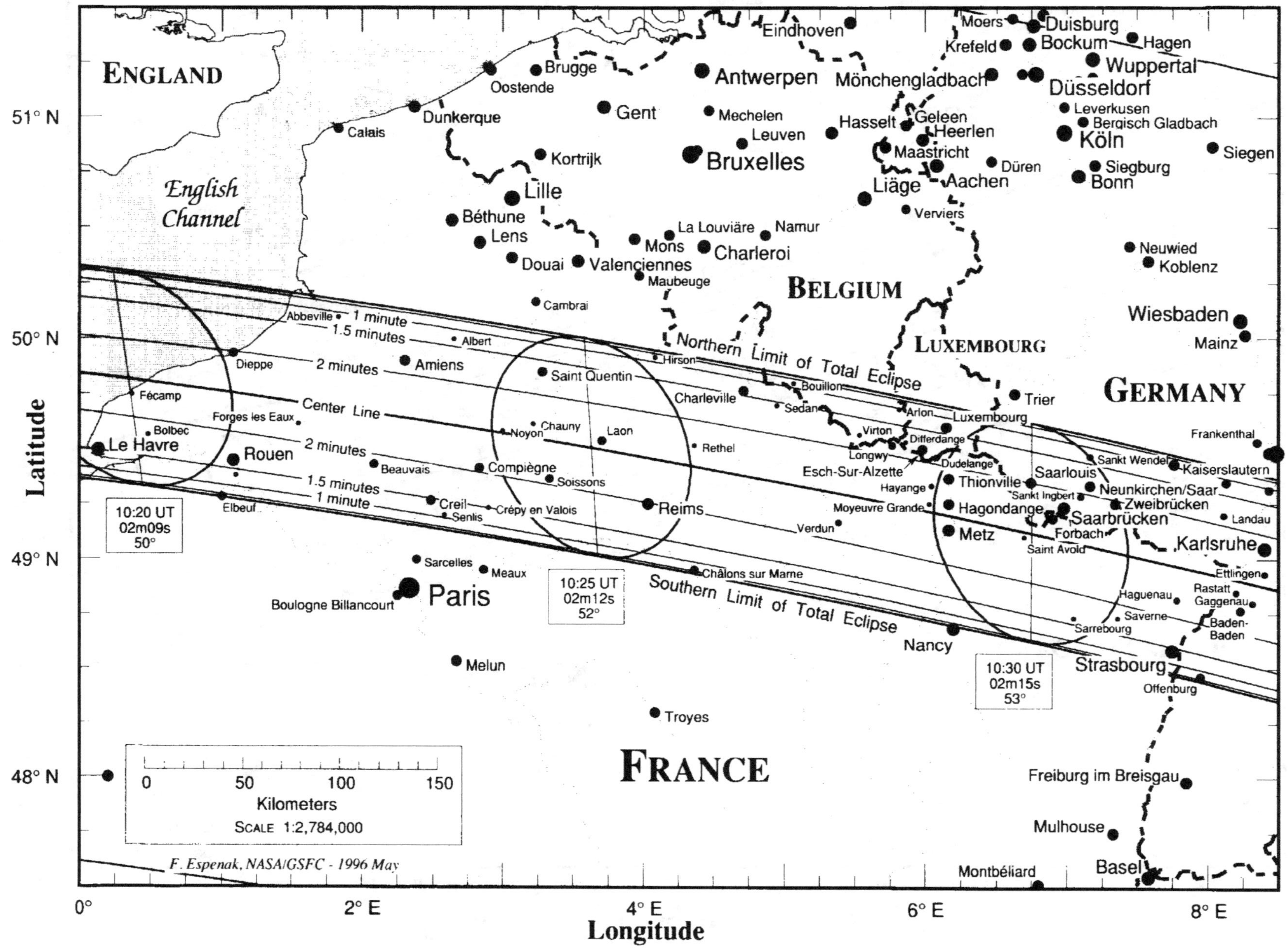

Total Solar Eclipse of 1999 August 11

FIGURE 8: THE ECLIPSE PATH THROUGH GERMANY AND AUSTRIA

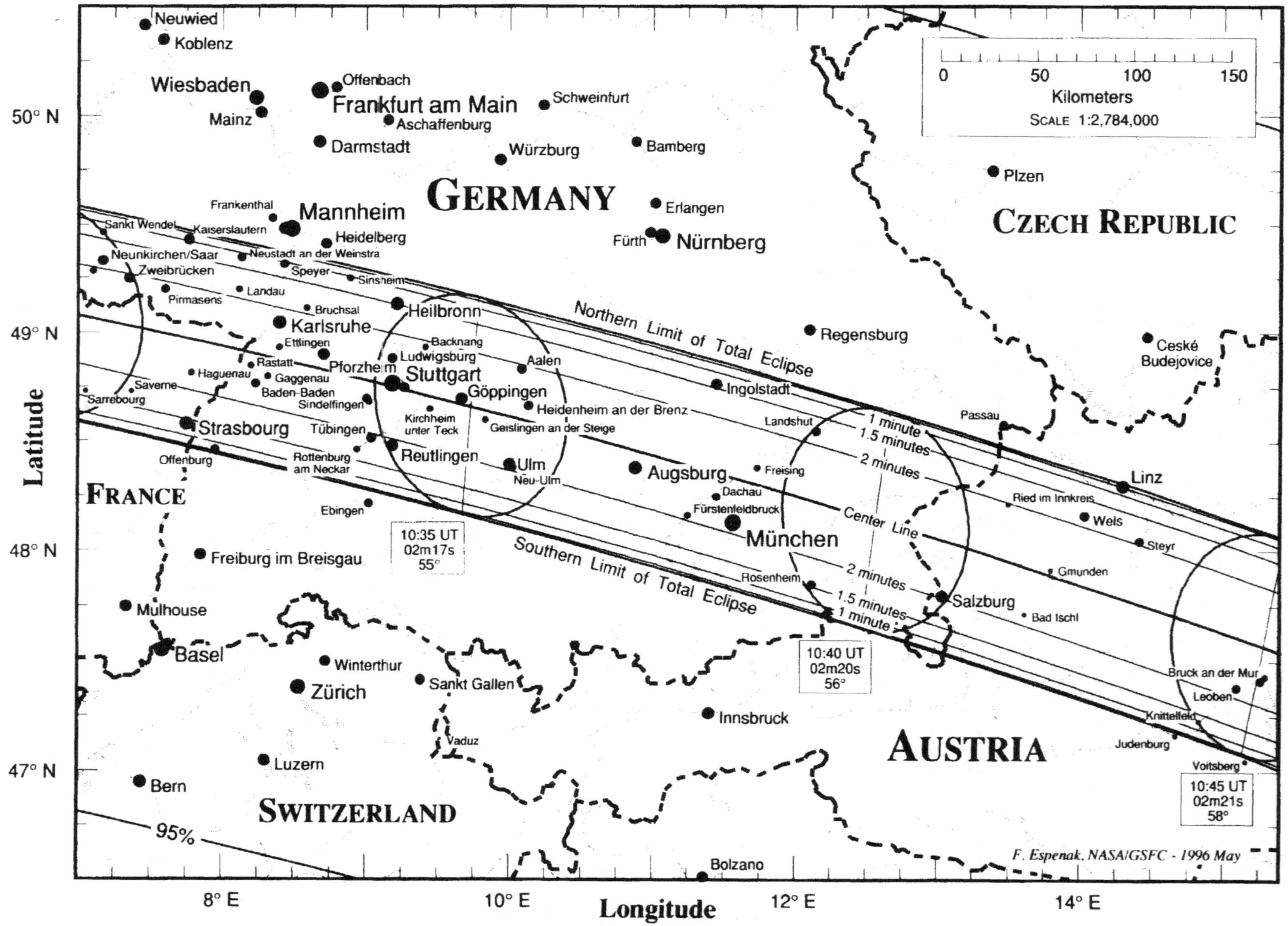

Total Solar Eclipse of 1999 August 11

FIGURE 9: THE ECLIPSE PATH THROUGH AUSTRIA, HUNGARY AND ROMANIA

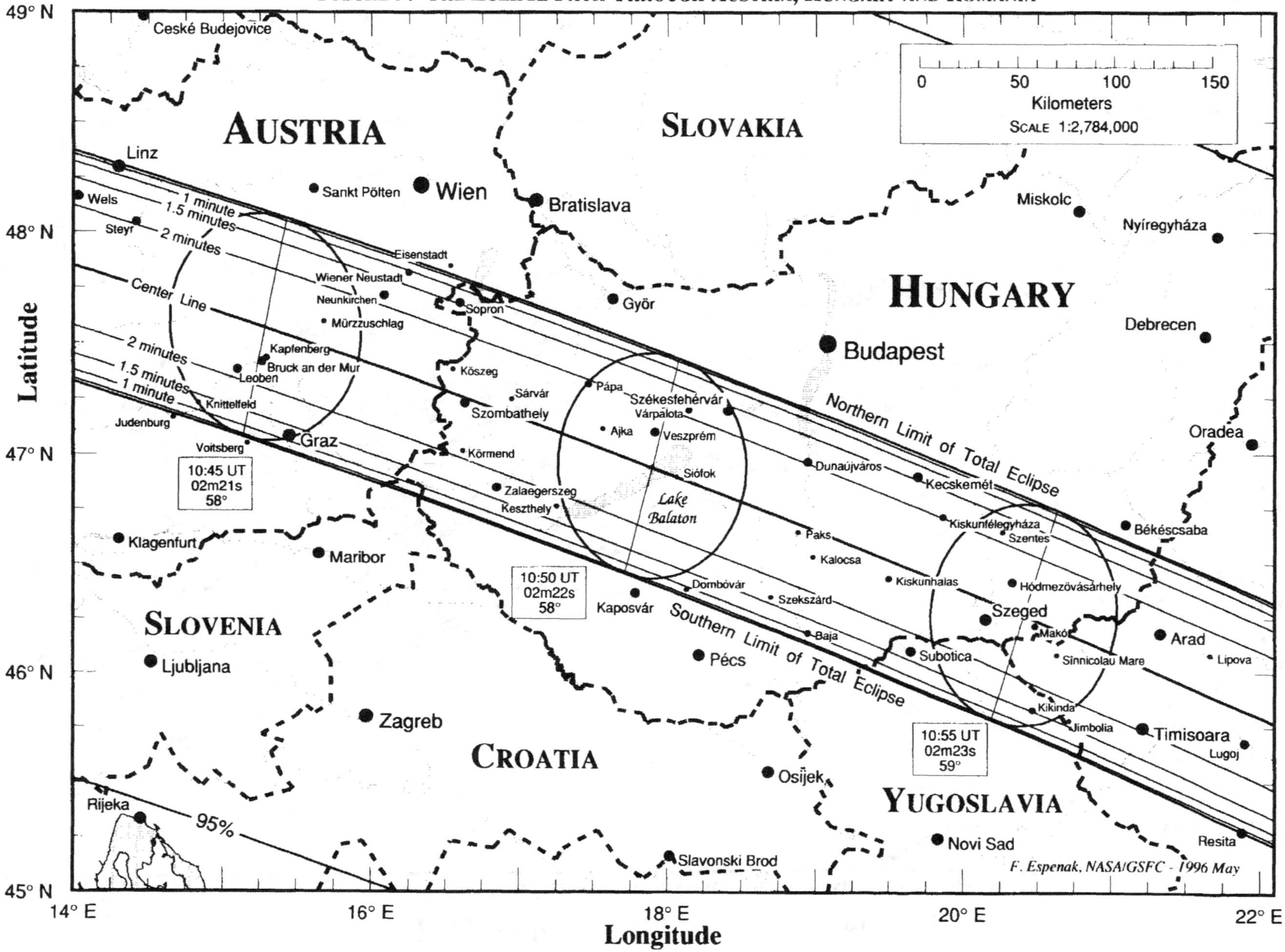

Total Solar Eclipse of 1999 August 11

FIGURE 10: THE ECLIPSE PATH THROUGH ROMANIA AND BULGARIA

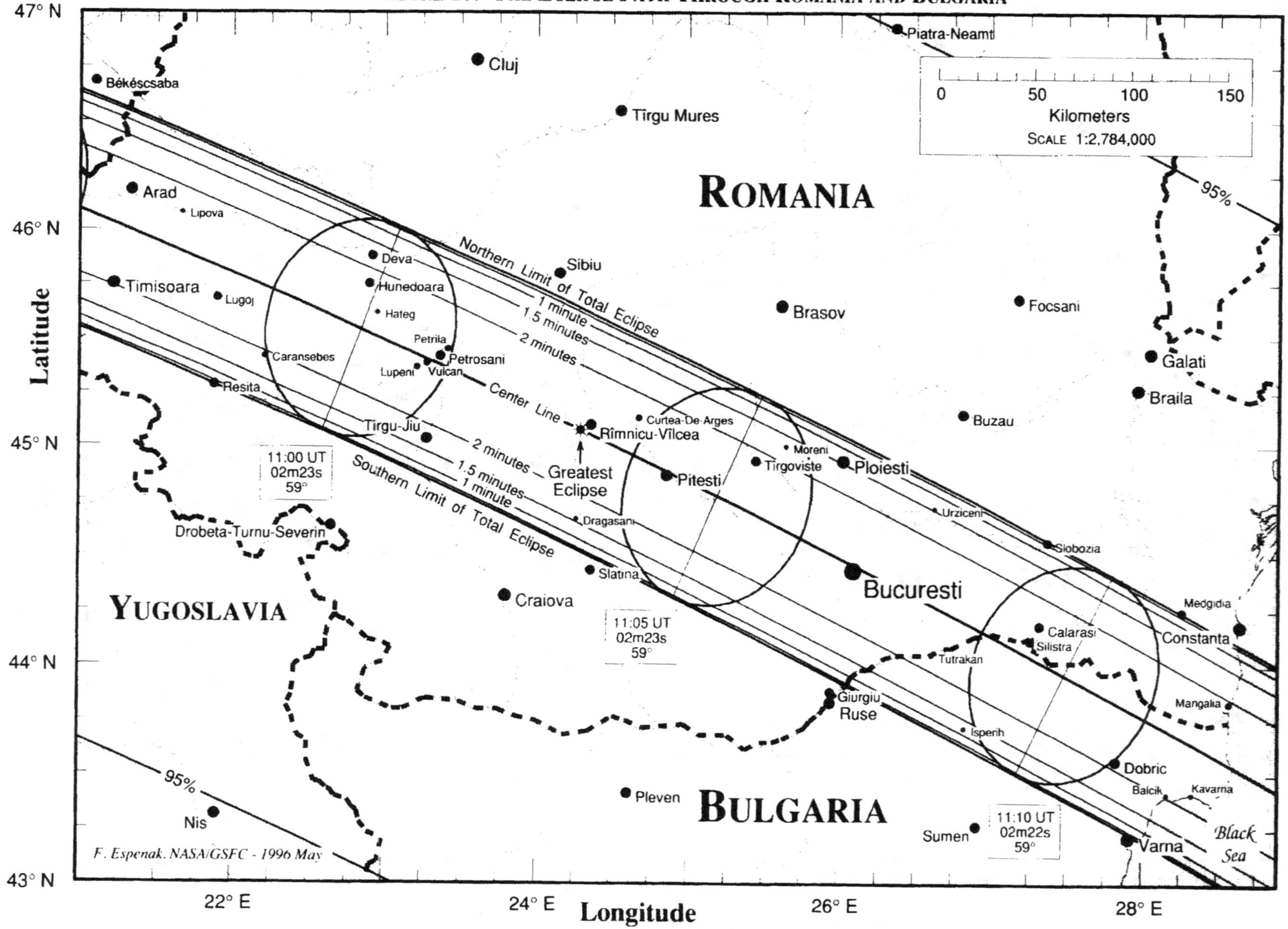

Total Solar Eclipse of 1999 August 11

FIGURE 11: THE ECLIPSE PATH THROUGH TURKEY

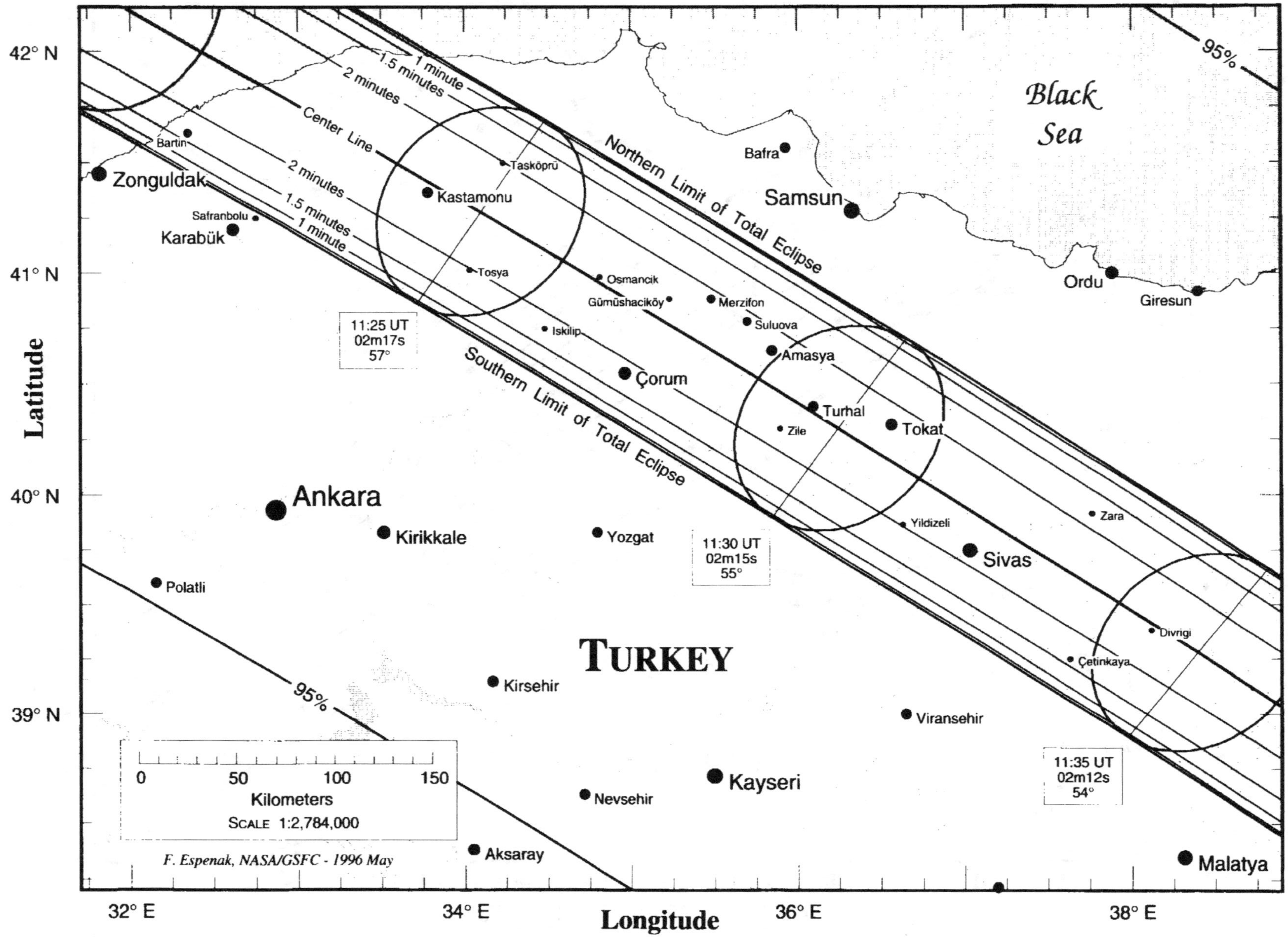

Total Solar Eclipse of 1999 August 11

Figure 12: The Eclipse Path Through Turkey, Syria and Iraq

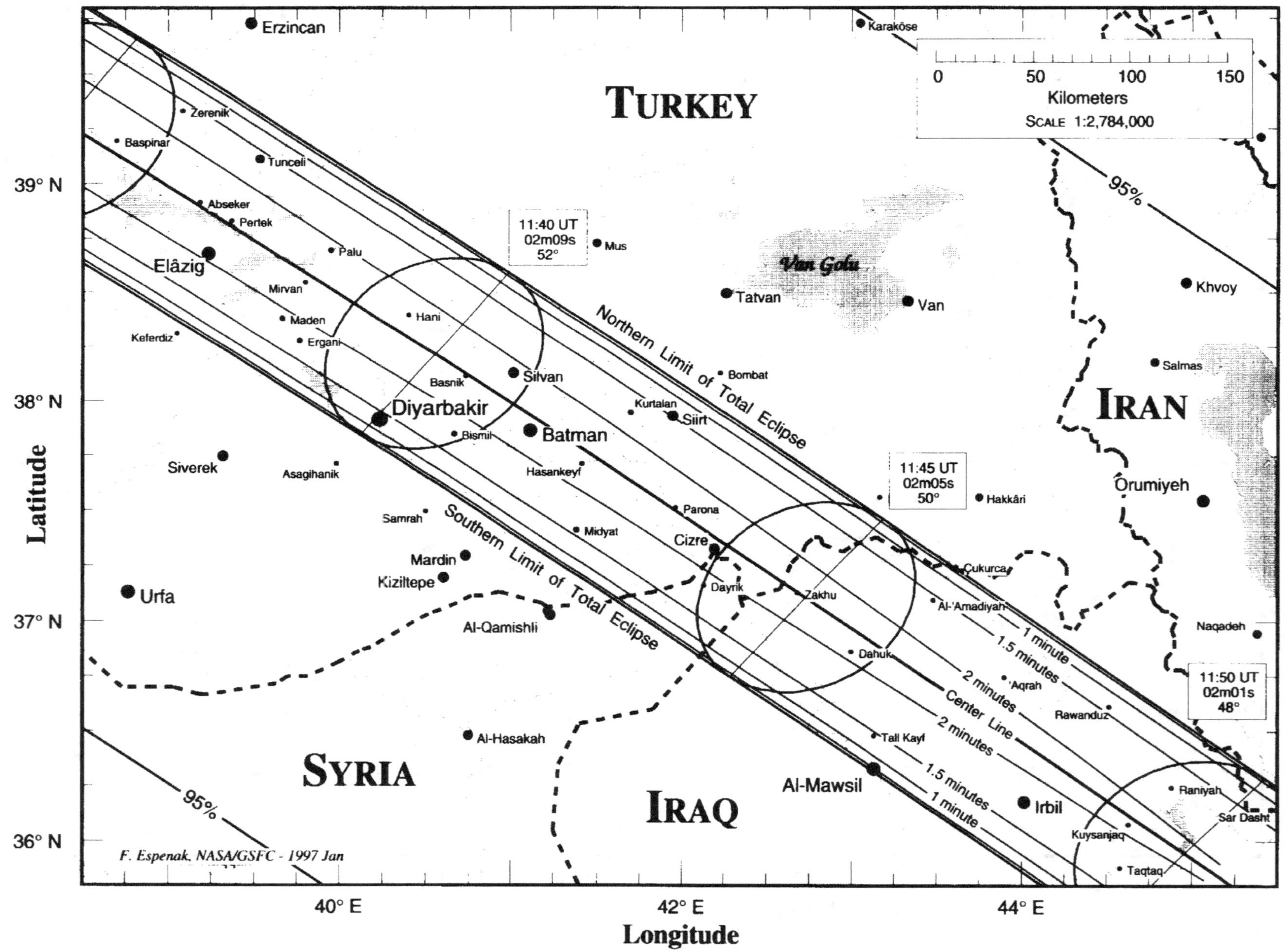

Total Solar Eclipse of 1999 August 11

Figure 13: The Eclipse Path Through Iraq and Iran

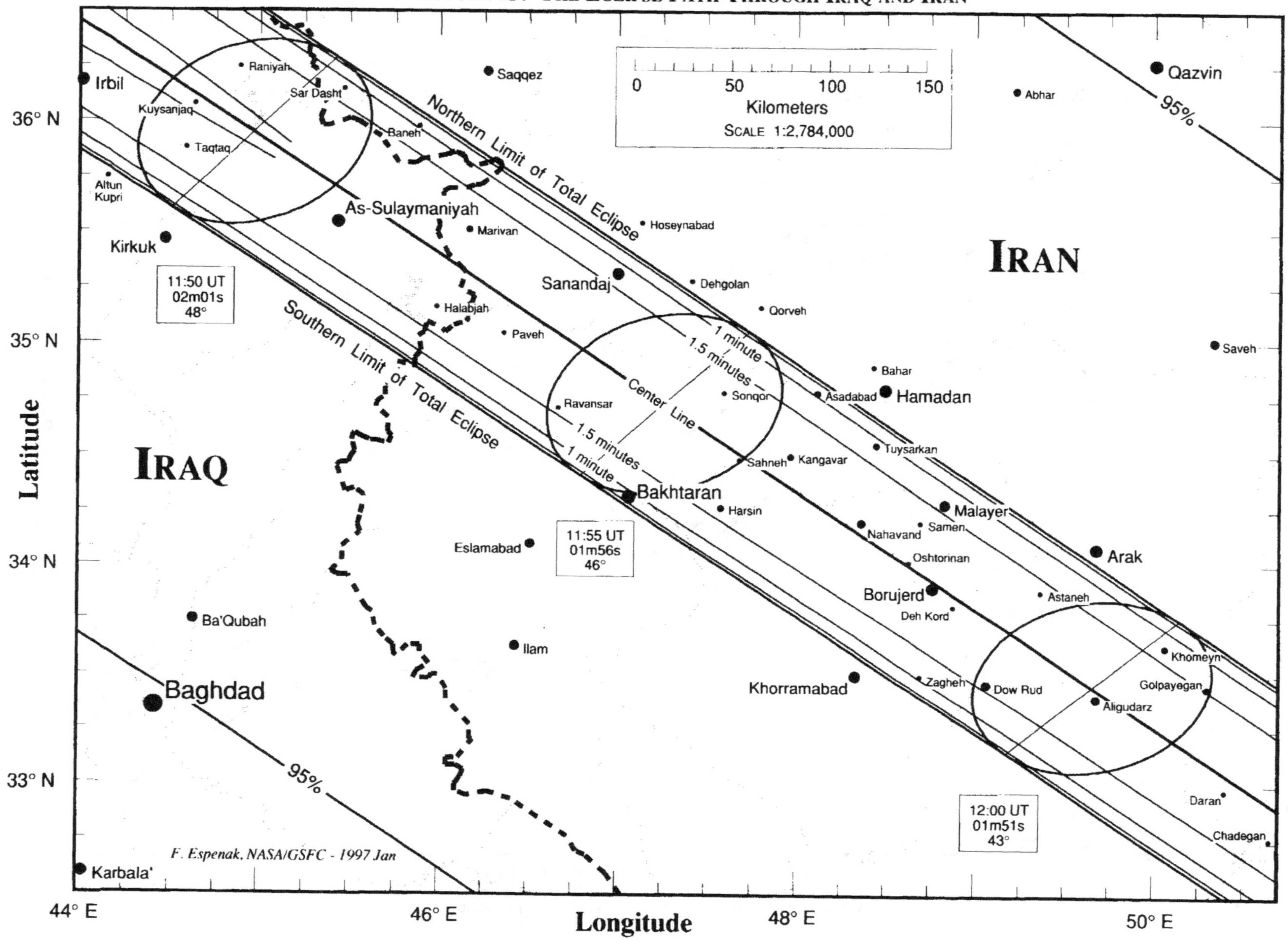

Total Solar Eclipse of 1999 August 11

Figure 14: The Eclipse Path Through Iran

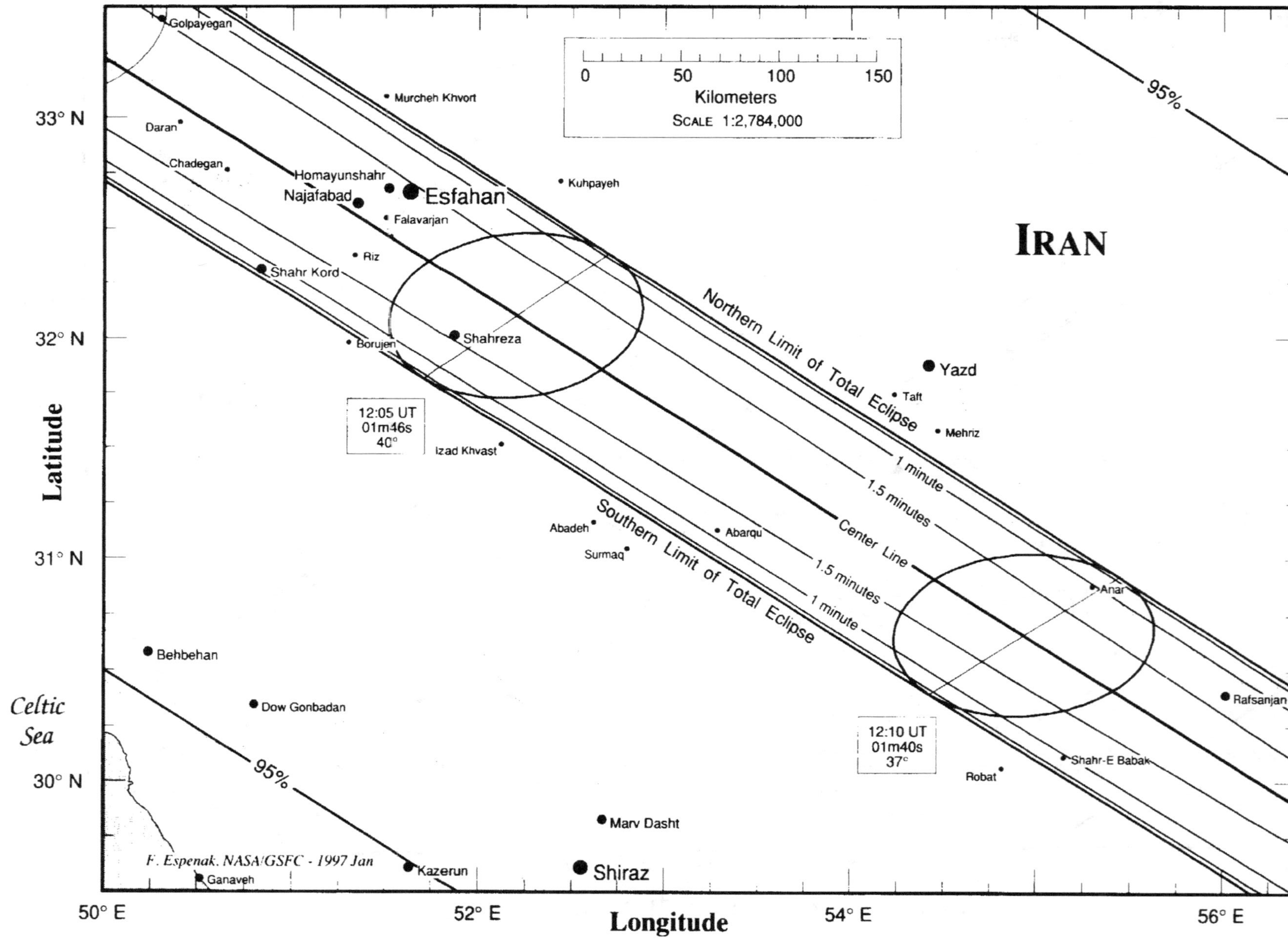

Total Solar Eclipse of 1999 August 11

FIGURE 15: THE ECLIPSE PATH THROUGH SOUTHERN IRAN

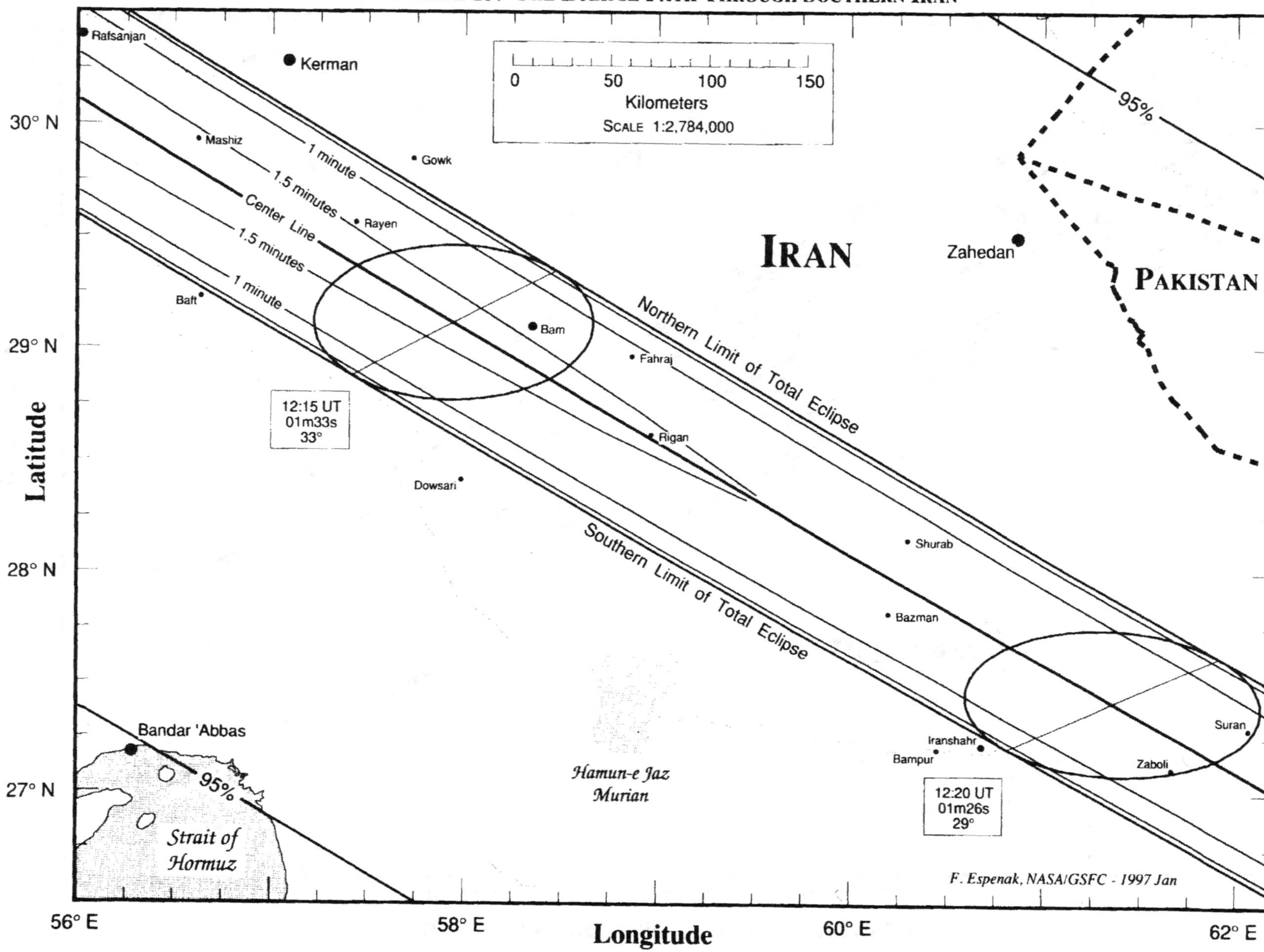

Total Solar Eclipse of 1999 August 11

FIGURE 16: THE ECLIPSE PATH THROUGH PAKISTAN

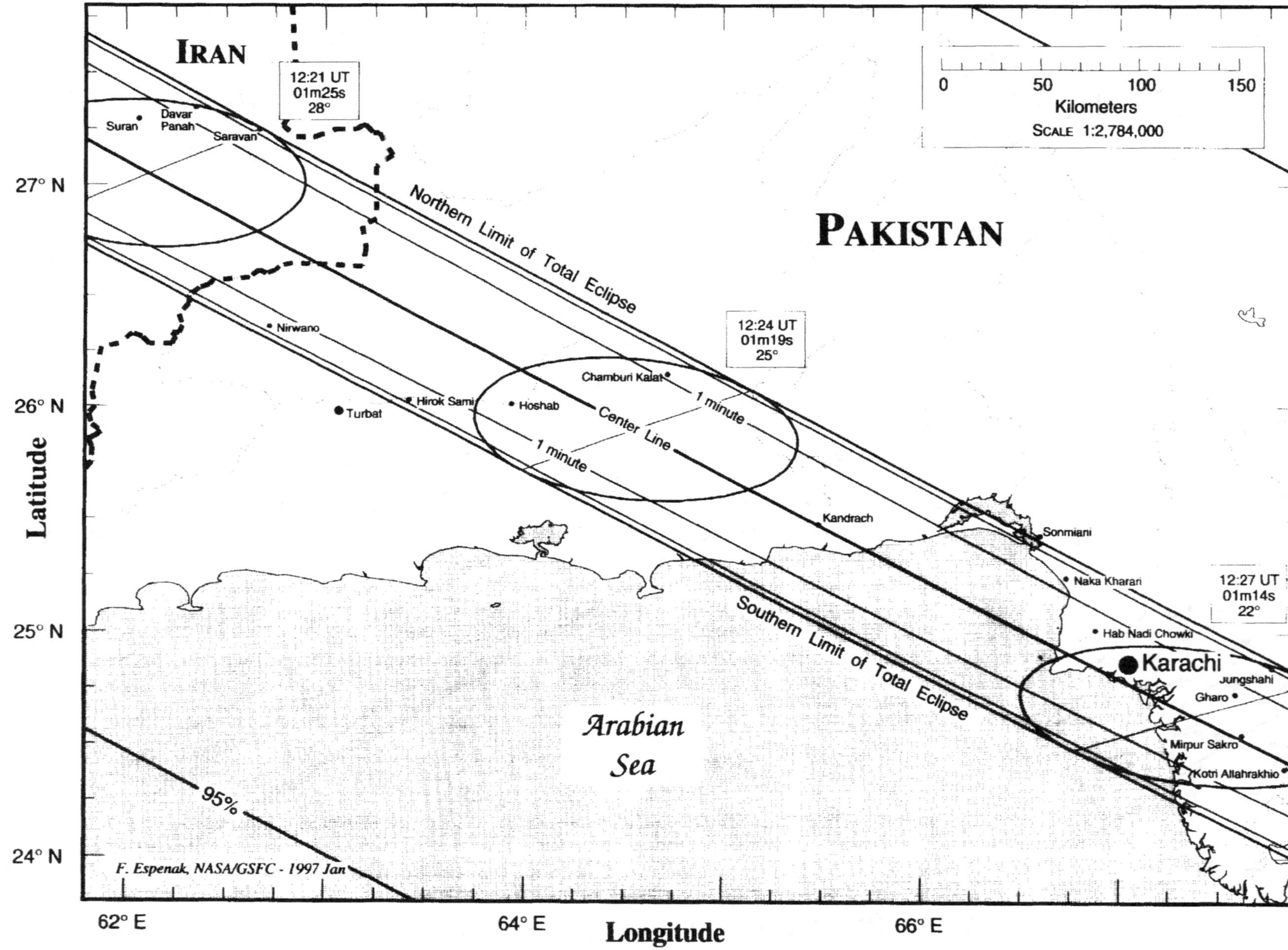

Total Solar Eclipse of 1999 August 11

FIGURE 17: THE ECLIPSE PATH THROUGH PAKISTAN AND INDIA

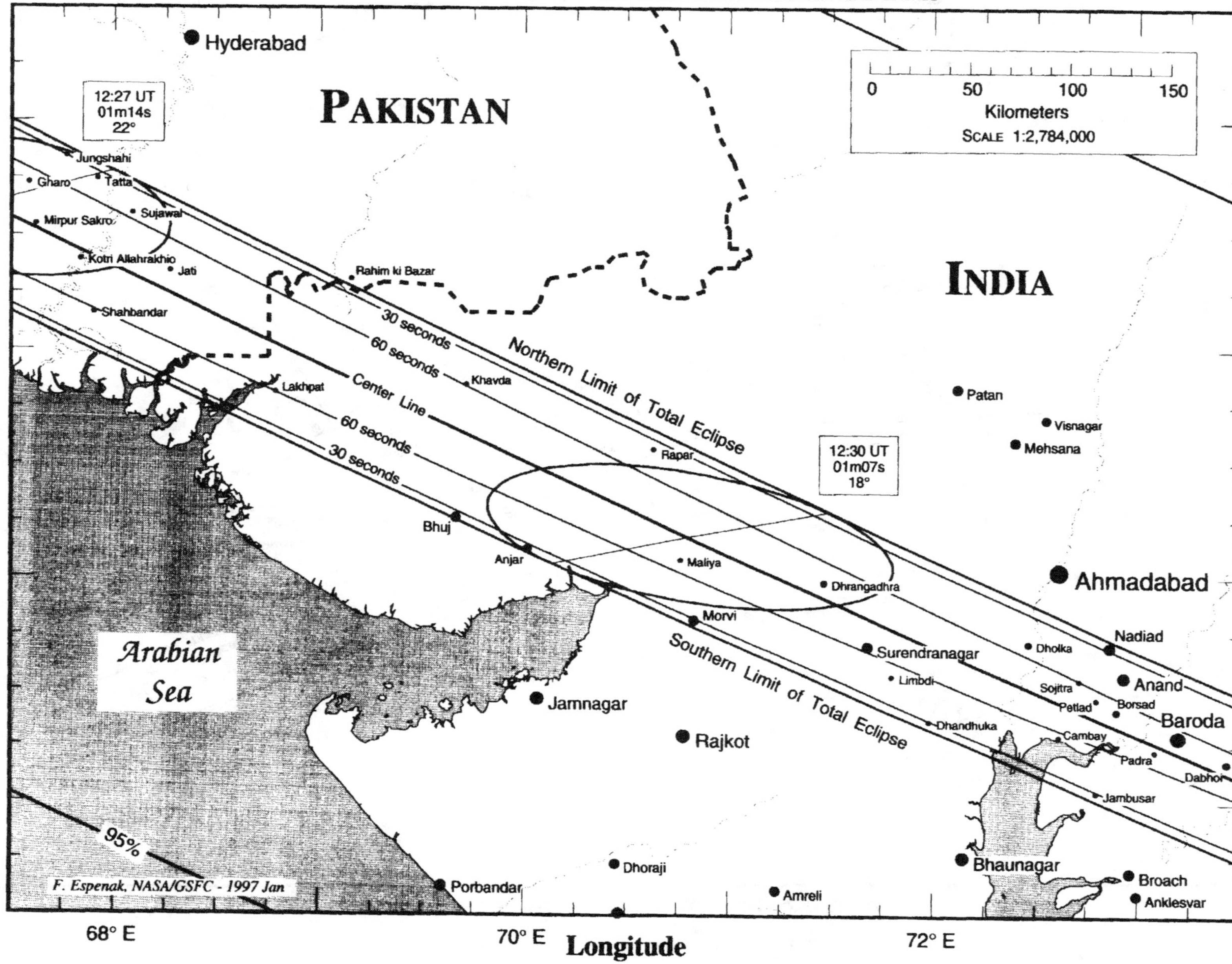

Total Solar Eclipse of 1999 August 11

FIGURE 18: THE ECLIPSE PATH THROUGH CENTRAL INDIA

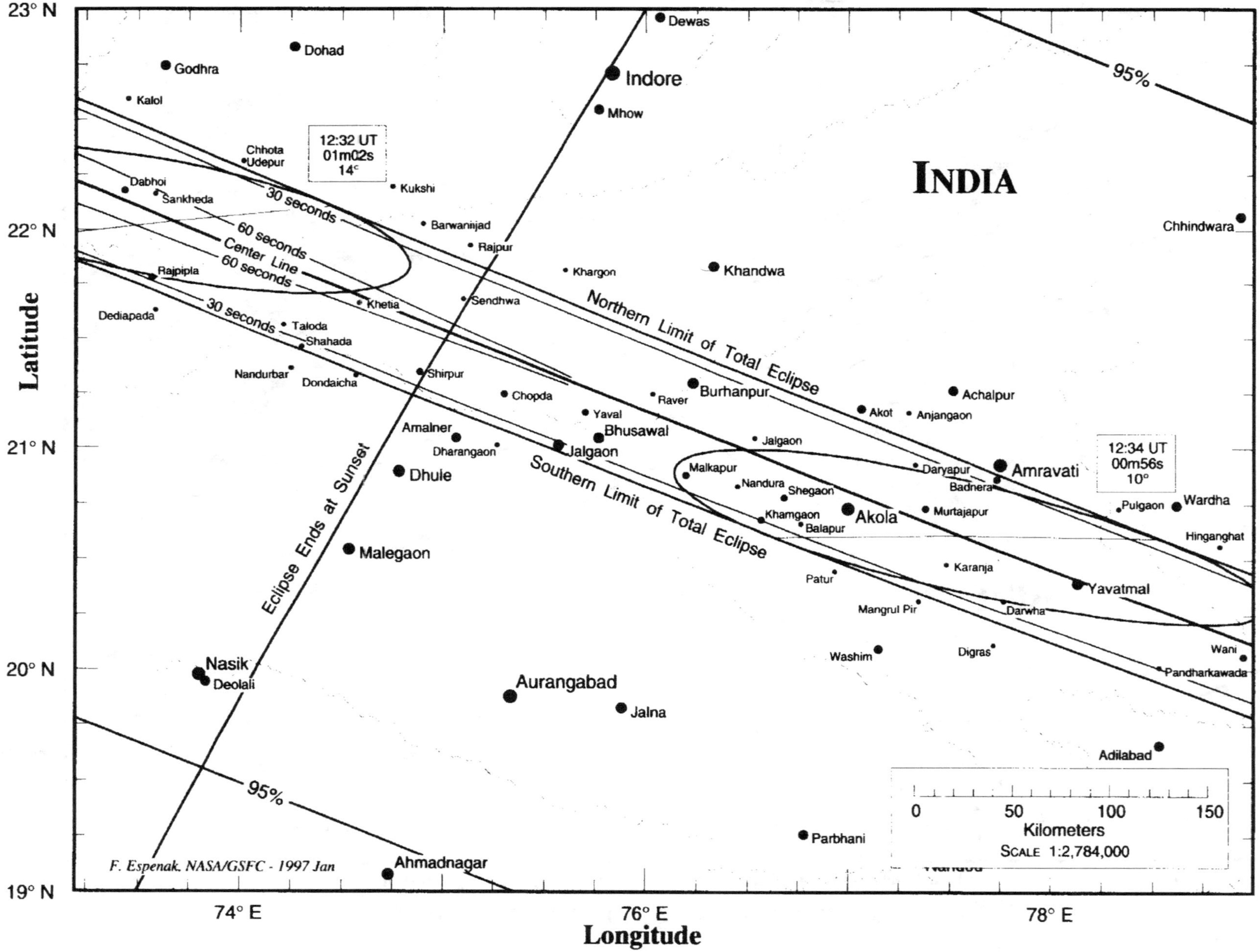

Total Solar Eclipse of 1999 August 11

FIGURE 19: THE ECLIPSE PATH THROUGH EASTERN INDIA

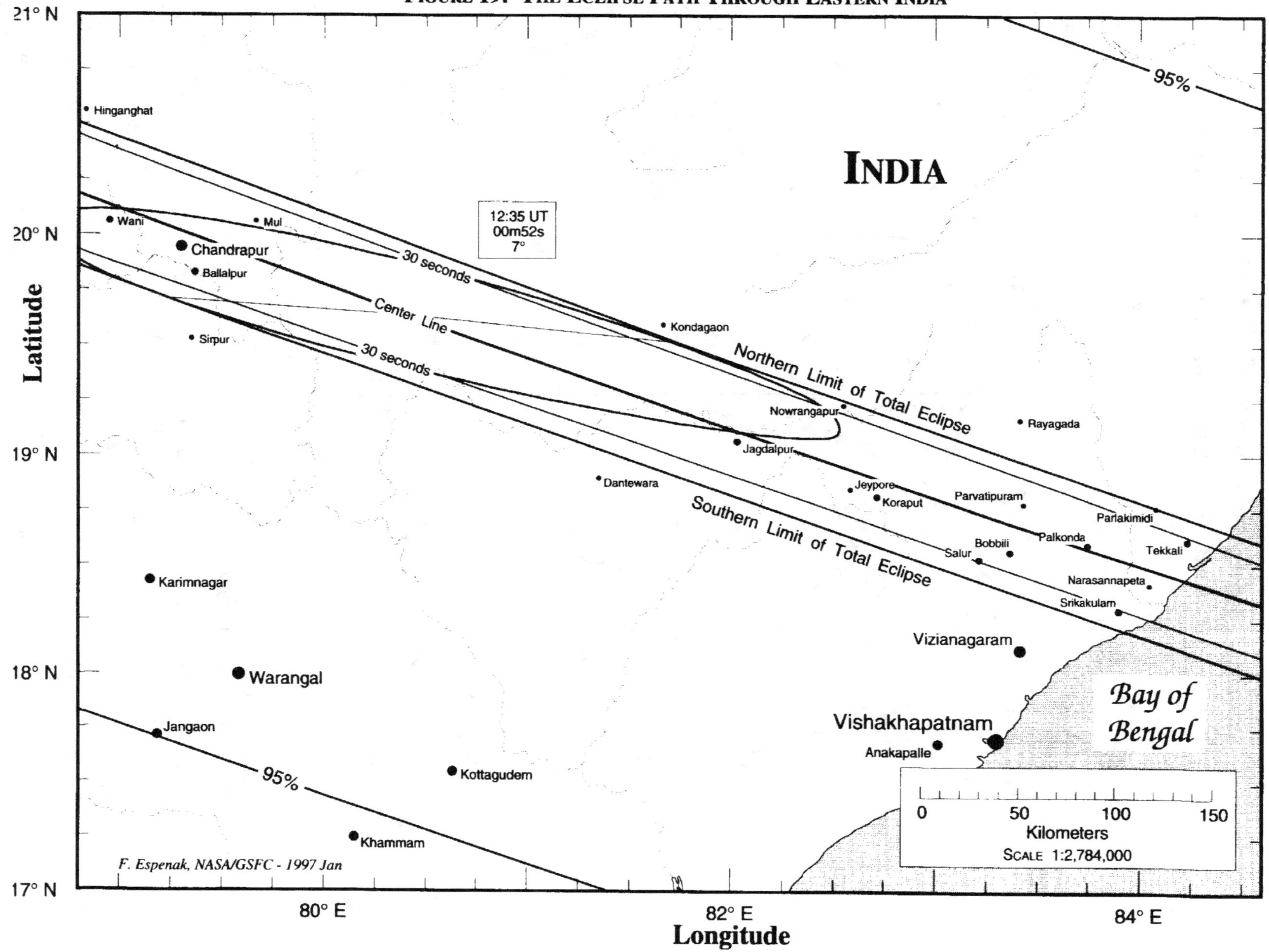

Total Solar Eclipse of 1999 August 11

FIGURE 20: THE LUNAR LIMB PROFILE AT 11:00 UT

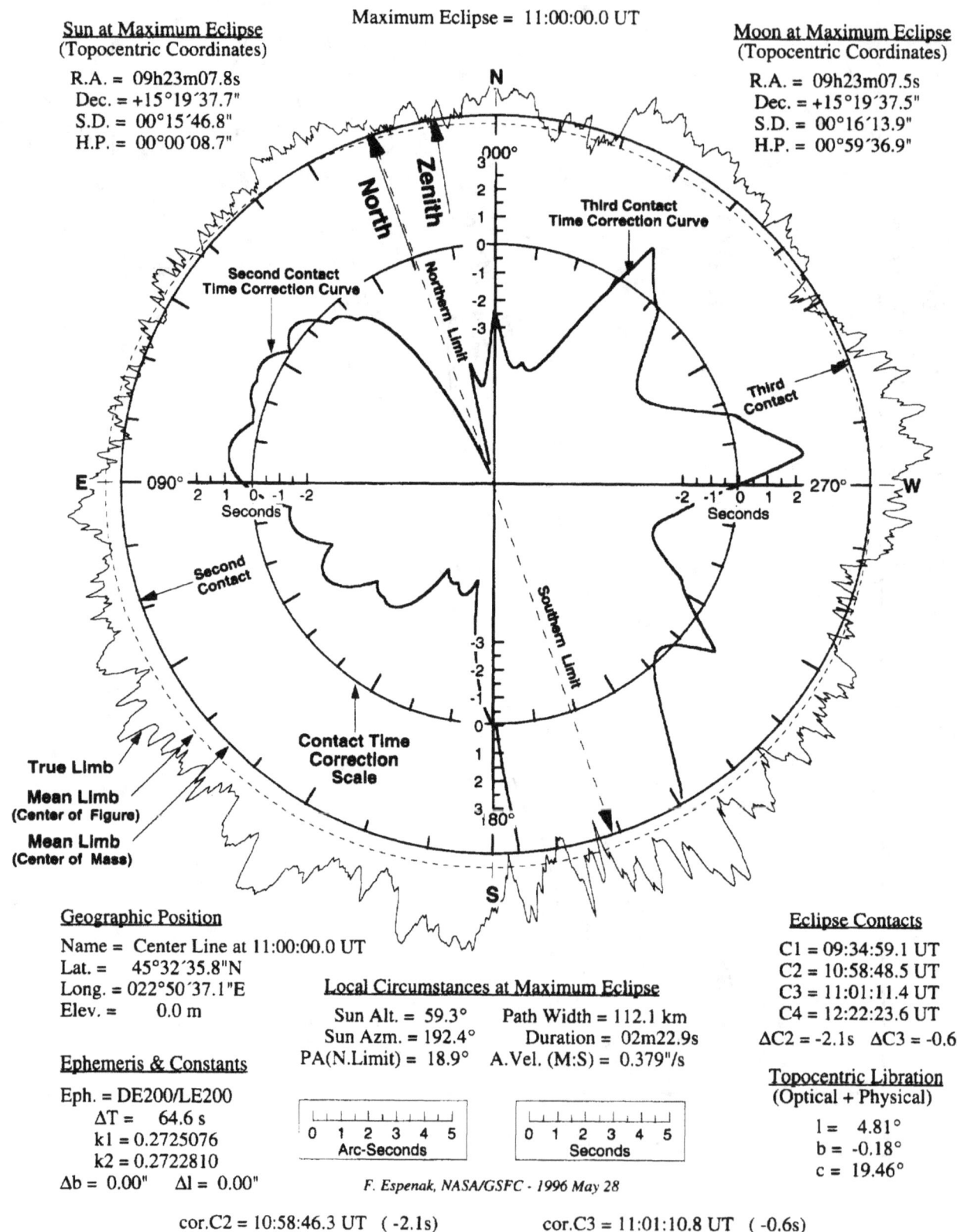

Maximum Eclipse = 11:00:00.0 UT

Sun at Maximum Eclipse
(Topocentric Coordinates)

R.A. = 09h23m07.8s
Dec. = +15°19′37.7"
S.D. = 00°15′46.8"
H.P. = 00°00′08.7"

Moon at Maximum Eclipse
(Topocentric Coordinates)

R.A. = 09h23m07.5s
Dec. = +15°19′37.5"
S.D. = 00°16′13.9"
H.P. = 00°59′36.9"

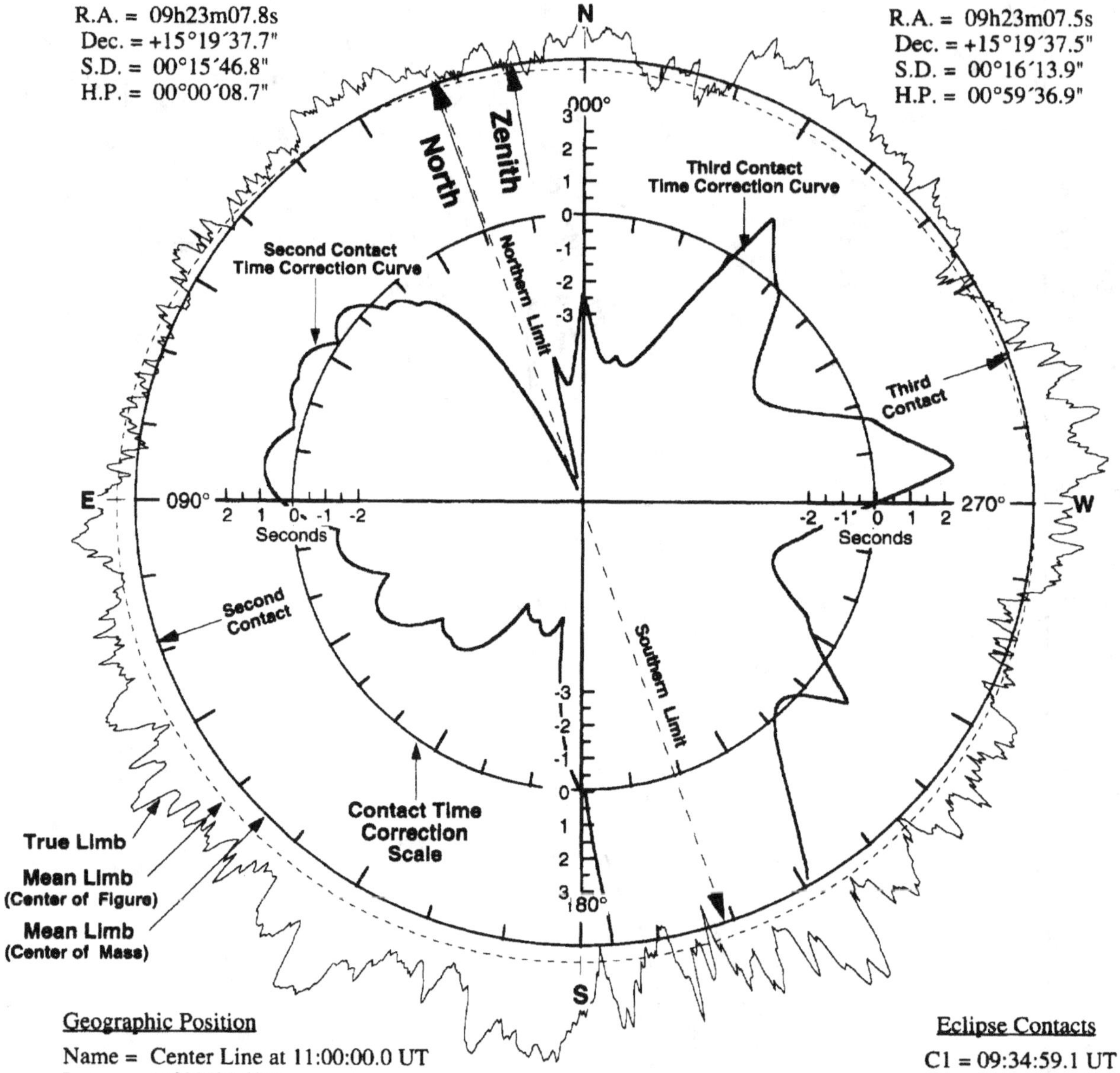

Geographic Position

Name = Center Line at 11:00:00.0 UT
Lat. = 45°32′35.8"N
Long. = 022°50′37.1"E
Elev. = 0.0 m

Local Circumstances at Maximum Eclipse

Sun Alt. = 59.3° Path Width = 112.1 km
Sun Azm. = 192.4° Duration = 02m22.9s
PA(N.Limit) = 18.9° A.Vel. (M:S) = 0.379"/s

Eclipse Contacts

C1 = 09:34:59.1 UT
C2 = 10:58:48.5 UT
C3 = 11:01:11.4 UT
C4 = 12:22:23.6 UT
ΔC2 = -2.1s ΔC3 = -0.6s

Ephemeris & Constants

Eph. = DE200/LE200
ΔT = 64.6 s
k1 = 0.2725076
k2 = 0.2722810
Δb = 0.00" Δl = 0.00"

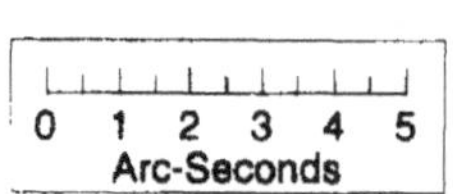

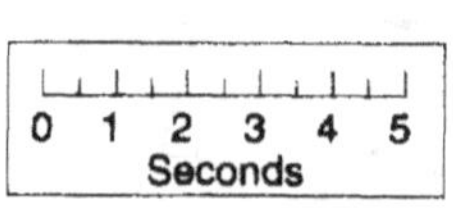

Topocentric Libration
(Optical + Physical)

l = 4.81°
b = -0.18°
c = 19.46°

F. Espenak, NASA/GSFC - 1996 May 28

cor.C2 = 10:58:46.3 UT (-2.1s) cor.C3 = 11:01:10.8 UT (-0.6s)

Total Solar Eclipse of 1999 August 11

Figure 21: Limb Profile Effects on the Duration of Totality

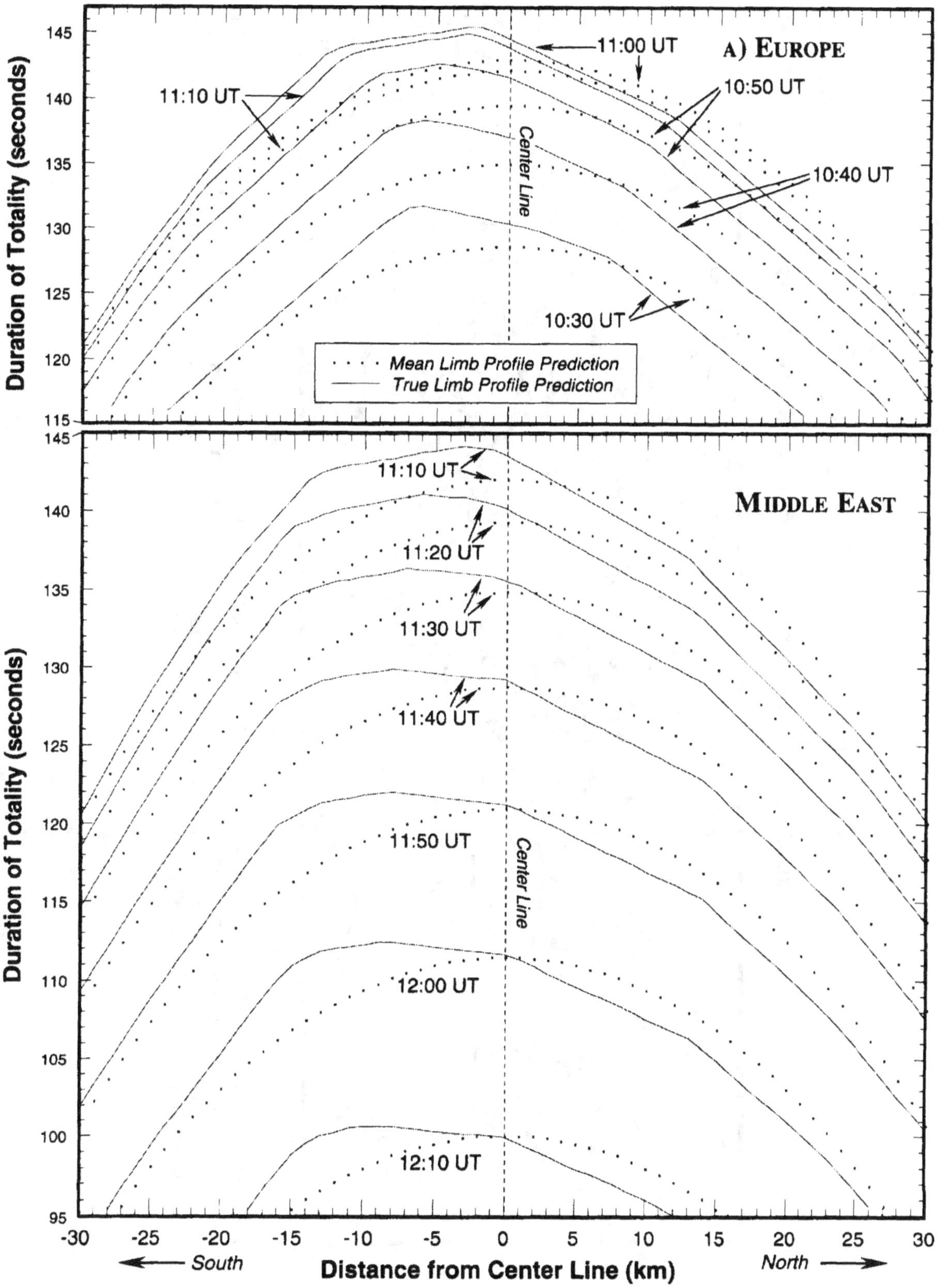

F. Espenak - 1997 Jan 31

FIGURE 22: MEAN CLOUD COVER IN AUGUST ALONG THE ECLIPSE PATH

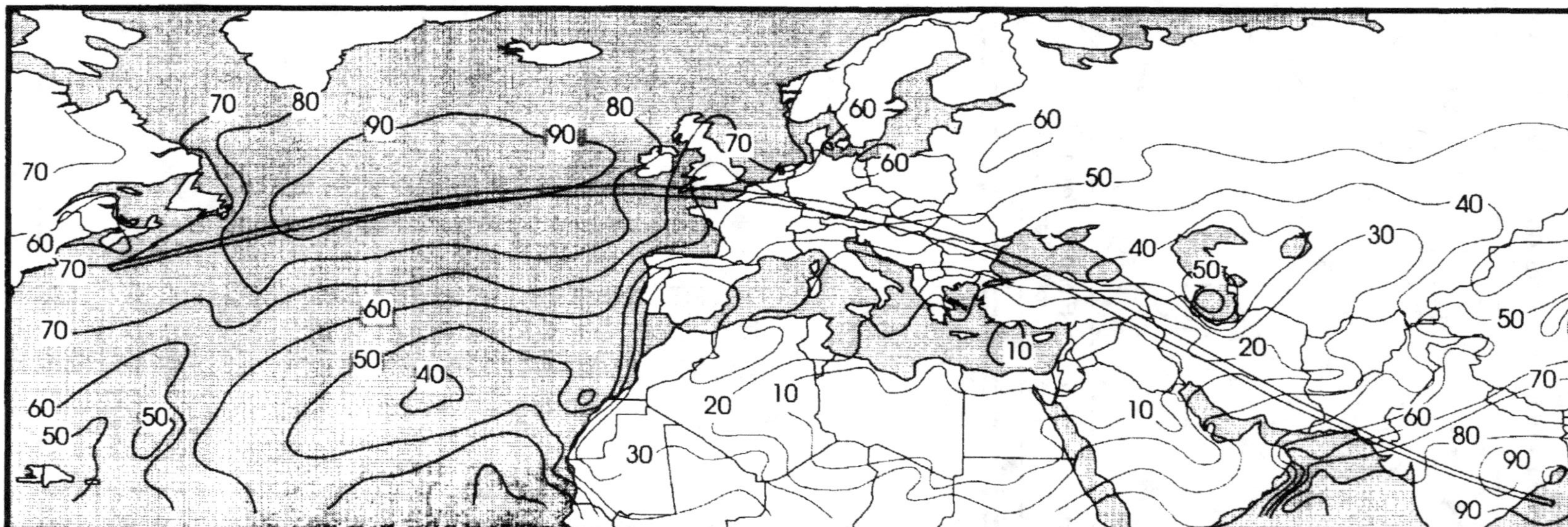

J. Anderson, Environment Canada - 1997 Jan

Figure 22: Mean cloud cover (in percent) along the eclipse track as determined by satellite measurements from eight years of analysis (1983-1990). These data are collected and analyzed globally from a number of satellites (International Satellite Cloud Climatology Project), and processed by computer. It provides an excellent comparative database for different locations around the globe. Statistics are collected over a 5° by 5° latitude/longitude area, and represent the large scale cloud characteristics of an area. Small scale variations are smoothed out.

Total Solar Eclipse of 1999 August 11

Figure 23: Probability of Seeing the Eclipse Along the Path

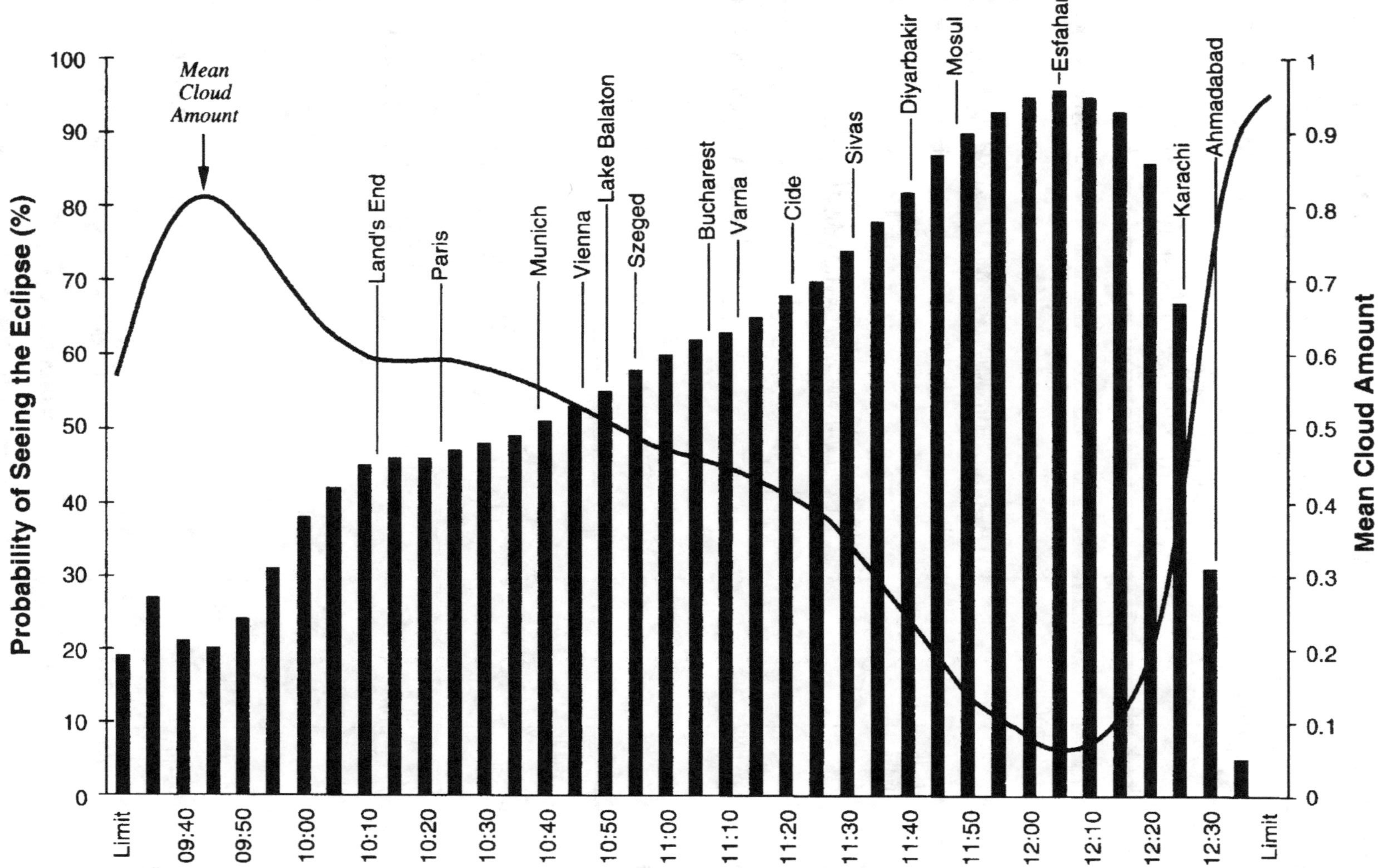

J. Anderson, Environment Canada - 1997 Jan

Figure 23: Probability of seeing the eclipse, adjusted for Sun altitude, time of day, and the actual date of the eclipse. Values are smoothed by the computer modeling so that fine details of the cloud cover are lost. Also included is a graph of the mean cloud cover along the track. The data are presented at 5 minute intervals along the eclipse track, with the locations of important cities and landmarks.

FIGURE 24: SPECTRAL RESPONSE OF SOME COMMONLY AVAILABLE SOLAR FILTERS

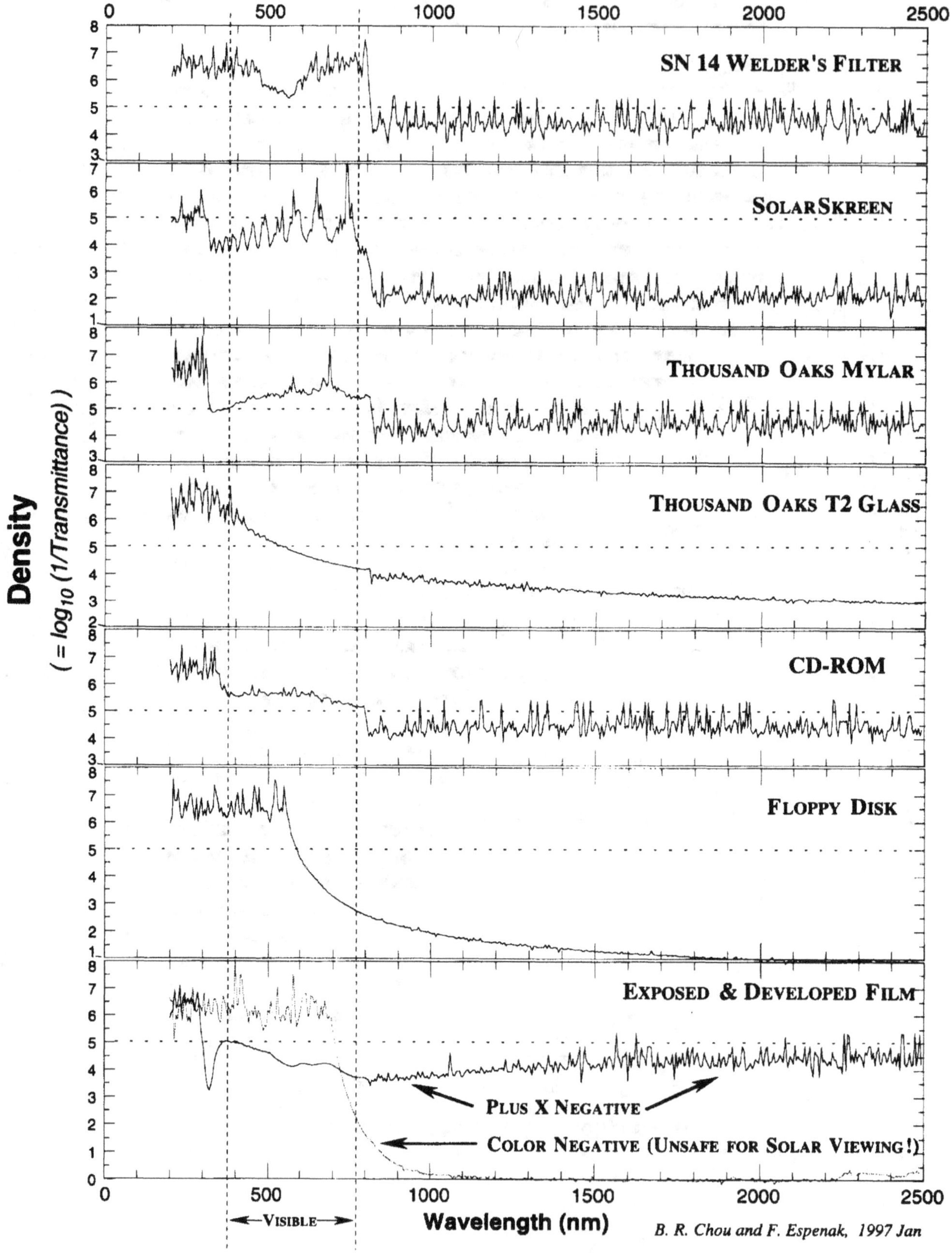

B. R. Chou and F. Espenak, 1997 Jan

Total Solar Eclipse of 1999 August 11

FIGURE 25: THE SKY DURING TOTALITY AS SEEN FROM CENTER LINE AT 11:00 UT

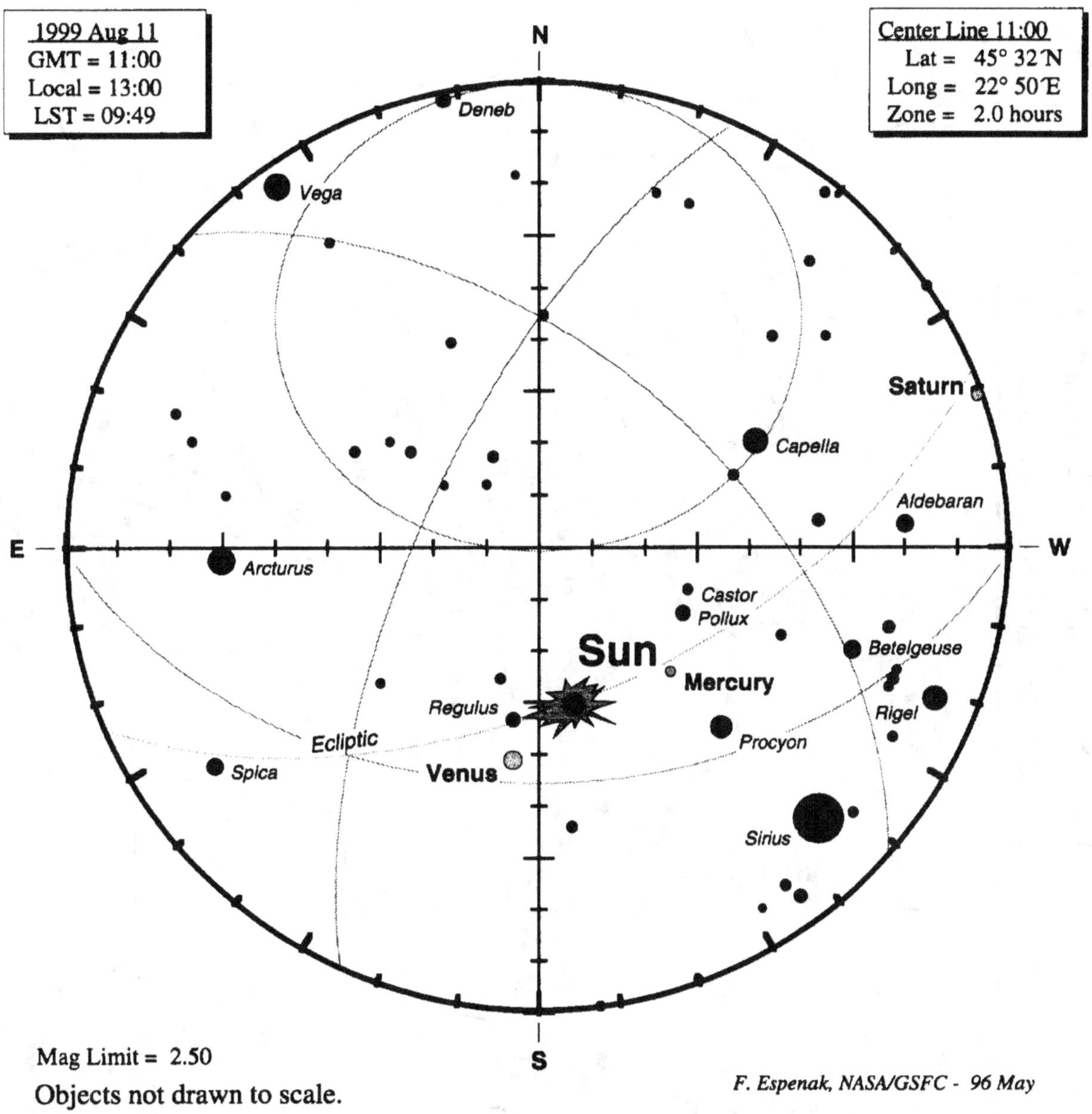

Figure 25: The sky during totality as seen from the center line in Romania at 11:00 UT. Venus (m=–3.5) will be the most conspicuous planet located 15° east of the Sun. Mercury (m=+0.7) should also be visible 18° west of the Sun. The southwestern sky will be dominated by the bright stars of winter, including Capella (m=+0.08), Albebaran (m=+0.85v), Procyon (+0.38), Betelgeuse(+0.5v), and Sirius (m=-1.46). Other bright stars which may also be visible include Spica (m=+1.0v), Arcturus (m=-0.04), and Regulus (m=+1.35).

For sky maps from other locations along the path of totality, see the special 1999 eclipse web site:
http://planets.gsfc.nasa.gov/eclipse/TSE1999/TSE1999.html

Total Solar Eclipse of 2001 June 21

FIGURE 26: THE ECLIPSE PATH THROUGH AFRICA

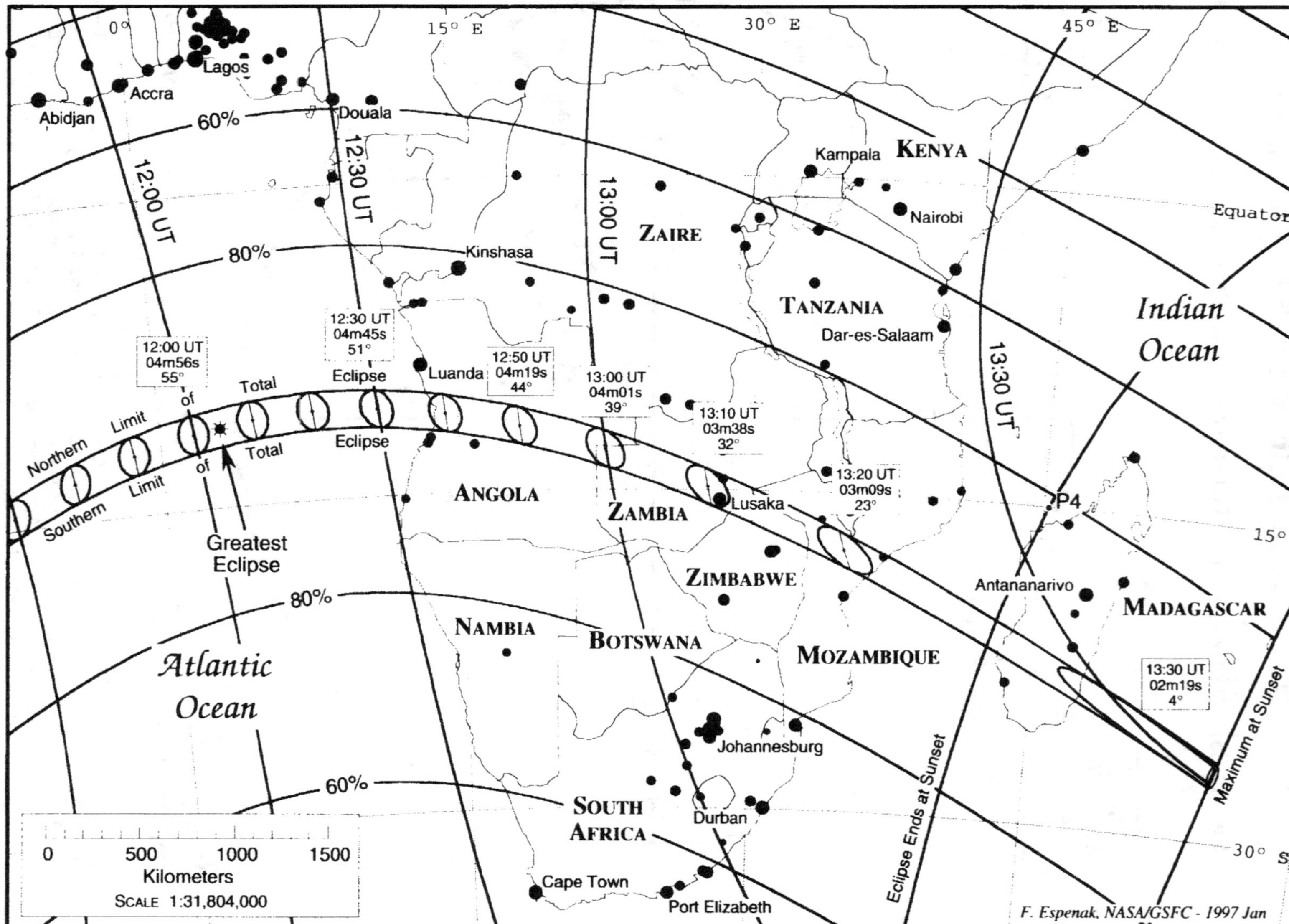

Total Solar Eclipse of 1999 August 11

Tables

TABLE 1

ELEMENTS OF THE TOTAL SOLAR ECLIPSE OF 1999 AUGUST 11

Geocentric Conjunction of Sun & Moon in R.A.: 10:52:16.66 TDT (=10:51:12.06 UT) J.D. = 2451401.952971

Instant of Greatest Eclipse: 11:04:09.01 TDT (=11:03:04.41 UT) J.D. = 2451401.961215

Geocentric Coordinates of Sun & Moon at Greatest Eclipse (DE200/LE200):

	Sun:		Moon:
R.A. =	09h23m08.297s	R.A. =	09h23m34.531s
Dec. =	+15°19′39.72"	Dec. =	+15°48′38.51"
Semi-Diameter =	15′46.77"	Semi-Diameter =	16′00.34"
Eq.Hor.Par. =	8.68"	Eq.Hor.Par. =	0°58′44.24"
Δ R.A. =	9.467s/h	Δ R.A. =	142.037s/h
Δ Dec. =	-44.35"/h	Δ Dec. =	-462.21"/h

Lunar Radius Constants: k1 = 0.2725076 (Penumbra) k2 = 0.2722810 (Umbra)

Shift in Lunar Position: Δb = 0.00" Δl = 0.00"

Geocentric Libration: (Optical + Physical) l = 4.8° b = -0.8° c = 19.7°

Brown Lun. No. = 1232
Saros Series = 145 (21/77)
Ephemeris = (DE200/LE200)

Eclipse Magnitude = 1.02859 Gamma = 0.50623 ΔT = 64.6 s

Polynomial Besselian Elements for: 1999 Aug 11 11:00:00.0 TDT ($=t_0$)

n	x	y	d	l_1	l_2	μ
0	0.0700559	0.5028388	15.3273401	0.5424893	-0.0036496	343.690308
1	0.5443045	-0.1184931	-0.0120348	0.0001168	0.0001163	15.002983
2	-0.0000406	-0.0001158	-0.0000033	-0.0000117	-0.0000116	0.000002
3	-0.0000081	0.0000017	0.0000000	0.0000000	0.0000000	0.000000

Tan f_1 = 0.0046129 Tan f_2 = 0.0045900

At time 't_1' (decimal hours), each Besselian element is evaluated by:

$$a = a_0 + a_1*t + a_2*t^2 + a_3*t^3 \quad (\text{or } a = \Sigma\ [a_n*t^n];\ n = 0 \text{ to } 3)$$

where: $a = x, y, d, l_1, l_2$, or μ
$t = t_1 - t_0$ (decimal hours) and t_0 = 11.000 TDT

The Besselian elements were derived from a least-squares fit to elements calculated at five separate times over a six hour period centered at t_0. Thus the Besselian elements are valid over the period $8.00 \le t_0 \le 14.00$ TDT.

Saros Series 145: Member 21 of 77 eclipses in series.

TABLE 2

SHADOW CONTACTS AND CIRCUMSTANCES
TOTAL SOLAR ECLIPSE OF 1999 AUGUST 11

ΔT = 64.6 s
=000°16′09.6"E

		Terrestrial Dynamical Time h m s	Latitude	Ephemeris Longitude†	True Longitude*
External/Internal Contacts of Penumbra:	P1	08:27:18.9	30°20.1′N	044°45.6′W	044°29.4′W
	P4	13:41:10.2	06°38.3′N	067°50.0′E	068°06.2′E
Extreme South Limits of Penumbral Path:	S1	09:09:07.2	10°07.7′N	048°46.6′W	048°30.4′W
	S2	12:59:11.4	13°43.1′S	072°40.6′E	072°56.8′E
External/Internal Contacts of Umbra:	U1	09:30:56.5	40°54.7′N	065°10.7′W	064°54.5′W
	U2	09:31:54.3	41°09.4′N	065°32.4′W	065°16.2′W
	U3	12:36:35.4	17°40.4′N	087°09.6′E	087°25.8′E
	U4	12:37:27.8	17°26.7′N	086°52.4′E	087°08.6′E
Extreme North/South Limits of Umbral Path:	N1	09:31:42.6	41°16.4′N	065°33.0′W	065°16.8′W
	S1	09:31:08.3	40°47.7′N	065°10.2′W	064°54.0′W
	N2	12:36:46.0	17°46.9′N	087°09.0′E	087°25.2′E
	S2	12:37:17.0	17°20.1′N	086°53.1′E	087°09.3′E
Extreme Limits of Center Line:	C1	09:31:25.4	41°02.0′N	065°21.5′W	065°05.4′W
	C2	12:37:01.6	17°33.5′N	087°01.0′E	087°17.2′E
Instant of Greatest Eclipse:	G0	11:04:09.0	45°04.5′N	024°01.8′E	024°18.0′E

Circumstances at Greatest Eclipse: Sun's Altitude = 59.3° Path Width = 112.3 km
Sun's Azimuth = 196.7° Central Duration = 02m22.9s

---

† Ephemeris Longitude is the terrestrial dynamical longitude assuming a uniformly rotating Earth.

* True Longitude is calculated by correcting the Ephemeris Longitude for the non-uniform rotation of Earth.
(T.L. = E.L. - 1.002738*ΔT/240, where ΔT(in seconds) = TDT - UT)

Note: Longitude is measured positive to the East.

Since ΔT is not known in advance, the value used in the predictions is an extrapolation based on pre-1997 measurements. Nevertheless, the actual value is expected to fall within ±0.3 seconds of the estimated ΔT used here.

Table 3

Path of the Umbral Shadow
Total Solar Eclipse of 1999 August 11

Universal	Northern Limit		Southern Limit		Center Line		Sun	Path	Central
Time	Latitude	Longitude	Latitude	Longitude	Latitude	Longitude	Alt °	Width km	Durat.
Limits	41°16.4′N	065°16.8′W	40°47.7′N	064°54.0′W	41°02.0′N	065°05.4′W	0	61	00m46.5s
09:35	46°08.0′N	046°39.6′W	45°48.4′N	044°45.1′W	45°58.6′N	045°41.4′W	16	79	01m09.1s
09:40	47°57.4′N	037°45.4′W	47°25.2′N	036°16.6′W	47°41.6′N	037°00.3′W	22	86	01m20.4s
09:45	49°04.2′N	030°56.5′W	48°24.6′N	029°42.0′W	48°44.6′N	030°18.7′W	28	91	01m29.4s
09:50	49°48.1′N	025°09.5′W	49°03.2′N	024°06.1′W	49°25.8′N	024°37.3′W	32	94	01m37.0s
09:55	50°16.4′N	020°01.7′W	49°27.5′N	019°08.1′W	49°52.0′N	019°34.4′W	36	97	01m43.8s
10:00	50°32.9′N	015°22.0′W	49°40.9′N	014°37.4′W	50°07.0′N	014°59.3′W	39	100	01m49.9s
10:05	50°40.1′N	011°04.0′W	49°45.6′N	010°27.9′W	50°12.9′N	010°45.6′W	42	102	01m55.4s
10:10	50°39.4′N	007°03.7′W	49°43.0′N	006°35.7′W	50°11.3′N	006°49.4′W	45	103	02m00.3s
10:15	50°32.0′N	003°18.4′W	49°34.2′N	002°58.2′W	50°03.1′N	003°08.0′W	47	105	02m04.7s
10:20	50°18.8′N	000°13.9′E	49°20.0′N	000°26.8′E	49°49.4′N	000°20.6′E	50	106	02m08.6s
10:25	50°00.3′N	003°34.8′E	49°00.9′N	003°40.8′E	49°30.6′N	003°37.9′E	52	107	02m12.1s
10:30	49°37.1′N	006°45.4′E	48°37.5′N	006°44.9′E	49°07.3′N	006°45.3′E	53	108	02m15.0s
10:35	49°09.8′N	009°46.9′E	48°10.1′N	009°40.3′E	48°39.9′N	009°43.7′E	55	109	02m17.5s
10:40	48°38.6′N	012°40.2′E	47°39.1′N	012°27.9′E	48°08.9′N	012°34.1′E	56	110	02m19.5s
10:45	48°03.9′N	015°26.0′E	47°04.9′N	015°08.4′E	47°34.4′N	015°17.2′E	58	111	02m21.1s
10:50	47°25.9′N	018°05.1′E	46°27.5′N	017°42.7′E	46°56.7′N	017°53.9′E	58	111	02m22.2s
10:55	46°44.8′N	020°38.3′E	45°47.3′N	020°11.5′E	46°16.1′N	020°24.8′E	59	112	02m22.8s
11:00	46°00.9′N	023°06.2′E	45°04.3′N	022°35.3′E	45°32.6′N	022°50.6′E	59	112	02m23.0s
11:05	45°14.2′N	025°29.4′E	44°18.7′N	024°54.8′E	44°46.5′N	025°12.0′E	59	112	02m22.7s
11:10	44°24.8′N	027°48.7′E	43°30.6′N	027°10.7′E	43°57.8′N	027°29.6′E	59	112	02m22.0s
11:15	43°32.9′N	030°04.6′E	42°40.1′N	029°23.6′E	43°06.5′N	029°44.0′E	59	113	02m20.9s
11:20	42°38.5′N	032°17.9′E	41°47.1′N	031°34.2′E	42°12.9′N	031°55.9′E	58	112	02m19.3s
11:25	41°41.6′N	034°29.3′E	40°51.7′N	033°43.1′E	41°16.7′N	034°06.0′E	57	112	02m17.3s
11:30	40°42.2′N	036°39.4′E	39°53.9′N	035°51.1′E	40°18.1′N	036°15.1′E	55	112	02m14.9s
11:35	39°40.2′N	038°49.2′E	38°53.6′N	037°58.8′E	39°17.0′N	038°23.8′E	54	111	02m12.0s
11:40	38°35.5′N	040°59.4′E	37°50.7′N	040°07.3′E	38°13.2′N	040°33.2′E	52	111	02m08.8s
11:45	37°28.0′N	043°11.2′E	36°45.1′N	042°17.4′E	37°06.6′N	042°44.1′E	50	110	02m05.1s
11:50	36°17.4′N	045°25.7′E	35°36.4′N	044°30.3′E	35°57.0′N	044°57.8′E	48	109	02m01.0s
11:55	35°03.4′N	047°44.3′E	34°24.5′N	046°47.5′E	34°44.0′N	047°15.7′E	46	107	01m56.5s
12:00	33°45.5′N	050°08.9′E	33°08.9′N	049°10.7′E	33°27.3′N	049°39.6′E	43	105	01m51.5s
12:05	32°23.2′N	052°41.9′E	31°49.0′N	051°42.3′E	32°06.2′N	052°11.9′E	40	103	01m46.0s
12:10	30°55.5′N	055°26.7′E	30°24.0′N	054°25.4′E	30°39.9′N	054°55.9′E	37	100	01m40.0s
12:15	29°21.0′N	058°28.3′E	28°52.4′N	057°25.1′E	29°06.9′N	057°56.5′E	33	97	01m33.4s
12:20	27°37.3′N	061°55.1′E	27°12.2′N	060°49.0′E	27°24.9′N	061°21.8′E	29	93	01m26.1s
12:25	25°39.8′N	066°02.6′E	25°19.3′N	064°51.8′E	25°29.7′N	065°27.0′E	24	87	01m17.7s
12:30	23°17.4′N	071°29.4′E	23°03.9′N	070°08.0′E	23°10.9′N	070°48.3′E	18	79	01m07.4s
12:35	19°31.1′N	081°43.2′E	19°42.9′N	079°13.7′E	19°38.5′N	080°24.0′E	7	65	00m51.6s
Limits	17°46.9′N	087°25.2′E	17°20.1′N	087°09.3′E	17°33.5′N	087°17.2′E	0	55	00m42.3s

Table 4

Physical Ephemeris of the Umbral Shadow
Total Solar Eclipse of 1999 August 11

Universal Time	Center Line Latitude	Center Line Longitude	Diameter Ratio	Eclipse Obscur.	Sun Alt °	Sun Azm °	Path Width km	Major Axis km	Minor Axis km	Umbra Veloc. km/s	Central Durat.
09:30.3	41°02.0´N	065°05.4´W	1.0143	1.0287	0.0	69.5	60.8	-	49.0	-	00m46.5s
09:35	45°58.6´N	045°41.4´W	1.0189	1.0382	15.6	83.9	78.9	241.0	64.6	2.856	01m09.1s
09:40	47°41.6´N	037°00.3´W	1.0208	1.0421	22.5	91.6	85.8	186.4	71.1	1.922	01m20.4s
09:45	48°44.6´N	030°18.7´W	1.0222	1.0449	27.6	98.3	90.5	163.4	75.7	1.531	01m29.4s
09:50	49°25.8´N	024°37.3´W	1.0233	1.0472	32.0	104.4	94.1	150.2	79.4	1.305	01m37.0s
09:55	49°52.0´N	019°34.4´W	1.0243	1.0491	35.7	110.3	97.1	141.3	82.5	1.156	01m43.8s
10:00	50°07.0´N	014°59.3´W	1.0250	1.0507	39.1	116.1	99.5	135.0	85.0	1.049	01m49.9s
10:05	50°12.9´N	010°45.6´W	1.0257	1.0520	42.1	121.8	101.6	130.2	87.3	0.969	01m55.4s
10:10	50°11.3´N	006°49.4´W	1.0263	1.0532	44.8	127.6	103.3	126.4	89.1	0.906	02m00.3s
10:15	50°03.1´N	003°08.0´W	1.0268	1.0542	47.3	133.4	104.9	123.4	90.8	0.857	02m04.7s
10:20	49°49.4´N	000°20.6´E	1.0272	1.0551	49.6	139.3	106.2	121.0	92.2	0.816	02m08.6s
10:25	49°30.6´N	003°37.9´E	1.0275	1.0558	51.6	145.4	107.4	119.0	93.4	0.784	02m12.1s
10:30	49°07.3´N	006°45.3´E	1.0278	1.0565	53.4	151.6	108.4	117.4	94.4	0.757	02m15.0s
10:35	48°39.9´N	009°43.7´E	1.0281	1.0570	55.0	158.0	109.3	116.1	95.2	0.735	02m17.5s
10:40	48°08.9´N	012°34.1´E	1.0283	1.0574	56.4	164.6	110.1	115.0	95.8	0.718	02m19.5s
10:45	47°34.4´N	015°17.2´E	1.0284	1.0577	57.5	171.4	110.7	114.1	96.3	0.704	02m21.1s
10:50	46°56.7´N	017°53.9´E	1.0285	1.0579	58.4	178.3	111.3	113.5	96.6	0.693	02m22.2s
10:55	46°16.1´N	020°24.8´E	1.0286	1.0580	59.0	185.3	111.7	113.0	96.8	0.686	02m22.8s
11:00	45°32.6´N	022°50.6´E	1.0286	1.0580	59.3	192.4	112.1	112.7	96.9	0.681	02m23.0s
11:05	44°46.5´N	025°12.0´E	1.0286	1.0580	59.3	199.4	112.3	112.5	96.8	0.679	02m22.7s
11:10	43°57.8´N	027°29.6´E	1.0285	1.0578	59.1	206.4	112.5	112.5	96.5	0.680	02m22.0s
11:15	43°06.5´N	029°44.0´E	1.0284	1.0576	58.6	213.1	112.5	112.6	96.1	0.683	02m20.9s
11:20	42°12.9´N	031°55.9´E	1.0282	1.0572	57.8	219.6	112.5	113.0	95.6	0.690	02m19.3s
11:25	41°16.7´N	034°06.0´E	1.0280	1.0568	56.7	225.8	112.2	113.5	94.9	0.699	02m17.3s
11:30	40°18.1´N	036°15.1´E	1.0278	1.0563	55.5	231.6	111.9	114.2	94.1	0.712	02m14.9s
11:35	39°17.0´N	038°23.8´E	1.0274	1.0556	53.9	237.1	111.4	115.1	93.0	0.728	02m12.0s
11:40	38°13.2´N	040°33.2´E	1.0271	1.0549	52.2	242.2	110.7	116.3	91.8	0.750	02m08.8s
11:45	37°06.6´N	042°44.1´E	1.0267	1.0540	50.2	247.0	109.8	117.8	90.5	0.776	02m05.1s
11:50	35°57.0´N	044°57.8´E	1.0262	1.0531	48.0	251.5	108.6	119.7	88.9	0.810	02m01.0s
11:55	34°44.0´N	047°15.7´E	1.0256	1.0519	45.5	255.7	107.2	122.0	87.1	0.852	01m56.5s
12:00	33°27.3´N	049°39.6´E	1.0250	1.0507	42.9	259.5	105.3	125.0	85.0	0.905	01m51.5s
12:05	32°06.2´N	052°11.9´E	1.0243	1.0492	39.9	263.2	103.1	128.9	82.7	0.974	01m46.0s
12:10	30°39.9´N	054°55.9´E	1.0235	1.0475	36.7	266.6	100.3	134.0	79.9	1.067	01m40.0s
12:15	29°06.9´N	057°56.5´E	1.0225	1.0456	33.0	269.9	96.9	141.0	76.8	1.196	01m33.4s
12:20	27°24.9´N	061°21.8´E	1.0214	1.0433	28.9	273.1	92.5	151.5	73.1	1.390	01m26.1s
12:25	25°29.7´N	065°27.0´E	1.0200	1.0405	23.9	276.2	86.9	168.9	68.5	1.714	01m17.7s
12:30	23°10.9´N	070°48.3´E	1.0182	1.0368	17.7	279.5	79.0	206.0	62.4	2.412	01m07.4s
12:35	19°38.5´N	080°24.0´E	1.0151	1.0304	7.1	283.8	64.8	422.7	51.7	6.535	00m51.6s
12:35.9	17°33.5´N	087°17.2´E	1.0130	1.0261	0.0	286.1	55.2	-	44.6	-	00m42.3s

TABLE 5

LOCAL CIRCUMSTANCES ON THE CENTER LINE
TOTAL SOLAR ECLIPSE OF 1999 AUGUST 11

Center Line Maximum Eclipse			First Contact				Second Contact			Third Contact			Fourth Contact			
U.T.	Durat.	Alt °	U.T.	P °	V °	Alt °	U.T.	P °	V °	U.T.	P °	V °	U.T.	P °	V °	Alt °
09:35	01m09.1s	16	08:36:00	277	321	6	09:34:26	98	144	09:35:35	278	324	10:39:27	98	144	27
09:40	01m20.4s	22	08:37:49	278	322	12	09:39:20	98	143	09:40:40	278	323	10:48:07	99	142	34
09:45	01m29.4s	28	08:40:13	279	322	17	09:44:15	99	142	09:45:45	279	322	10:55:56	100	139	39
09:50	01m37.0s	32	08:42:56	279	322	21	09:49:12	100	141	09:50:49	280	321	11:03:17	101	136	43
09:55	01m43.8s	36	08:45:52	280	322	25	09:54:08	101	140	09:55:52	281	320	11:10:16	102	133	46
10:00	01m49.9s	39	08:48:58	280	321	28	09:59:05	101	138	10:00:55	281	318	11:16:58	103	129	49
10:05	01m55.4s	42	08:52:11	281	321	31	10:04:02	102	137	10:05:58	282	317	11:23:24	103	125	51
10:10	02m00.3s	45	08:55:32	282	321	34	10:09:00	103	135	10:11:00	283	315	11:29:38	104	121	53
10:15	02m04.7s	47	08:59:00	282	320	37	10:13:58	103	133	10:16:02	283	312	11:35:38	105	117	54
10:20	02m08.6s	50	09:02:34	283	320	40	10:18:56	104	130	10:21:04	284	310	11:41:28	106	112	55
10:25	02m12.1s	52	09:06:14	283	319	42	10:23:54	105	128	10:26:06	285	307	11:47:06	106	107	56
10:30	02m15.0s	53	09:10:00	284	318	45	10:28:53	105	125	10:31:08	285	304	11:52:34	107	103	56
10:35	02m17.5s	55	09:13:53	285	317	47	10:33:51	106	121	10:36:09	286	301	11:57:53	107	98	56
10:40	02m19.5s	56	09:17:52	285	315	49	10:38:50	107	118	10:41:10	287	297	12:03:03	108	94	56
10:45	02m21.1s	58	09:21:58	286	314	51	10:43:49	107	114	10:46:11	287	293	12:08:05	108	89	55
10:50	02m22.2s	58	09:26:11	286	312	54	10:48:49	108	109	10:51:11	288	289	12:12:58	109	85	54
10:55	02m22.8s	59	09:30:31	287	310	56	10:53:49	108	105	10:56:11	288	284	12:17:44	109	82	53
11:00	02m23.0s	59	09:34:59	288	307	57	10:58:48	109	100	11:01:11	289	280	12:22:24	110	78	52
11:05	02m22.7s	59	09:39:35	288	304	59	11:03:49	109	96	11:06:11	289	275	12:26:56	110	75	51
11:10	02m22.0s	59	09:44:18	289	300	61	11:08:49	110	91	11:11:11	290	270	12:31:22	110	72	49
11:15	02m20.9s	59	09:49:10	289	296	62	11:13:49	110	86	11:16:10	290	265	12:35:43	111	69	48
11:20	02m19.3s	58	09:54:11	290	291	63	11:18:50	111	81	11:21:10	291	261	12:39:57	111	67	46
11:25	02m17.3s	57	09:59:21	290	286	64	11:23:51	111	77	11:26:09	291	257	12:44:07	111	64	44
11:30	02m14.9s	55	10:04:40	291	280	64	11:28:52	111	73	11:31:07	291	253	12:48:11	111	62	42
11:35	02m12.0s	54	10:10:08	291	273	65	11:33:54	112	69	11:36:06	292	249	12:52:10	111	60	40
11:40	02m08.8s	52	10:15:46	291	267	64	11:38:55	112	66	11:41:04	292	245	12:56:04	112	58	38
11:45	02m05.1s	50	10:21:35	292	260	64	11:43:57	112	62	11:46:02	292	242	12:59:53	112	56	36
11:50	02m01.0s	48	10:27:34	292	254	62	11:48:59	112	59	11:51:00	292	239	13:03:37	112	55	33
11:55	01m56.5s	46	10:33:44	292	248	61	11:54:02	112	56	11:55:58	292	236	13:07:17	112	53	31
12:00	01m51.5s	43	10:40:05	292	243	59	11:59:04	112	54	12:00:56	292	234	13:10:50	111	51	28
12:05	01m46.0s	40	10:46:39	293	238	56	12:04:07	112	51	12:05:53	292	231	13:14:18	111	50	25
12:10	01m40.0s	37	10:53:26	293	233	53	12:09:10	112	49	12:10:50	292	229	13:17:39	111	48	22
12:15	01m33.4s	33	11:00:28	292	229	49	12:14:13	112	47	12:15:47	292	227	13:20:52	111	47	19
12:20	01m26.1s	29	11:07:48	292	226	45	12:19:17	111	44	12:20:43	291	224	13:23:54	110	45	15
12:25	01m17.7s	24	11:15:32	292	222	40	12:24:21	111	42	12:25:39	291	222	13:26:41	110	44	10
12:30	01m07.4s	18	11:23:57	291	219	33	12:29:26	110	40	12:30:34	290	220	13:29:01	109	42	5
12:35	00m51.6s	7	11:34:19	290	215	21	12:34:34	109	37	12:35:26	289	217	-	-	-	-
12:35	00m42.3s	7	11:34:19	290	215	21	12:34:34	109	37	12:35:26	289	217	-	-	-	-
12:35	00m42.3s	7	11:34:19	290	215	21	12:34:34	109	37	12:35:26	289	217	-	-	-	-

TABLE 6

TOPOCENTRIC DATA AND PATH CORRECTIONS DUE TO LUNAR LIMB PROFILE
TOTAL SOLAR ECLIPSE OF 1999 AUGUST 11

Universal Time	Moon Topo H.P.	Moon Topo S.D.	Moon Rel. Ang.V	Topo Lib. Long	Sun Alt.	Sun Az.	Path Az.	North Limit P.A.	North Limit Int.	North Limit Ext.	South Limit Int.	South Limit Ext.	Central Durat. Cor.
	"	"	"/s	°	°	°	°	°	'	'	'	'	s
09:35	3542.7	964.7	0.518	5.54	15.6	83.9	72.6	7.7	-0.5	0.6	0.7	-2.4	1.3
09:40	3549.5	966.5	0.491	5.49	22.5	91.6	75.4	8.5	-0.5	0.8	0.7	-1.7	1.7
09:45	3554.4	967.8	0.471	5.45	27.6	98.3	78.2	9.2	-0.5	0.7	0.7	-0.8	1.7
09:50	3558.2	968.9	0.455	5.41	32.0	104.4	81.0	10.0	-0.4	0.4	0.7	-0.9	1.7
09:55	3561.5	969.7	0.442	5.37	35.7	110.3	83.8	10.7	-0.4	0.3	0.7	-1.7	1.6
10:00	3564.2	970.5	0.431	5.32	39.1	116.1	86.6	11.4	-0.4	0.6	0.7	-2.2	1.6
10:05	3566.5	971.1	0.422	5.28	42.1	121.8	89.3	12.1	-0.4	0.8	0.8	-2.6	1.7
10:10	3568.6	971.7	0.413	5.24	44.8	127.6	92.0	12.8	-0.3	0.8	0.8	-2.8	1.7
10:15	3570.3	972.1	0.406	5.20	47.3	133.4	94.6	13.5	-0.3	0.7	0.8	-2.9	1.7
10:20	3571.8	972.5	0.400	5.16	49.6	139.3	97.1	14.2	-0.3	0.6	0.8	-2.8	1.7
10:25	3573.1	972.9	0.395	5.11	51.6	145.4	99.6	14.8	-0.2	0.5	0.8	-2.6	1.7
10:30	3574.1	973.2	0.390	5.07	53.4	151.6	102.0	15.5	-0.2	0.4	0.8	-2.2	2.0
10:35	3575.0	973.4	0.387	5.03	55.0	158.0	104.3	16.1	-0.2	0.3	0.8	-1.7	2.0
10:40	3575.7	973.6	0.384	4.99	56.4	164.6	106.5	16.7	-0.2	0.2	0.9	-1.5	2.1
10:45	3576.2	973.7	0.382	4.94	57.5	171.4	108.5	17.3	-0.2	0.2	0.9	-1.8	2.0
10:50	3576.6	973.8	0.380	4.90	58.4	178.3	110.5	17.9	-0.2	0.3	0.8	-2.2	1.7
10:55	3576.8	973.9	0.379	4.86	59.0	185.3	112.3	18.4	-0.2	0.5	0.9	-2.4	1.6
11:00	3576.9	973.9	0.379	4.81	59.3	192.4	114.0	18.9	-0.2	0.7	0.9	-2.6	1.6
11:05	3576.8	973.9	0.379	4.77	59.3	199.4	115.6	19.4	-0.2	0.7	0.9	-2.6	1.5
11:10	3576.5	973.8	0.380	4.73	59.1	206.4	117.1	19.9	-0.1	0.6	0.9	-2.6	1.5
11:15	3576.1	973.7	0.382	4.69	58.6	213.1	118.4	20.3	-0.1	0.6	0.8	-2.7	0.9
11:20	3575.5	973.5	0.384	4.64	57.8	219.6	119.5	20.7	-0.1	0.7	0.8	-2.9	0.9
11:25	3574.8	973.3	0.386	4.60	56.7	225.8	120.5	21.0	-0.1	0.8	0.8	-3.0	0.8
11:30	3573.9	973.1	0.390	4.56	55.5	231.6	121.4	21.3	-0.1	0.9	0.8	-3.2	0.7
11:35	3572.8	972.8	0.394	4.52	53.9	237.1	122.0	21.6	-0.1	0.9	0.8	-3.3	0.6
11:40	3571.5	972.5	0.398	4.47	52.2	242.2	122.5	21.8	-0.0	1.0	0.8	-3.5	0.4
11:45	3570.1	972.1	0.404	4.43	50.2	247.0	122.9	21.9	-0.0	1.0	0.8	-3.5	0.3
11:50	3568.4	971.6	0.410	4.39	48.0	251.5	123.0	22.1	-0.0	1.0	0.8	-3.6	0.3
11:55	3566.5	971.1	0.417	4.35	45.5	255.7	122.9	22.1	-0.0	1.0	0.8	-3.6	0.2
12:00	3564.3	970.5	0.425	4.30	42.9	259.5	122.6	22.1	-0.0	1.0	0.8	-3.6	0.2
12:05	3561.8	969.8	0.434	4.26	39.9	263.2	122.1	22.0	-0.0	1.0	0.8	-3.7	-0.1
12:10	3558.9	969.0	0.445	4.22	36.7	266.6	121.2	21.9	-0.0	1.0	0.8	-3.6	-0.1
12:15	3555.6	968.1	0.457	4.18	33.0	269.9	120.1	21.7	-0.0	1.0	0.8	-3.5	-0.2
12:20	3551.6	967.1	0.471	4.13	28.9	273.1	118.7	21.4	-0.1	1.0	0.8	-3.4	-0.3
12:25	3546.8	965.8	0.489	4.09	23.9	276.2	116.7	20.9	-0.1	0.9	0.8	-3.1	-0.5
12:30	3540.5	964.1	0.512	4.05	17.7	279.5	114.0	20.3	-0.1	0.7	0.8	-2.7	-0.7
12:35	3529.5	961.1	0.553	4.01	7.1	283.8	109.2	19.0	-0.1	0.7	0.8	-2.7	-0.8

Table 7
Mapping Coordinates for the Umbral Path
Total Solar Eclipse of 1999 August 11

Longitude	Latitude of:			Universal Time at:			Circumstances on the Center Line .			
	Northern Limit	Southern Limit	Center Line	Northern Limit	Southern Limit	Center Line	Sun Alt	Sun Az.	Path Width	Center Durat.
				h m s	h m s	h m s	°	°	km	
065°00.0´W	41°21.17´N	–	41°03.56´N	09:30:20	–	09:30:14	0	–	–	–
064°00.0´W	41°38.43´N	41°02.72´N	41°20.47´N	09:30:39	09:30:04	09:30:22	1	70	62	00m47.6s
063°00.0´W	41°55.51´N	41°19.26´N	41°37.28´N	09:30:42	09:30:06	09:30:24	2	71	63	00m48.7s
062°00.0´W	42°12.41´N	41°35.66´N	41°53.93´N	09:30:47	09:30:09	09:30:28	2	72	64	00m49.8s
061°00.0´W	42°29.17´N	41°51.84´N	42°10.44´N	09:30:53	09:30:14	09:30:33	3	72	65	00m50.9s
060°00.0´W	42°45.79´N	42°07.92´N	42°26.78´N	09:31:00	09:30:20	09:30:40	4	73	66	00m52.0s
059°00.0´W	43°02.23´N	42°23.87´N	42°42.96´N	09:31:09	09:30:28	09:30:48	5	74	67	00m53.1s
058°00.0´W	43°18.49´N	42°39.60´N	42°58.96´N	09:31:20	09:30:37	09:30:58	6	74	68	00m54.3s
057°00.0´W	43°34.57´N	42°55.16´N	43°14.78´N	09:31:31	09:30:48	09:31:10	7	75	69	00m55.4s
056°00.0´W	43°50.44´N	43°10.53´N	43°30.40´N	09:31:45	09:31:00	09:31:22	7	76	70	00m56.6s
055°00.0´W	44°06.11´N	43°25.71´N	43°45.83´N	09:32:00	09:31:14	09:31:37	8	77	71	00m57.7s
054°00.0´W	44°21.57´N	43°40.69´N	44°01.05´N	09:32:16	09:31:29	09:31:52	9	77	72	00m58.9s
053°00.0´W	44°36.81´N	43°55.47´N	44°16.06´N	09:32:34	09:31:46	09:32:10	10	78	72	01m00.1s
052°00.0´W	44°51.81´N	44°10.08´N	44°30.85´N	09:32:53	09:32:04	09:32:28	11	79	73	01m01.3s
051°00.0´W	45°06.71´N	44°24.31´N	44°45.41´N	09:33:14	09:32:24	09:32:49	11	80	74	01m02.5s
050°00.0´W	45°21.22´N	44°38.40´N	44°59.73´N	09:33:36	09:32:45	09:33:10	12	80	75	01m03.7s
049°00.0´W	45°35.52´N	44°52.27´N	45°13.82´N	09:33:59	09:33:08	09:33:33	13	81	76	01m05.0s
048°00.0´W	45°49.57´N	45°05.89´N	45°27.65´N	09:34:24	09:33:32	09:33:58	14	82	77	01m06.2s
047°00.0´W	46°03.36´N	45°19.26´N	45°41.23´N	09:34:51	09:33:57	09:34:24	15	83	78	01m07.5s
046°00.0´W	46°16.89´N	45°32.37´N	45°54.55´N	09:35:19	09:34:24	09:34:51	15	84	79	01m08.7s
045°00.0´W	46°30.14´N	45°45.22´N	46°07.60´N	09:35:48	09:34:53	09:35:20	16	84	79	01m10.0s
044°00.0´W	46°43.11´N	45°57.80´N	46°20.38´N	09:36:18	09:35:23	09:35:50	17	85	80	01m11.3s
043°00.0´W	46°55.80´N	46°10.10´N	46°32.88´N	09:36:50	09:35:54	09:36:22	18	86	81	01m12.6s
042°00.0´W	47°08.20´N	46°22.12´N	46°45.09´N	09:37:24	09:36:26	09:36:55	19	87	82	01m13.9s
041°00.0´W	47°20.30´N	46°33.85´N	46°57.00´N	09:37:58	09:37:00	09:37:29	19	88	83	01m15.1s
040°00.0´W	47°32.10´N	46°45.28´N	47°08.62´N	09:38:34	09:37:36	09:38:05	20	89	83	01m16.5s
039°00.0´W	47°43.58´N	46°56.42´N	47°19.93´N	09:39:12	09:38:13	09:38:42	21	90	84	01m17.8s
038°00.0´W	47°54.76´N	47°07.25´N	47°30.94´N	09:39:50	09:38:51	09:39:20	22	91	85	01m19.1s
037°00.0´W	48°05.61´N	47°17.77´N	47°41.62´N	09:40:30	09:39:31	09:40:00	22	92	86	01m20.4s
036°00.0´W	48°16.14´N	47°27.97´N	47°51.99´N	09:41:12	09:40:11	09:40:41	23	93	87	01m21.7s
035°00.0´W	48°26.34´N	47°37.84´N	48°02.03´N	09:41:54	09:40:54	09:41:24	24	94	87	01m23.1s
034°00.0´W	48°36.20´N	47°47.39´N	48°11.74´N	09:42:38	09:41:37	09:42:08	25	95	88	01m24.4s
033°00.0´W	48°45.72´N	47°56.61´N	48°21.11´N	09:43:23	09:42:22	09:42:53	26	96	89	01m25.7s
032°00.0´W	48°54.90´N	48°05.49´N	48°30.14´N	09:44:10	09:43:09	09:43:39	26	97	89	01m27.1s
031°00.0´W	49°03.73´N	48°14.03´N	48°38.82´N	09:44:57	09:43:56	09:44:26	27	98	90	01m28.4s
030°00.0´W	49°12.20´N	48°22.22´N	48°47.16´N	09:45:46	09:44:45	09:45:15	28	99	91	01m29.8s
029°00.0´W	49°20.31´N	48°30.05´N	48°55.14´N	09:46:36	09:45:35	09:46:06	29	100	91	01m31.1s
028°00.0´W	49°28.07´N	48°37.53´N	49°02.75´N	09:47:28	09:46:27	09:46:57	29	101	92	01m32.5s
027°00.0´W	49°35.45´N	48°44.65´N	49°10.01´N	09:48:20	09:47:20	09:47:50	30	102	93	01m33.8s
026°00.0´W	49°42.46´N	48°51.40´N	49°16.89´N	09:49:14	09:48:14	09:48:44	31	103	93	01m35.2s
025°00.0´W	49°49.10´N	48°57.78´N	49°23.40´N	09:50:09	09:49:09	09:49:39	32	104	94	01m36.5s
024°00.0´W	49°55.36´N	49°03.78´N	49°29.53´N	09:51:05	09:50:06	09:50:35	32	105	95	01m37.9s
023°00.0´W	50°01.23´N	49°09.40´N	49°35.28´N	09:52:02	09:51:04	09:51:33	33	106	95	01m39.2s
022°00.0´W	50°06.71´N	49°14.63´N	49°40.64´N	09:53:01	09:52:03	09:52:32	34	107	96	01m40.6s
021°00.0´W	50°11.80´N	49°19.48´N	49°45.61´N	09:54:01	09:53:04	09:53:32	35	109	96	01m41.9s
020°00.0´W	50°16.50´N	49°23.93´N	49°50.18´N	09:55:02	09:54:05	09:54:33	35	110	97	01m43.3s

TABLE 7 - continued
MAPPING COORDINATES FOR THE UMBRAL PATH
TOTAL SOLAR ECLIPSE OF 1999 AUGUST 11

Longitude	Latitude of:			Universal Time at:			Circumstances on the Center Line .			
	Northern Limit	Southern Limit	Center Line	Northern Limit h m s	Southern Limit h m s	Center Line h m s	Sun Alt °	Sun Az. °	Path Width km	Center Durat.
019°00.0´W	50°20.79´N	49°27.98´N	49°54.36´N	09:56:04	09:55:09	09:55:36	36	111	97	01m44.6s
018°00.0´W	50°24.68´N	49°31.63´N	49°58.13´N	09:57:07	09:56:13	09:56:40	37	112	98	01m45.9s
017°00.0´W	50°28.16´N	49°34.87´N	50°01.49´N	09:58:12	09:57:19	09:57:45	38	114	98	01m47.3s
016°00.0´W	50°31.22´N	49°37.69´N	50°04.43´N	09:59:18	09:58:26	09:58:52	38	115	99	01m48.6s
015°00.0´W	50°33.86´N	49°40.10´N	50°06.96´N	10:00:25	09:59:34	09:59:59	39	116	100	01m49.9s
014°00.0´W	50°36.08´N	49°42.08´N	50°09.06´N	10:01:33	10:00:43	10:01:08	40	117	100	01m51.2s
013°00.0´W	50°37.88´N	49°43.63´N	50°10.74´N	10:02:42	10:01:54	10:02:18	40	119	100	01m52.5s
012°00.0´W	50°39.24´N	49°44.75´N	50°11.98´N	10:03:53	10:03:07	10:03:30	41	120	101	01m53.8s
011°00.0´W	50°40.16´N	49°45.43´N	50°12.78´N	10:05:05	10:04:20	10:04:42	42	122	101	01m55.1s
010°00.0´W	50°40.64´N	49°45.67´N	50°13.14´N	10:06:18	10:05:35	10:05:56	43	123	102	01m56.4s
009°00.0´W	50°40.67´N	49°45.45´N	50°13.05´N	10:07:32	10:06:51	10:07:12	43	124	102	01m57.6s
008°00.0´W	50°40.24´N	49°44.78´N	50°12.50´N	10:08:48	10:08:09	10:08:28	44	126	103	01m58.9s
007°00.0´W	50°39.36´N	49°43.64´N	50°11.49´N	10:10:05	10:09:28	10:09:46	45	127	103	02m00.1s
006°00.0´W	50°38.01´N	49°42.03´N	50°10.02´N	10:11:23	10:10:48	10:11:05	45	129	104	02m01.3s
005°00.0´W	50°36.20´N	49°39.95´N	50°08.07´N	10:12:42	10:12:10	10:12:26	46	130	104	02m02.5s
004°00.0´W	50°33.90´N	49°37.39´N	50°05.64´N	10:14:03	10:13:33	10:13:48	47	132	105	02m03.7s
003°00.0´W	50°31.12´N	49°34.34´N	50°02.73´N	10:15:25	10:14:57	10:15:11	47	134	105	02m04.9s
002°00.0´W	50°27.85´N	49°30.80´N	49°59.32´N	10:16:49	10:16:23	10:16:36	48	135	105	02m06.0s
001°00.0´W	50°24.09´N	49°26.75´N	49°55.42´N	10:18:14	10:17:51	10:18:02	49	137	106	02m07.2s
000°00.0´E	50°19.82´N	49°22.19´N	49°51.00´N	10:19:40	10:19:20	10:19:30	49	139	106	02m08.3s
001°00.0´E	50°15.03´N	49°17.11´N	49°46.07´N	10:21:07	10:20:50	10:20:59	50	141	106	02m09.3s
002°00.0´E	50°09.73´N	49°11.50´N	49°40.62´N	10:22:36	10:22:22	10:22:29	51	142	107	02m10.4s
003°00.0´E	50°03.90´N	49°05.37´N	49°34.64´N	10:24:07	10:23:56	10:24:01	51	144	107	02m11.4s
004°00.0´E	49°57.54´N	48°58.68´N	49°28.12´N	10:25:39	10:25:31	10:25:35	52	146	108	02m12.4s
005°00.0´E	49°50.63´N	48°51.45´N	49°21.05´N	10:27:12	10:27:07	10:27:10	52	148	108	02m13.4s
006°00.0´E	49°43.17´N	48°43.66´N	49°13.43´N	10:28:47	10:28:45	10:28:46	53	150	108	02m14.3s
007°00.0´E	49°35.15´N	48°35.30´N	49°05.23´N	10:30:24	10:30:25	10:30:24	54	152	108	02m15.2s
008°00.0´E	49°26.56´N	48°26.36´N	48°56.47´N	10:32:02	10:32:07	10:32:04	54	154	109	02m16.1s
009°00.0´E	49°17.39´N	48°16.83´N	48°47.12´N	10:33:41	10:33:50	10:33:45	55	156	109	02m16.9s
010°00.0´E	49°07.63´N	48°06.70´N	48°37.18´N	10:35:22	10:35:35	10:35:28	55	159	109	02m17.7s
011°00.0´E	48°57.27´N	47°55.97´N	48°26.64´N	10:37:05	10:37:21	10:37:13	56	161	110	02m18.4s
012°00.0´E	48°46.30´N	47°44.62´N	48°15.48´N	10:38:49	10:39:09	10:38:59	56	163	110	02m19.1s
013°00.0´E	48°34.72´N	47°32.65´N	48°03.70´N	10:40:35	10:40:59	10:40:47	57	166	110	02m19.8s
014°00.0´E	48°22.51´N	47°20.04´N	47°51.29´N	10:42:23	10:42:51	10:42:36	57	168	110	02m20.4s
015°00.0´E	48°09.66´N	47°06.78´N	47°38.24´N	10:44:12	10:44:44	10:44:28	57	171	111	02m20.9s
016°00.0´E	47°56.16´N	46°52.87´N	47°24.54´N	10:46:03	10:46:39	10:46:21	58	173	111	02m21.4s
017°00.0´E	47°42.01´N	46°38.30´N	47°10.17´N	10:47:56	10:48:36	10:48:15	58	176	111	02m21.8s
018°00.0´E	47°27.19´N	46°23.06´N	46°55.14´N	10:49:50	10:50:34	10:50:12	58	179	111	02m22.2s
019°00.0´E	47°11.69´N	46°07.13´N	46°39.43´N	10:51:46	10:52:35	10:52:10	59	181	111	02m22.5s
020°00.0´E	46°55.51´N	45°50.52´N	46°23.03´N	10:53:44	10:54:37	10:54:10	59	184	112	02m22.7s

Table 7 - continued
Mapping Coordinates for the Umbral Path
Total Solar Eclipse of 1999 August 11

Longitude	Latitude of: Northern Limit	Latitude of: Southern Limit	Latitude of: Center Line	Universal Time at: Northern Limit h m s	Universal Time at: Southern Limit h m s	Universal Time at: Center Line h m s	Circumstances on the Center Line: Sun Alt °	Sun Az. °	Path Width km	Center Durat.
021°00.0´E	46°38.64´N	45°33.22´N	46°05.94´N	10:55:43	10:56:40	10:56:12	59	187	112	02m22.9s
022°00.0´E	46°21.07´N	45°15.21´N	45°48.15´N	10:57:45	10:58:46	10:58:15	59	190	112	02m23.0s
023°00.0´E	46°02.79´N	44°56.50´N	45°29.66´N	10:59:47	11:00:53	11:00:20	59	193	112	02m23.0s
024°00.0´E	45°43.80´N	44°37.08´N	45°10.45´N	11:01:52	11:03:01	11:02:26	59	196	112	02m22.9s
025°00.0´E	45°24.09´N	44°16.95´N	44°50.53´N	11:03:58	11:05:11	11:04:34	59	199	112	02m22.8s
026°00.0´E	45°03.67´N	43°56.10´N	44°29.89´N	11:06:05	11:07:23	11:06:44	59	202	112	02m22.5s
027°00.0´E	44°42.52´N	43°34.54´N	44°08.54´N	11:08:14	11:09:36	11:08:55	59	205	112	02m22.2s
028°00.0´E	44°20.65´N	43°12.28´N	43°46.47´N	11:10:25	11:11:51	11:11:07	59	208	113	02m21.8s
029°00.0´E	43°58.07´N	42°49.31´N	43°23.69´N	11:12:37	11:14:06	11:13:21	59	211	113	02m21.3s
030°00.0´E	43°34.77´N	42°25.65´N	43°00.20´N	11:14:50	11:16:23	11:15:36	58	214	113	02m20.7s
031°00.0´E	43°10.76´N	42°01.29´N	42°36.02´N	11:17:04	11:18:41	11:17:52	58	217	112	02m20.0s
032°00.0´E	42°46.05´N	41°36.27´N	42°11.15´N	11:19:19	11:21:00	11:20:09	58	220	112	02m19.2s
033°00.0´E	42°20.66´N	41°10.58´N	41°45.60´N	11:21:36	11:23:19	11:22:27	57	223	112	02m18.4s
034°00.0´E	41°54.59´N	40°44.26´N	41°19.40´N	11:23:53	11:25:39	11:24:46	57	225	112	02m17.4s
035°00.0´E	41°27.87´N	40°17.32´N	40°52.57´N	11:26:11	11:28:00	11:27:05	56	228	112	02m16.3s
036°00.0´E	41°00.52´N	39°49.79´N	40°25.12´N	11:28:29	11:30:21	11:29:25	56	231	112	02m15.2s
037°00.0´E	40°32.55´N	39°21.69´N	39°57.08´N	11:30:48	11:32:42	11:31:45	55	234	112	02m13.9s
038°00.0´E	40°04.01´N	38°53.06´N	39°28.49´N	11:33:06	11:35:03	11:34:05	54	236	111	02m12.6s
039°00.0´E	39°34.91´N	38°23.93´N	38°59.37´N	11:35:25	11:37:23	11:36:24	53	239	111	02m11.1s
040°00.0´E	39°05.31´N	37°54.34´N	38°29.76´N	11:37:43	11:39:43	11:38:43	53	241	111	02m09.6s
041°00.0´E	38°35.22´N	37°24.32´N	37°59.70´N	11:40:01	11:42:02	11:41:02	52	243	111	02m08.0s
042°00.0´E	38°04.69´N	36°53.94´N	37°29.23´N	11:42:18	11:44:20	11:43:19	51	245	110	02m06.4s
043°00.0´E	37°33.77´N	36°23.21´N	36°58.40´N	11:44:35	11:46:37	11:45:36	50	248	110	02m04.6s
044°00.0´E	37°02.50´N	35°52.20´N	36°27.26´N	11:46:50	11:48:52	11:47:51	49	250	109	02m02.8s
045°00.0´E	36°30.93´N	35°20.96´N	35°55.84´N	11:49:03	11:51:06	11:50:05	48	252	109	02m00.9s
046°00.0´E	35°59.11´N	34°49.52´N	35°24.21´N	11:51:15	11:53:17	11:52:17	47	253	108	01m59.0s
047°00.0´E	35°27.08´N	34°17.95´N	34°52.40´N	11:53:25	11:55:27	11:54:26	46	255	107	01m57.0s
048°00.0´E	34°54.90´N	33°46.29´N	34°20.47´N	11:55:33	11:57:34	11:56:34	45	257	107	01m54.9s
049°00.0´E	34°22.62´N	33°14.59´N	33°48.48´N	11:57:39	11:59:38	11:58:39	44	259	106	01m52.9s
050°00.0´E	33°50.30´N	32°42.90´N	33°16.46´N	11:59:42	12:01:40	12:00:41	42	260	105	01m50.8s
051°00.0´E	33°17.97´N	32°11.26´N	32°44.48´N	12:01:42	12:03:38	12:02:41	41	262	104	01m48.6s
052°00.0´E	32°45.69´N	31°39.73´N	32°12.56´N	12:03:40	12:05:34	12:04:37	40	263	103	01m46.5s
053°00.0´E	32°13.51´N	31°08.34´N	31°40.78´N	12:05:34	12:07:26	12:06:31	39	264	102	01m44.3s
054°00.0´E	31°41.47´N	30°37.15´N	31°09.15´N	12:07:25	12:09:15	12:08:21	38	266	101	01m42.1s
055°00.0´E	31°09.61´N	30°06.17´N	30°37.74´N	12:09:13	12:11:00	12:10:07	37	267	100	01m39.9s
056°00.0´E	30°37.98´N	29°35.46´N	30°06.56´N	12:10:58	12:12:42	12:11:50	35	268	99	01m37.7s
057°00.0´E	30°06.62´N	29°05.05´N	29°35.67´N	12:12:38	12:14:20	12:13:30	34	269	98	01m35.5s
058°00.0´E	29°35.54´N	28°34.97´N	29°05.10´N	12:14:15	12:15:55	12:15:05	33	270	97	01m33.3s
059°00.0´E	29°04.81´N	28°05.24´N	28°34.86´N	12:15:49	12:17:25	12:16:38	32	271	96	01m31.1s
060°00.0´E	28°34.43´N	27°35.90´N	28°05.00´N	12:17:19	12:18:52	12:18:06	31	272	94	01m29.0s

TABLE 7 - continued
MAPPING COORDINATES FOR THE UMBRAL PATH
TOTAL SOLAR ECLIPSE OF 1999 AUGUST 11

Longitude	Latitude of: Northern Limit	Southern Limit	Center Line	Universal Time at: Northern Limit h m s	Southern Limit h m s	Center Line h m s	Circumstances on the Center Line: Sun Alt °	Sun Az. °	Path Width km	Center Durat.
061°00.0´E	28°04.43´N	27°06.95´N	27°35.53´N	12:18:45	12:20:15	12:19:30	29	273	93	01m26.9s
062°00.0´E	27°34.85´N	26°38.43´N	27°06.49´N	12:20:07	12:21:34	12:20:51	28	274	92	01m24.7s
063°00.0´E	27°05.70´N	26°10.36´N	26°37.87´N	12:21:25	12:22:50	12:22:08	27	274	90	01m22.7s
064°00.0´E	26°36.99´N	25°42.74´N	26°09.71´N	12:22:39	12:24:01	12:23:21	26	275	89	01m20.6s
065°00.0´E	26°08.75´N	25°15.58´N	25°42.01´N	12:23:50	12:25:09	12:24:30	24	276	88	01m18.6s
066°00.0´E	25°40.98´N	24°48.91´N	25°14.79´N	12:24:57	12:26:13	12:25:36	23	277	86	01m16.6s
067°00.0´E	25°13.70´N	24°22.72´N	24°48.07´N	12:26:00	12:27:14	12:26:37	22	277	85	01m14.6s
068°00.0´E	24°46.92´N	23°57.02´N	24°21.83´N	12:27:00	12:28:11	12:27:36	21	278	83	01m12.7s
069°00.0´E	24°20.65´N	23°31.83´N	23°56.10´N	12:27:56	12:29:04	12:28:30	20	278	82	01m10.8s
070°00.0´E	23°54.88´N	23°07.13´N	23°30.87´N	12:28:48	12:29:54	12:29:21	19	279	80	01m08.9s
071°00.0´E	23°29.62´N	22°42.94´N	23°06.15´N	12:29:37	12:30:40	12:30:09	17	280	79	01m07.1s
072°00.0´E	23°04.88´N	22°19.26´N	22°41.94´N	12:30:23	12:31:23	12:30:53	16	280	77	01m05.3s
073°00.0´E	22°40.65´N	21°56.08´N	22°18.24´N	12:31:05	12:32:03	12:31:34	15	281	76	01m03.5s
074°00.0´E	22°16.94´N	21°33.40´N	21°55.05´N	12:31:44	12:32:39	12:32:12	14	281	74	01m01.8s
075°00.0´E	21°53.74´N	21°11.22´N	21°32.36´N	12:32:19	12:33:12	12:32:46	13	282	73	01m00.1s
076°00.0´E	21°31.06´N	20°49.54´N	21°10.19´N	12:32:52	12:33:43	12:33:17	12	282	71	00m58.5s
077°00.0´E	21°08.88´N	20°28.36´N	20°48.51´N	12:33:21	12:34:10	12:33:46	11	282	70	00m56.8s
078°00.0´E	20°47.22´N	20°07.66´N	20°27.33´N	12:33:47	12:34:34	12:34:11	10	283	68	00m55.3s
079°00.0´E	20°26.05´N	19°47.45´N	20°06.65´N	12:34:11	12:34:56	12:34:33	9	283	67	00m53.7s
080°00.0´E	20°05.39´N	19°27.70´N	19°46.45´N	12:34:31	12:35:14	12:34:53	7	284	65	00m52.2s
081°00.0´E	19°45.25´N	19°08.40´N	19°26.74´N	12:34:49	12:35:30	12:35:10	6	284	64	00m50.7s
082°00.0´E	19°25.83´N	18°49.03´N	19°07.51´N	12:35:04	12:35:44	12:35:24	5	284	63	00m49.3s
083°00.0´E	19°05.99´N	18°31.43´N	18°48.75´N	12:35:17	12:35:54	12:35:35	4	285	61	00m47.9s
084°00.0´E	18°50.16´N	18°13.57´N	18°30.45´N	12:35:25	12:36:02	12:35:44	3	285	60	00m46.5s
085°00.0´E	19°09.38´N	17°56.30´N	18°12.62´N	12:35:14	12:36:08	12:35:51	2	285	58	00m45.2s
086°00.0´E	18°11.49´N	17°39.16´N	17°55.23´N	12:35:38	12:36:11	12:35:55	1	286	57	00m43.9s

TABLE 8

MAPPING COORDINATES FOR THE ZONES OF GRAZING ECLIPSE
TOTAL SOLAR ECLIPSE OF 1999 AUGUST 11

Longitude	North Graze Zone Latitudes		Northern Limit	South Graze Zone Latitudes		Southern Limit			
	Northern Limit	Southern Limit	Universal Time	Northern Limit	Southern Limit	Universal Time	Path Azm	Elev Fact	Scale Fact
° ′	° ′	° ′	h m s	° ′	° ′	h m s	°		km/"
005 00.0W	50 36.96N	50 35.87N	10:12:42	49 40.76N	49 37.09N	10:12:10	93.3	-0.58	2.09
004 00.0W	50 34.64N	50 33.59N	10:14:03	49 38.20N	49 34.52N	10:13:33	94.0	-0.58	2.09
003 00.0W	50 31.86N	50 30.82N	10:15:25	49 35.16N	49 31.46N	10:14:57	94.7	-0.58	2.09
002 00.0W	50 28.57N	50 27.56N	10:16:49	49 31.61N	49 27.93N	10:16:23	95.4	-0.58	2.09
001 00.0W	50 24.78N	50 23.81N	10:18:14	49 27.57N	49 23.90N	10:17:51	96.1	-0.57	2.09
000 00.0E	50 20.47N	50 19.55N	10:19:40	49 23.01N	49 19.37N	10:19:20	96.9	-0.57	2.08
001 00.0E	50 15.65N	50 14.78N	10:21:07	49 17.94N	49 14.35N	10:20:50	97.6	-0.57	2.08
002 00.0E	50 10.32N	50 09.49N	10:22:36	49 12.34N	49 08.80N	10:22:22	98.4	-0.57	2.08
003 00.0E	50 04.46N	50 03.67N	10:24:07	49 06.21N	49 02.74N	10:23:56	99.1	-0.57	2.08
004 00.0E	49 58.04N	49 57.30N	10:25:39	48 59.53N	48 56.16N	10:25:31	99.9	-0.57	2.08
005 00.0E	49 51.09N	49 50.41N	10:27:12	48 52.28N	48 49.02N	10:27:07	100.6	-0.57	2.08
006 00.0E	49 43.61N	49 42.95N	10:28:47	48 44.49N	48 41.35N	10:28:45	101.4	-0.57	2.08
007 00.0E	49 35.56N	49 34.92N	10:30:24	48 36.14N	48 33.13N	10:30:25	102.2	-0.56	2.08
008 00.0E	49 26.93N	49 26.33N	10:32:02	48 27.20N	48 24.35N	10:32:07	102.9	-0.56	2.08
009 00.0E	49 17.70N	49 17.16N	10:33:41	48 17.67N	48 14.99N	10:33:50	103.7	-0.56	2.08
010 00.0E	49 07.86N	49 07.40N	10:35:22	48 07.55N	48 05.05N	10:35:35	104.5	-0.56	2.08
011 00.0E	48 57.49N	48 57.04N	10:37:05	47 56.82N	47 54.37N	10:37:21	105.2	-0.56	2.08
012 00.0E	48 46.54N	48 46.07N	10:38:49	47 45.48N	47 43.08N	10:39:09	106.0	-0.56	2.08
013 00.0E	48 34.95N	48 34.49N	10:40:35	47 33.51N	47 31.17N	10:40:59	106.8	-0.56	2.08
014 00.0E	48 22.73N	48 22.28N	10:42:23	47 20.90N	47 18.39N	10:42:51	107.5	-0.57	2.08
015 00.0E	48 09.90N	48 09.43N	10:44:12	47 07.65N	47 04.97N	10:44:44	108.3	-0.57	2.08
016 00.0E	47 56.42N	47 55.93N	10:46:03	46 53.74N	46 50.91N	10:46:39	109.1	-0.57	2.08
017 00.0E	47 42.27N	47 41.78N	10:47:56	46 39.18N	46 36.20N	10:48:36	109.8	-0.57	2.08
018 00.0E	47 27.51N	47 26.98N	10:49:50	46 23.91N	46 20.83N	10:50:34	110.6	-0.57	2.08
019 00.0E	47 12.11N	47 11.50N	10:51:46	46 07.99N	46 04.81N	10:52:35	111.3	-0.57	2.08
020 00.0E	46 56.01N	46 55.33N	10:53:44	45 51.38N	45 48.11N	10:54:37	112.0	-0.57	2.09
021 00.0E	46 39.21N	46 38.46N	10:55:43	45 34.07N	45 30.73N	10:56:40	112.7	-0.58	2.09
022 00.0E	46 21.69N	46 20.89N	10:57:45	45 16.06N	45 12.67N	10:58:46	113.4	-0.58	2.09
023 00.0E	46 03.44N	46 02.61N	10:59:47	44 57.35N	44 53.92N	11:00:53	114.1	-0.58	2.09
024 00.0E	45 44.47N	45 43.63N	11:01:52	44 37.93N	44 34.47N	11:03:01	114.8	-0.59	2.10
025 00.0E	45 24.77N	45 23.93N	11:03:58	44 17.80N	44 14.33N	11:05:11	115.5	-0.59	2.10
026 00.0E	45 04.33N	45 03.51N	11:06:05	43 56.96N	43 53.49N	11:07:23	116.1	-0.59	2.10
027 00.0E	44 43.16N	44 42.37N	11:08:14	43 35.40N	43 31.96N	11:09:36	116.8	-0.60	2.11
028 00.0E	44 21.25N	44 20.51N	11:10:25	43 13.13N	43 09.73N	11:11:51	117.4	-0.60	2.11
029 00.0E	43 58.65N	43 57.93N	11:12:37	42 50.12N	42 46.67N	11:14:06	118.0	-0.61	2.12
030 00.0E	43 35.36N	43 34.63N	11:14:50	42 26.46N	42 22.90N	11:16:23	118.5	-0.61	2.12
031 00.0E	43 11.38N	43 10.63N	11:17:04	42 02.11N	41 58.45N	11:18:41	119.0	-0.62	2.12
032 00.0E	42 46.70N	42 45.93N	11:19:19	41 37.08N	41 33.34N	11:21:00	119.6	-0.62	2.13
033 00.0E	42 21.37N	42 20.54N	11:21:36	41 11.40N	41 07.59N	11:23:19	120.0	-0.63	2.14
034 00.0E	41 55.37N	41 54.49N	11:23:53	40 45.07N	40 41.20N	11:25:40	120.5	-0.63	2.14
035 00.0E	41 28.70N	41 27.77N	11:26:11	40 18.13N	40 14.21N	11:28:00	120.9	-0.64	2.15
036 00.0E	41 01.38N	41 00.42N	11:28:29	39 50.60N	39 46.59N	11:30:21	121.3	-0.64	2.15
037 00.0E	40 33.45N	40 32.46N	11:30:48	39 22.50N	39 18.42N	11:32:42	121.6	-0.65	2.16
038 00.0E	40 04.94N	40 03.93N	11:33:06	38 53.87N	38 49.72N	11:35:03	121.9	-0.66	2.17
039 00.0E	39 35.86N	39 34.84N	11:35:25	38 24.72N	38 20.53N	11:37:23	122.2	-0.66	2.17
040 00.0E	39 06.28N	39 05.26N	11:37:43	37 55.12N	37 50.89N	11:39:43	122.4	-0.67	2.18

TABLE 8 - continued

MAPPING COORDINATES FOR THE ZONES OF GRAZING ECLIPSE
TOTAL SOLAR ECLIPSE OF 1999 AUGUST 11

Longitude	North Graze Zone Latitudes		Northern Limit	South Graze Zone Latitudes		Southern Limit			
	Northern Limit	Southern Limit	Universal Time	Northern Limit	Southern Limit	Universal Time	Path Azm	Elev Fact	Scale Fact
° ′	° ′	° ′	h m s	° ′	° ′	h m s	°		km/"
041 00.0E	38 36.20N	38 35.18N	11:40:01	37 25.11N	37 20.84N	11:42:02	122.6	-0.68	2.19
042 00.0E	38 05.68N	38 04.66N	11:42:18	36 54.72N	36 50.42N	11:44:20	122.8	-0.68	2.19
043 00.0E	37 34.76N	37 33.74N	11:44:35	36 24.00N	36 19.67N	11:46:37	122.9	-0.69	2.20
044 00.0E	37 03.49N	37 02.47N	11:46:50	35 52.99N	35 48.64N	11:48:52	123.0	-0.70	2.21
045 00.0E	36 31.92N	36 30.90N	11:49:03	35 21.75N	35 17.37N	11:51:06	123.0	-0.71	2.22
046 00.0E	36 00.10N	35 59.08N	11:51:15	34 50.32N	34 45.93N	11:53:17	123.0	-0.71	2.22
047 00.0E	35 28.07N	35 27.05N	11:53:25	34 18.75N	34 14.35N	11:55:27	122.9	-0.72	2.23
048 00.0E	34 55.90N	34 54.87N	11:55:33	33 47.09N	33 42.68N	11:57:34	122.8	-0.73	2.24
049 00.0E	34 23.62N	34 22.59N	11:57:39	33 15.40N	33 10.98N	11:59:38	122.7	-0.73	2.24
050 00.0E	33 51.29N	33 50.26N	11:59:42	32 43.66N	32 39.23N	12:01:40	122.5	-0.74	2.25
051 00.0E	33 19.00N	33 17.96N	12:01:42	32 12.03N	32 07.61N	12:03:38	122.3	-0.74	2.26
052 00.0E	32 46.72N	32 45.68N	12:03:40	31 40.50N	31 36.08N	12:05:34	122.1	-0.75	2.26
053 00.0E	32 14.54N	32 13.49N	12:05:34	31 09.12N	31 04.70N	12:07:26	121.8	-0.75	2.27
054 00.0E	31 42.50N	31 41.45N	12:07:25	30 37.93N	30 33.52N	12:09:15	121.6	-0.76	2.27
055 00.0E	31 10.64N	31 09.59N	12:09:13	30 06.96N	30 02.57N	12:11:00	121.2	-0.76	2.28
056 00.0E	30 39.00N	30 37.95N	12:10:58	29 36.26N	29 31.89N	12:12:42	120.9	-0.77	2.28
057 00.0E	30 07.63N	30 06.58N	12:12:38	29 05.85N	29 01.50N	12:14:20	120.5	-0.77	2.29
058 00.0E	29 36.55N	29 35.50N	12:14:15	28 35.77N	28 31.46N	12:15:55	120.1	-0.77	2.29
059 00.0E	29 05.80N	29 04.76N	12:15:49	28 06.05N	28 01.77N	12:17:25	119.7	-0.78	2.29
060 00.0E	28 35.41N	28 34.37N	12:17:19	27 36.71N	27 32.47N	12:18:52	119.3	-0.78	2.30
061 00.0E	28 05.41N	28 04.37N	12:18:45	27 07.77N	27 03.58N	12:20:15	118.8	-0.78	2.30
062 00.0E	27 35.80N	27 34.78N	12:20:07	26 39.25N	26 35.13N	12:21:34	118.4	-0.78	2.30
063 00.0E	27 06.62N	27 05.62N	12:21:25	26 11.18N	26 07.12N	12:22:50	117.9	-0.79	2.30
064 00.0E	26 37.88N	26 36.91N	12:22:39	25 43.56N	25 39.57N	12:24:01	117.4	-0.79	2.30
065 00.0E	26 09.60N	26 08.66N	12:23:50	25 16.38N	25 12.49N	12:25:09	116.9	-0.79	2.31
066 00.0E	25 41.85N	25 40.92N	12:24:57	24 49.71N	24 45.87N	12:26:13	116.4	-0.79	2.31
067 00.0E	25 14.56N	25 13.64N	12:26:00	24 23.52N	24 19.74N	12:27:14	115.9	-0.79	2.31
068 00.0E	24 47.76N	24 46.85N	12:27:00	23 57.83N	23 54.12N	12:28:11	115.4	-0.79	2.31
069 00.0E	24 21.46N	24 20.57N	12:27:56	23 32.64N	23 28.99N	12:29:04	114.9	-0.79	2.31
070 00.0E	23 55.66N	23 54.80N	12:28:48	23 07.94N	23 04.38N	12:29:54	114.4	-0.79	2.30
071 00.0E	23 30.37N	23 29.53N	12:29:37	22 43.75N	22 40.28N	12:30:40	113.9	-0.79	2.30
072 00.0E	23 05.59N	23 04.79N	12:30:23	22 20.07N	22 16.57N	12:31:23	113.4	-0.78	2.30
073 00.0E	22 41.31N	22 40.56N	12:31:05	21 56.89N	21 53.37N	12:32:03	112.9	-0.78	2.30
074 00.0E	22 17.61N	22 16.84N	12:31:44	21 34.21N	21 30.67N	12:32:39	112.4	-0.78	2.30
075 00.0E	21 54.45N	21 53.64N	12:32:19	21 12.03N	21 08.48N	12:33:12	111.9	-0.78	2.30
076 00.0E	21 31.79N	21 30.95N	12:32:52	20 50.35N	20 46.80N	12:33:43	111.4	-0.78	2.29
077 00.0E	21 09.63N	21 08.77N	12:33:21	20 29.17N	20 25.62N	12:34:10	110.9	-0.78	2.29
078 00.0E	20 47.97N	20 47.10N	12:33:47	20 08.47N	20 04.94N	12:34:34	110.4	-0.77	2.29
079 00.0E	20 26.81N	20 25.94N	12:34:11	19 48.25N	19 44.75N	12:34:56	109.9	-0.77	2.28
080 00.0E	20 06.15N	20 05.28N	12:34:31	19 28.51N	19 25.05N	12:35:14	109.4	-0.76	2.28
081 00.0E	19 45.99N	19 45.13N	12:34:49	19 09.17N	19 05.88N	12:35:30	108.9	-0.75	2.27
082 00.0E	19 26.62N	19 25.78N	12:35:04	18 49.92N	18 46.62N	12:35:44	108.3	-0.73	2.24
083 00.0E	19 06.49N	19 05.66N	12:35:17	18 32.36N	18 28.56N	12:35:54	107.6	-0.66	2.17
084 00.0E	18 47.98N	18 47.18N	12:35:26	18 13.57N	18 13.57N	12:36:02	107.4	-0.72	2.23

TABLE 9
LOCAL CIRCUMSTANCES FOR THE UNITED STATES OF AMERICA
TOTAL SOLAR ECLIPSE OF 1999 AUGUST 11

Location Name	Latitude	Longitude	Elev.	First Contact				Second Contact			Third Contact			Fourth Contact				Maximum Eclipse					Eclip. Mag.	Eclip. Obs.	Umbral Durat.
			m	U.T. h m s	P °	V °	Alt °	U.T. h m s	P °	V °	U.T. h m s	P °	V °	U.T. h m s	P °	V °	Alt °	U.T. h m s	P °	V °	Alt °	Azm °			
CONNECTICUT																									
Bridgeport	41°11′N	073°11′W	3	–				–			–			10:27:42.5	100	149	5	09:59 Rise	–	–	0	69	0.526	0.421	
Hartford	41°46′N	072°41′W	52	–				–			–			10:28:11.9	101	149	5	09:54 Rise	–	–	0	68	0.611	0.520	
New Haven	41°18′N	072°56′W	12	–				–			–			10:27:47.0	100	149	5	09:57 Rise	–	–	0	69	0.556	0.455	
Stamford	41°03′N	073°32′W	11	–				–			–			10:27:37.2	100	149	4	10:00 Rise	–	–	0	69	0.500	0.393	
Waterbury	41°30′N	073°00′W	79	–				–			–			10:27:58.8	100	149	5	09:56 Rise	–	–	0	68	0.581	0.485	
DELAWARE																									
Dover	39°09′N	075°31′W	–	–				–			–			10:26:10.9	98	148	2	10:12 Rise	–	–	0	70	0.252	0.147	
DISTRICT OF COLUMBIA																									
Washington	38°53′N	077°02′W	4	–				–			–			10:26:16.3	98	148	1	10:19 Rise	–	–	0	70	0.142	0.063	
MAINE																									
Augusta	44°18′N	069°46′W	14	–				–			–			10:30:29.7	103	149	8	09:38 Rise	–	–	0	68	0.887	0.860	
Portland	43°40′N	070°17′W	14	–				–			–			10:29:53.1	102	149	8	09:41 Rise	–	–	0	68	0.848	0.811	
MARYLAND																									
Annapolis	38°58′N	076°29′W	–	–				–			–			10:26:13.3	98	148	2	10:17 Rise	–	–	0	70	0.177	0.087	
Baltimore	39°17′N	076°36′W	45	–				–			–			10:26:31.9	99	148	2	10:15 Rise	–	–	0	69	0.209	0.111	
MASSACHUSETTS																									
Boston	42°21′N	071°03′W	5	–				–			–			10:28:37.7	101	149	7	09:48 Rise	–	–	0	68	0.734	0.669	
Springfield	42°07′N	072°33′W	26	–				–			–			10:28:30.8	101	149	6	09:53 Rise	–	–	0	68	0.630	0.543	
Worcester	42°16′N	071°49′W	145	–				–			–			10:28:36.1	101	149	6	09:49 Rise	–	–	0	68	0.714	0.644	
MICHIGAN																									
Sault Ste. Marie	46°28′N	084°22′W	220	–				–			–			10:34:11.6	111	153	0	10:28 Rise	–	–	0	66	0.106	0.041	
NEW HAMPSHIRE																									
Concord	43°12′N	071°32′W	104	–				–			–			10:29:28.7	102	149	7	09:46 Rise	–	–	0	68	0.772	0.715	
NEW JERSEY																									
Elizabeth	40°40′N	074°13′W	6	–				–			–			10:27:21.4	100	148	4	10:04 Rise	–	–	0	69	0.430	0.317	
Jersey City	40°43′N	074°05′W	6	–				–			–			10:27:23.0	100	148	4	10:03 Rise	–	–	0	69	0.442	0.330	
Newark	40°44′N	074°10′W	2	–				–			–			10:27:24.7	100	148	4	10:04 Rise	–	–	0	69	0.434	0.321	
Paterson	40°55′N	074°10′W	30	–				–			–			10:27:35.0	100	149	4	10:02 Rise	–	–	0	69	0.458	0.347	
Trenton	40°13′N	074°44′W	11	–				–			–			10:27:01.3	99	148	3	10:07 Rise	–	–	0	69	0.374	0.259	
NEW YORK																									
Albany	42°39′N	073°45′W	84	–				–			–			10:29:08.6	102	149	5	09:56 Rise	–	–	0	68	0.592	0.497	
Buffalo	42°53′N	078°52′W	215	–				–			–			10:30:08.5	104	150	2	10:15 Rise	–	–	0	68	0.275	0.167	
New York	40°43′N	074°01′W	40	–				–			–			10:27:22.4	100	148	4	10:02 Rise	–	–	0	69	0.461	0.349	
Rochester	43°09′N	077°36′W	167	–				–			–			10:30:08.1	104	150	3	10:10 Rise	–	–	0	68	0.370	0.256	
Syracuse	43°02′N	076°08′W	125	–				–			–			10:29:47.9	104	150	4	10:05 Rise	–	–	0	68	0.457	0.346	
Yonkers	40°57′N	073°54′W	3	–				–			–			10:27:34.6	100	149	4	10:02 Rise	–	–	0	69	0.464	0.354	
OHIO																									
Youngstown	41°05′N	080°38′W	256	–				–			–			10:29:02.7	103	150	0	10:26 Rise	–	–	0	68	0.061	0.018	
PENNSYLVANIA																									
Allentown	40°35′N	075°30′W	78	–				–			–			10:27:29.4	100	149	3	10:08 Rise	–	–	0	69	0.360	0.246	
Erie	42°07′N	080°05′W	209	–				–			–			10:29:45.9	104	150	1	10:22 Rise	–	–	0	68	0.149	0.068	
Harrisburg	40°16′N	076°53′W	111	–				–			–			10:27:28.2	100	149	2	10:14 Rise	–	–	0	69	0.255	0.149	
Philadelphia	39°57′N	075°09′W	2	–				–			–			10:26:50.9	99	148	3	10:09 Rise	–	–	0	69	0.324	0.211	
Pittsburgh	40°26′N	079°59′W	228	–				–			–			10:28:20.3	102	149	0	10:25 Rise	–	–	0	69	0.067	0.021	
Scranton	41°24′N	075°39′W	221	–				–			–			10:28:15.3	101	149	3	10:05 Rise	–	–	0	68	0.416	0.302	
RHODE ISLAND																									
Providence	41°49′N	071°24′W	16	–				–			–			10:28:07.9	100	149	6	09:50 Rise	–	–	0	68	0.690	0.614	
VERMONT																									
Montpelier	44°15′N	072°34′W	148	–				–			–			10:30:32.3	104	150	6	09:47 Rise	–	–	0	67	0.759	0.699	
VIRGINIA																									
Alexandria	38°49′N	077°05′W	–	–				–			–			10:26:13.6	98	148	1	10:19 Rise	–	–	0	70	0.128	0.054	
Chesapeake	38°48′N	076°16′W	–	–				–			–			10:26:01.3	98	148	2	10:16 Rise	–	–	0	70	0.183	0.092	
Hampton	37°02′N	076°21′W	–	–				–			–			10:24:25.7	95	147	1	10:20 Rise	–	–	0	70	0.087	0.031	
Newport News	37°03′N	076°29′W	–	–				–			–			10:24:28.6	95	147	1	10:20 Rise	–	–	0	70	0.079	0.027	
Norfolk	36°50′N	076°17′W	3	–				–			–			10:24:13.7	95	147	1	10:19 Rise	–	–	0	70	0.088	0.031	
Portsmouth	36°50′N	076°19′W	3	–				–			–			10:24:14.2	95	147	1	10:20 Rise	–	–	0	70	0.085	0.030	
Richmond	37°33′N	077°27′W	50	–				–			–			10:25:11.0	96	147	0	10:22 Rise	–	–	0	70	0.062	0.019	
Virginia Beach	36°50′N	075°58′W	–	–				–			–			10:24:08.8	95	147	1	10:19 Rise	–	–	0	70	0.103	0.039	

TABLE 10
LOCAL CIRCUMSTANCES FOR CANADA
TOTAL SOLAR ECLIPSE OF 1999 AUGUST 11

Location Name	Latitude	Longitude	Elev.	First Contact				Second Contact			Third Contact			Fourth Contact				Maximum Eclipse					Eclip.	Eclip.	Umbral
				U.T.	P	V	Alt	U.T.	P	V	U.T.	P	V	U.T.	P	V	Alt	U.T.	P	V	Alt	Azm	Mag.	Obs.	Durat.
			m	h m s	°	°	°	h m s	°	°	h m s	°	°	h m s	°	°	°	h m s	°	°	°	°			
NEW BRUNSWICK																									
Fredericton, NB	45°58′N	066°39′W	9	–				–			–			10:32:17.2	104	150	11	09:35:49.8	187	230	2	69	0.881	0.852	
NEWFOUNDLAND																									
Saint John's, NF	47°34′N	052°43′W	64	08:39:18.1	273	314	2	–			–			10:37:34.3	102	147	22	09:36:05.1	187	231	11	80	0.930	0.916	
NOVA SCOTIA																									
Halifax, NS	44°39′N	063°36′W	25	–				–			–			10:31:14.9	101	148	13	09:33:41.4	187	231	3	71	0.931	0.916	
ONTARIO																									
Etobicoke, ON	43°39′N	079°34′W	–	–				–			–			10:30:55.6	106	151	2	10:19 Rise	–	–	0	68	0.215	0.116	
Hamilton, ON	43°15′N	079°51′W	100	–				–			–			10:30:39.1	105	151	1	10:19 Rise	–	–	0	68	0.211	0.113	
Kitchener, ON	43°27′N	080°29′W	335	–				–			–			10:30:56.9	106	151	1	10:19 Rise	–	–	0	67	0.210	0.112	
London, ON	42°59′N	081°14′W	251	–				–			–			10:30:44.4	106	151	1	10:24 Rise	–	–	0	68	0.123	0.051	
Mississauga, ON	43°35′N	079°37′W	–	–				–			–			10:30:52.9	106	151	2	10:19 Rise	–	–	0	68	0.208	0.110	
North York, ON	43°46′N	079°25′W	–	–				–			–			10:30:59.7	106	151	2	10:18 Rise	–	–	0	68	0.231	0.129	
Saint Catharine…	43°10′N	079°15′W	110	–				–			–			10:30:27.5	105	151	2	10:17 Rise	–	–	0	68	0.249	0.144	
Scarborough, ON	43°44′N	079°16′W	–	–				–			–			10:30:56.3	106	151	2	10:18 Rise	–	–	0	68	0.240	0.136	
Toronto, ON	43°39′N	079°23′W	116	–							–			10:30:53.4	106	151	2	10:16 Rise	–	–	0	68	0.268	0.160	
PRINCE EDWARD ISLAND																									
Charlottetown, …	46°14′N	063°08′W	55	–				–			–			10:33:00.0	104	149	14	09:35:19.8	187	230	4	72	0.897	0.874	
QUEBEC																									
Laval, QC	45°35′N	073°45′W	–	–				–			–			10:31:51.3	106	151	6	09:51 Rise	–	–	0	67	0.702	0.629	
Montréal, QC	45°31′N	073°34′W	57	–				–			–			10:31:46.9	106	151	6	09:49 Rise	–	–	0	67	0.735	0.670	
Ottawa, QC	45°25′N	075°42′W	114	–				–			–			10:31:51.9	107	151	5	09:57 Rise	–	–	0	67	0.609	0.517	
Québec, QC	46°49′N	071°14′W	73	–				–			–			10:32:55.7	107	151	8	09:38:05.3	187	228	0	67	0.832	0.791	

TABLE 11
LOCAL CIRCUMSTANCES FOR THE NORTH ATLANTIC
TOTAL SOLAR ECLIPSE OF 1999 AUGUST 11

Location Name	Latitude	Longitude	Elev.	First Contact				Second Contact			Third Contact			Fourth Contact				Maximum Eclipse					Eclip.	Eclip.	Umbral
				U.T.	P	V	Alt	U.T.	P	V	U.T.	P	V	U.T.	P	V	Alt	U.T.	P	V	Alt	Azm	Mag.	Obs.	Durat.
			m	h m s	°	°	°	h m s	°	°	h m s	°	°	h m s	°	°	°	h m s	°	°	°	°			
AZORES																									
Ponta Delgada	37°44′N	025°40′W	36	08:32:25.7	297	352	18	–			–			10:48:16.2	81	133	45	09:36:29.5	9	64	31	94	0.682	0.606	
BERMUDA																									
Hamilton	32°17′N	064°46′W	46	–				–			–			10:17:15.1	83	141	7	09:40 Rise	–	–	0	71	0.608	0.517	
CANARY ISLANDS																									
Las Palmas G.Ca…	28°07′N	015°28′W	6	08:44:06.8	320	26	28	–			–			10:47:01.2	60	123	55	09:42:07.1	10	76	41	94	0.348	0.235	
Santa Cruz Tene…	28°25′N	016°16′W	–	08:42:46.3	319	25	27	–			–			10:46:06.6	61	124	54	09:40:57.0	10	76	40	94	0.360	0.246	
FAEROE ISLANDS																									
Tórshavn	62°01′N	006°46′W	–	09:12:16.8	264	289	32	–			–			11:27:23.1	122	132	42	10:18:34.0	193	212	38	138	0.701	0.630	
GREENLAND																									
Godthåb	64°11′N	051°44′W	20	09:06:07.8	254	280	11	–						10:54:45.7	125	152	23	09:59:10.7	189	216	17	89	0.574	0.477	
ICELAND																									
Reykjavik	64°09′N	021°51′W	28	09:08:34.7	260	266	24	–			–			11:13:30.2	123	141	36	10:09:38.4	191	215	30	120	0.646	0.563	
ST. PIERRE & MIQUELON																									
Saint-Pierre Is…	46°47′N	056°11′W	–	08:39:26.4	272	313	0	–			–			10:35:20.7	102	148	19	09:35:08.7	187	231	9	77	0.928	0.913	

TABLE 12
LOCAL CIRCUMSTANCES FOR AFRICA
TOTAL SOLAR ECLIPSE OF 1999 AUGUST 11

Location Name	Latitude	Longitude	Elev.	First Contact U.T.				Second Contact U.T.			Third Contact U.T.			Fourth Contact U.T.				Maximum Eclipse U.T.					Eclip. Mag.	Eclip. Obs.	Umbral Durat.
			m	h m s	P °	V °	Alt °	h m s	P °	V °	h m s	P °	V °	h m s	P °	V °	Alt °	h m s	P °	V °	Alt °	Azm °			
ALGERIA																									
Alger (Algiers)	36°47′N	003°03′E	59	09:03:12.1	306	358	47	–			–			11:48:41.0	86	88	69	10:23:30.0	16	54	61	131	0.634	0.549	
Annaba (Bône)	36°54′N	007°46′E	20	09:11:11.9	305	355	52	–			–			12:00:44.2	88	74	68	10:34:32.3	17	46	65	145	0.661	0.581	
Constantine	36°22′N	006°37′E	–	09:09:32.2	306	357	51	–			–			11:57:50.5	87	76	69	10:31:59.8	17	49	64	140	0.638	0.554	
Wahran (Oran)	35°43′N	000°43′W	–	08:57:49.0	307	3	43	–			–			11:38:17.2	82	99	69	10:14:56.7	14	61	58	122	0.588	0.495	
DJIBOUTI																									
Djibouti	11°36′N	043°09′E	7	11:22:39.3	333	241	58	–			–			13:25:13.2	75	350	29	12:27:39.8	24	296	43	280	0.384	0.270	
EGYPT																									
Alexandria	31°12′N	029°54′E	32	10:04:51.8	309	310	74	–			–			12:53:29.2	96	35	49	11:33:15.3	23	335	64	237	0.712	0.644	
Al-Jizah (Giza)	30°01′N	031°13′E	–	10:10:05.2	310	302	75	–			–			12:57:00.6	95	33	47	11:37:53.5	23	330	63	242	0.698	0.626	
Al-Mahallah al-…	30°58′N	031°10′E	–	10:08:02.6	309	303	74	–			–			12:55:49.4	96	35	47	11:36:11.7	23	332	63	240	0.722	0.656	
Al-Manshah	26°28′N	031°48′E	–	10:19:31.0	316	293	78	–			–			13:01:35.0	90	22	46	11:45:08.2	24	322	63	251	0.612	0.523	
Al-Qahirah (Cairo)	30°03′N	031°15′E	116	10:10:05.2	310	302	75	–			–			12:57:01.6	95	33	47	11:37:54.2	23	330	63	242	0.699	0.628	
As-Suways (Suez)	29°58′N	032°33′E	–	10:13:01.4	310	294	75	–			–			12:59:10.1	96	33	45	11:40:35.4	23	329	62	245	0.714	0.646	
Asyut	27°11′N	031°11′E	–	10:16:27.2	315	299	78	–			–			12:59:52.1	91	24	47	11:42:39.6	24	324	63	249	0.623	0.536	
Bahtim	30°08′N	031°17′E	–	10:09:59.1	310	302	75	–			–			12:56:59.1	95	33	47	11:37:49.7	23	330	63	242	0.702	0.631	
Bur Sa'id (Port…	31°16′N	032°18′E	–	10:09:52.4	308	297	74	–			–			12:57:12.5	97	37	46	11:37:54.2	23	331	62	241	0.744	0.684	
Shubra al-Khaym…	30°06′N	031°15′E	–	10:09:59.0	310	302	75	–			–			12:56:58.1	95	33	47	11:37:49.1	23	330	63	242	0.700	0.630	
Tanta	30°47′N	031°00′E	–	10:08:03.1	309	304	74	–			–			12:55:46.5	96	35	47	11:36:10.2	23	332	63	240	0.715	0.648	
ERITREA																									
Asmera	15°20′N	038°53′E	2325	11:05:21.8	331	244	67	–			–			13:19:06.6	78	356	35	12:16:27.8	25	300	50	276	0.421	0.309	
ETHIOPIA																									
Adis Abeba (Add…	09°02′N	038°42′E	2450	11:29:20.9	345	246	60	–			–			13:16:05.5	65	336	35	12:25:30.4	25	292	47	283	0.247	0.143	
LIBYA																									
Banghazi	32°07′N	020°04′E	25	09:41:47.3	312	350	68	–			–			12:33:02.7	89	38	60	11:09:28.6	21	4	72	200	0.621	0.534	
Tarabulus (Trip…	32°54′N	013°11′E	22	09:25:58.1	313	2	60	–			–			12:15:30.6	85	49	67	10:50:52.2	19	34	72	163	0.578	0.483	
MAURITANIA																									
Nouakchott	18°06′N	015°57′W	21	09:18:12.2	356	74	35	–			–			09:56:15.7	23	103	44	09:36:59.1	9	89	39	85	0.027	0.005	
MOROCCO																									
Casablanca (Dar…	33°39′N	007°35′W	50	08:49:08.2	310	10	36	–			–			11:17:55.7	75	115	65	09:59:42.5	12	68	50	108	0.514	0.409	
Fès	34°05′N	004°57′W	–	08:52:31.0	310	9	39	–			–			11:25:24.9	77	110	67	10:05:20.3	13	66	53	112	0.529	0.427	
Marrakech	31°38′N	008°00′W	460	08:50:07.1	314	16	36	–			–			11:13:30.3	70	116	65	09:58:03.5	12	71	50	105	0.449	0.339	
Meknès	33°53′N	005°37′W	–	08:51:41.9	310	9	38	–			–			11:23:23.3	76	111	66	10:03:51.6	13	67	53	111	0.522	0.419	
Oujda	34°41′N	001°45′W	–	08:56:52.1	309	6	42	–			–			11:34:40.6	79	101	69	10:12:30.6	14	63	57	118	0.553	0.454	
Rabat	34°02′N	006°51′W	65	08:49:53.7	310	9	37	–			–			11:20:22.0	76	114	65	10:01:20.9	13	67	51	109	0.526	0.423	
Salé	34°04′N	006°50′W	–	08:49:53.9	310	8	37	–			–			11:20:27.3	76	114	65	10:01:23.7	13	67	51	109	0.527	0.424	
Tanger (Tangier)	35°48′N	005°45′W	73	08:50:32.1	306	3	38	–			–			11:25:19.8	80	112	65	10:04:14.2	13	64	52	113	0.582	0.488	
QATAR																									
Ad-Dawhah (Doha)	25°17′N	051°32′E	–	10:58:07.4	303	235	56	–			–			13:23:52.4	102	34	23	12:15:49.6	23	313	39	270	0.847	0.812	
SOMALIA																									
Muqdisho (Mogad…	02°04′N	045°22′E	12	12:00:43.1	353	250	44	–			–			13:19:13.2	55	320	26	12:41:27.9	24	286	35	287	0.152	0.070	
SUDAN																									
Al-Khartum	15°36′N	032°32′E	390	10:56:14.8	337	250	75	–			–			13:06:45.6	72	349	44	12:05:14.0	25	300	59	274	0.331	0.219	
Al-Khartum Bahri	15°38′N	032°33′E	–	10:56:07.8	337	250	75	–			–			13:06:48.1	73	350	44	12:05:12.2	25	300	59	274	0.333	0.220	
Umm Durman (Omd…	15°38′N	032°30′E	–	10:56:04.0	337	250	75	–			–			13:06:41.2	72	349	44	12:05:06.5	25	300	59	274	0.332	0.219	
TUNISIA																									
Sfax	34°44′N	010°46′E	–	09:18:44.9	309	359	56	–			–			12:08:46.0	87	60	68	10:43:08.0	18	41	69	153	0.614	0.526	
Tunis	36°48′N	010°11′E	66	09:15:40.3	305	353	54	–			–			12:06:45.5	90	68	67	10:40:22.4	18	40	66	153	0.673	0.596	
WESTERN SAHARA																									
El Aaiun	27°09′N	013°12′W	–	08:48:30.9	323	31	31	–			–			10:49:27.4	58	122	58	09:45:42.5	11	78	44	95	0.311	0.200	

TABLE 13
LOCAL CIRCUMSTANCES FOR EUROPE - ALBANIA, AUSTRIA, BELARUS & BELGIUM
TOTAL SOLAR ECLIPSE OF 1999 AUGUST 11

Location Name	Latitude	Longitude	Elev.	First Contact U.T.	P	V	Alt	Second Contact U.T.	P	V	Third Contact U.T.	P	V	Fourth Contact U.T.	P	V	Alt	Maximum Eclipse U.T.	P	V	Alt	Azm	Eclip. Mag.	Eclip. Obs.	Umbral Durat.
			m	h m s	°	°	°	h m s	°	°	h m s	°	°	h m s	°	°	°	h m s	°	°	°	°			
ALBANIA																									
Tiranë	41°20'N	019°50'E	7	09:30:59.6	295	324	59	–			–			12:23:23.9	103	68	57	10:58:21.0	19	14	64	187	0.877	0.852	
AUSTRIA																									
Bad Ischl	47°43'N	013°37'E	–	09:19:21.3	286	316	50	10:40:56.1	132	141	10:43:03.6	262	271	12:05:13.6	108	91	56	10:41:59.8	17	26	57	167	1.028	1.000	02m08s
Bruck an der Mur	47°25'N	015°16'E	–	09:21:54.3	286	314	52	10:43:55.5	125	132	10:46:10.1	270	275	12:08:14.0	108	89	55	10:45:02.8	17	23	58	171	1.028	1.000	02m15s
Eisenstadt	47°51'N	016°32'E	–	09:24:02.3	285	311	52	–			–			12:09:41.1	109	89	54	10:47:01.6	197	201	57	175	0.999	1.000	
Gmunden	47°55'N	013°48'E	–	09:19:41.9	285	315	50	10:41:03.8	105	114	10:43:23.9	289	297	12:05:17.9	108	92	55	10:42:13.9	197	206	57	168	1.028	1.000	02m20s
Graz	47°05'N	015°27'E	–	09:22:08.6	287	315	52	10:44:56.9	167	173	10:46:09.2	228	234	12:08:55.7	108	88	55	10:45:33.1	17	23	58	172	1.029	1.000	01m12s
Innsbruck	47°16'N	011°24'E	582	09:15:47.0	287	319	49	–			–			12:02:00.7	106	93	57	10:38:14.9	17	30	57	161	0.984	0.985	
Judenburg	47°10'N	014°40'E	–	09:20:53.4	287	316	51	–			–			12:07:33.6	108	89	56	10:44:06.6	17	25	58	170	0.999	1.000	
Kapfenberg	47°26'N	015°18'E	–	09:21:57.8	286	314	52	10:43:57.6	122	129	10:46:14.0	272	278	12:08:16.0	108	89	55	10:45:05.8	17	23	58	171	1.028	1.000	02m16s
Klagenfurt	46°37'N	014°18'E	–	09:20:12.0	288	318	52	–			–			12:07:35.0	107	88	56	10:43:43.6	17	25	58	169	0.982	0.983	
Knittelfeld	47°14'N	014°50'E	–	09:21:10.3	286	315	51	10:43:44.2	165	172	10:45:00.6	230	237	12:07:45.2	108	89	56	10:44:22.4	17	24	58	170	1.028	1.000	01m16s
Leoben	47°23'N	015°06'E	–	09:21:37.8	286	315	52	10:43:42.7	134	140	10:45:49.6	261	267	12:08:00.4	108	89	55	10:44:46.2	17	24	58	171	1.028	1.000	02m07s
Linz	48°18'N	014°18'E	–	09:20:35.5	285	313	50	10:42:40.9	29	37	10:43:10.6	5	12	12:05:39.2	109	92	55	10:42:55.8	197	205	57	169	1.028	1.000	00m30s
Mürzzuschlag	47°36'N	015°41'E	–	09:22:36.8	286	313	52	10:44:32.2	94	100	10:46:49.7	300	305	12:08:40.4	109	89	55	10:45:41.0	197	203	58	172	1.028	1.000	02m17s
Neunkirchen	47°43'N	016°05'E	–	09:23:17.0	285	312	52	10:45:23.9	69	74	10:47:14.3	326	330	12:09:09.4	109	89	55	10:46:19.1	197	202	57	173	1.028	1.000	01m50s
Ried im Innkreis	48°13'N	013°30'E	–	09:19:19.4	285	314	50	10:40:34.5	77	87	10:42:35.7	317	325	12:04:28.7	108	93	55	10:41:35.1	197	206	57	167	1.028	1.000	02m01s
Salzburg	47°48'N	013°02'E	–	09:18:28.2	286	316	50	10:39:55.4	136	147	10:41:57.5	257	267	12:04:11.1	108	92	56	10:40:56.5	17	27	57	166	1.028	1.000	02m02s
Sankt Pölten	48°12'N	015°37'E	–	09:22:38.3	285	312	51	–			–			12:07:49.8	109	91	55	10:45:14.8	197	203	57	172	0.995	0.997	
Steyr	48°03'N	014°25'E	–	09:20:42.1	285	314	51	10:42:16.0	74	82	10:44:13.4	320	327	12:06:07.8	109	91	55	10:43:14.7	197	205	57	169	1.028	1.000	01m57s
Voitsberg	47°03'N	015°10'E	–	09:21:40.7	287	315	52	–			–			12:08:30.6	108	88	56	10:45:03.8	17	24	58	171	0.999	1.000	
Wels	48°10'N	014°02'E	–	09:20:08.1	285	314	50	10:41:36.5	69	78	10:43:26.9	325	333	12:05:23.2	109	92	55	10:42:31.7	197	205	57	168	1.028	1.000	01m50s
Wien (Vienna)	48°13'N	016°20'E	202	09:23:47.2	285	311	52	–			–			12:08:54.6	110	90	54	10:46:28.3	197	202	57	174	0.990	0.992	
Wiener Neustadt	47°49'N	016°15'E	–	09:23:34.3	285	312	52	10:46:00.3	45	50	10:47:06.1	350	354	12:09:17.4	109	89	55	10:46:33.2	197	202	57	174	1.028	1.000	01m06s
BELARUS																									
Baranovici	53°08'N	026°02'E	–	09:40:18.1	275	285	51	–			–			12:13:44.5	120	96	46	10:58:03.0	198	189	52	194	0.803	0.757	
Bobrujsk	53°09'N	029°14'E	–	09:45:20.4	274	280	52	–			–			12:16:55.9	122	95	45	11:02:26.4	198	185	51	201	0.780	0.728	
Borisov	54°15'N	028°30'E	–	09:44:18.5	273	279	51	–			–			12:14:10.3	123	98	44	11:00:22.9	198	187	50	199	0.761	0.704	
Brest	52°06'N	023°42'E	–	09:36:25.9	277	291	52	–			–			12:12:57.6	118	95	48	10:55:34.3	198	191	53	190	0.843	0.808	
Gomel'	52°25'N	031°00'E	–	09:48:10.5	275	278	53	–			–			12:20:00.2	122	93	44	11:05:36.5	198	183	51	205	0.784	0.733	
Grodno	53°41'N	023°50'E	–	09:37:03.2	275	287	50	–			–			12:10:26.2	120	98	47	10:54:31.5	198	192	51	190	0.804	0.759	
Minsk	53°54'N	027°34'E	225	09:42:49.0	273	281	51	–			–			12:13:55.0	122	97	45	10:59:27.9	198	187	50	197	0.775	0.722	
Mogil'ov	53°54'N	030°21'E	–	09:47:08.3	273	277	51	–			–			12:16:33.0	123	97	44	11:03:09.6	198	185	50	202	0.755	0.698	
Mozyr'	52°03'N	029°14'E	–	09:45:18.8	276	282	53	–			–			12:18:59.7	121	93	45	11:03:34.1	198	185	52	202	0.805	0.760	
Orsa	54°30'N	030°24'E	–	09:47:15.4	272	276	51	–			–			12:15:26.5	124	98	43	11:02:36.6	198	185	49	202	0.742	0.681	
Pinsk	52°07'N	026°04'E	–	09:40:11.8	276	287	52	–			–			12:15:34.7	119	94	47	10:59:00.3	198	188	52	195	0.827	0.787	
Vitebsk	55°12'N	030°11'E	–	09:47:00.2	271	275	50	–			–			12:13:54.3	125	100	43	11:01:38.0	198	185	49	201	0.728	0.664	
BELGIUM																									
Antwerpen (Anve…	51°13'N	004°25'E	–	09:08:33.5	281	314	42	–			–			11:47:17.3	109	109	54	10:26:28.4	195	215	51	148	0.964	0.961	
Arlon	49°41'N	005°49'E	–	09:09:07.9	283	317	44	10:27:57.7	44	64	10:29:01.2	347	6	11:50:36.4	107	105	56	10:28:29.4	195	215	53	150	1.028	1.000	01m03s
Bouillon	49°48'N	005°04'E	–	09:08:15.0	283	317	43	10:26:53.5	37	58	10:27:43.1	353	14	11:49:17.6	107	106	56	10:27:18.3	195	216	52	149	1.028	1.000	00m50s
Brugge (Bruges)	51°13'N	003°14'E	–	09:07:08.4	281	314	41	–			–			11:45:26.0	109	110	54	10:24:42.4	195	216	50	146	0.967	0.965	
Bruxelles (Brus…	50°50'N	004°20'E	–	09:08:08.2	281	315	42	–			–			11:47:25.0	108	109	54	10:26:17.4	195	215	51	148	0.975	0.974	
Charleroi	50°25'N	004°26'E	–	09:07:55.3	282	316	43	–			–			11:47:51.1	108	108	55	10:26:23.0	195	216	51	148	0.986	0.987	
Gent (Gand)	51°03'N	003°43'E	–	09:07:34.2	281	314	42	–			–			11:46:18.0	108	110	54	10:25:23.7	195	216	50	147	0.970	0.969	
Hasselt	50°56'N	005°20'E	–	09:09:27.2	281	314	43	–			–			11:48:55.5	109	108	54	10:27:49.6	195	214	51	150	0.969	0.967	
Kortrijk (Court…	50°50'N	003°16'E	–	09:06:50.8	281	315	42	–			–			11:45:43.2	108	110	54	10:24:40.6	195	216	50	146	0.978	0.978	
La Louviäre	50°28'N	004°11'E	–	09:07:39.2	282	316	42	–			–			11:47:25.0	108	108	55	10:26:00.4	195	216	51	147	0.985	0.987	
Leuven (Louvain)	50°53'N	004°42'E	–	09:08:37.7	281	314	42	–			–			11:47:57.7	109	108	54	10:26:51.3	195	215	51	149	0.972	0.971	
Liäge (Luik)	50°38'N	005°34'E	–	09:09:30.6	282	314	43	–			–			11:49:30.8	109	107	55	10:28:09.5	195	214	52	150	0.976	0.976	
Mechelen (Malin…	51°02'N	004°28'E	–	09:08:27.9	281	314	42	–			–			11:47:29.5	109	109	54	10:26:31.3	195	215	51	148	0.969	0.967	
Mons (Bergen)	50°27'N	003°56'E	–	09:07:20.0	282	316	42	–			–			11:47:01.5	108	108	55	10:25:37.2	195	216	51	147	0.986	0.988	
Namur	50°28'N	004°52'E	–	09:08:30.1	282	315	43	–			–			11:48:30.9	108	107	55	10:27:03.6	195	215	51	149	0.983	0.984	
Oostende (Osten…	51°13'N	002°55'E	–	09:06:46.0	281	314	41	–			–			11:44:56.2	108	111	54	10:24:14.2	195	216	50	145	0.968	0.966	
Schaerbeek	50°51'N	004°23'E	–	09:08:12.7	281	315	42	–			–			11:47:29.0	108	109	54	10:26:22.1	195	215	51	148	0.974	0.973	
Verviers	50°35'N	005°52'E	–	09:09:51.1	282	314	43	–			–			11:50:01.6	109	106	55	10:28:37.2	195	214	52	151	0.976	0.976	
Virton	49°34'N	005°32'E	–	09:08:41.0	283	318	44	10:27:07.1	70	91	10:28:56.6	320	340	11:50:13.4	107	105	56	10:28:01.8	195	215	53	149	1.028	1.000	01m49s

TABLE 14
LOCAL CIRCUMSTANCES FOR EUROPE -
BOSNIA & HERZEGOWINA, BULGARIA, CROATIA, CZECH REPUBLIC, DENMARK, ESTONIA & FINLAND
TOTAL SOLAR ECLIPSE OF 1999 AUGUST 11

Location Name	Latitude	Longitude	Elev.	First Contact				Second Contact			Third Contact			Fourth Contact				Maximum Eclipse					Eclip.	Eclip.	Umbral
			m	U.T. h m s	P °	V °	Alt °	U.T. h m s	P °	V °	U.T. h m s	P °	V °	U.T. h m s	P °	V °	Alt °	U.T. h m s	P °	V °	Alt °	Azm °	Mag.	Obs.	Durat.
BOSNIA & HERZEGOWINA																									
Banja Luka	44°46′N	017°11′E	–	09:25:00.7	290	319	55	–			–			12:14:35.5	106	80	56	10:50:12.6	18	20	61	177	0.951	0.945	
Mostar	43°20′N	017°49′E	–	09:26:24.7	292	322	56	–			–			12:17:25.0	104	76	57	10:52:31.3	19	19	62	179	0.916	0.901	
Prijedor	44°59′N	016°43′E	–	09:24:10.8	290	319	54	–			–			12:13:32.5	106	81	56	10:49:10.4	18	21	60	176	0.953	0.948	
Sarajevo	43°52′N	018°25′E	–	09:27:21.7	291	320	56	–			–			12:17:47.4	105	77	56	10:53:15.8	19	18	61	181	0.935	0.925	
Tuzla	44°32′N	018°41′E	–	09:27:41.9	290	317	56	–			–			12:17:23.1	106	78	55	10:53:14.1	18	17	61	181	0.955	0.950	
Zenica	44°12′N	017°55′E	–	09:26:23.6	291	319	56	–			–			12:16:31.7	105	78	56	10:52:02.1	18	19	61	179	0.940	0.932	
BULGARIA																									
Balcik	43°25′N	028°10′E	–	09:45:55.2	289	300	61	11:10:51.2	141	120	11:12:52.8	259	237	12:33:10.5	110	70	49	11:11:52.1	20	359	59	209	1.029	1.000	02m02s
Burgas	42°30′N	027°28′E	–	09:45:08.8	291	303	62	–			–			12:33:44.2	109	68	50	11:11:47.1	20	359	60	208	0.977	0.978	
Dobric	43°34′N	027°50′E	–	09:45:11.1	289	300	61	11:10:02.8	138	118	11:12:08.0	262	241	12:32:29.0	110	71	49	11:11:05.5	20	359	59	208	1.029	1.000	02m05s
Isperih	43°43′N	026°50′E	–	09:43:10.4	289	302	61	11:08:31.3	169	151	11:09:44.7	231	212	12:30:53.5	110	72	50	11:09:08.0	20	1	60	205	1.029	1.000	01m13s
Kavarna	43°25′N	028°20′E	–	09:46:14.6	289	299	61	11:11:04.7	134	113	11:13:14.4	266	244	12:33:23.6	110	70	49	11:12:09.6	20	358	59	209	1.028	1.000	02m10s
Pleven	43°25′N	024°37′E	–	09:39:04.8	291	308	60	–			–			12:28:13.8	108	71	52	11:05:27.7	20	5	61	199	0.974	0.974	
Plovdiv	42°09′N	024°45′E	–	09:40:02.9	293	311	61	–			–			12:30:21.0	107	68	52	11:07:10.3	20	4	62	201	0.943	0.935	
Ruse	43°50′N	025°57′E	–	09:41:25.2	289	304	60	–			–			12:29:29.1	109	72	51	11:07:24.8	20	3	60	203	0.997	0.998	
Silistra	44°07′N	027°16′E	–	09:43:47.4	289	301	60	11:08:14.4	103	84	11:10:35.2	297	278	12:30:49.1	111	72	49	11:09:24.9	200	181	59	206	1.029	1.000	02m21s
Sliven	42°40′N	026°19′E	–	09:42:47.2	291	306	62	–			–			12:31:51.4	108	69	51	11:09:28.7	20	1	61	205	0.971	0.970	
Sofija (Sofia)	42°41′N	023°19′E	550	09:36:57.5	292	313	60	–			–			12:27:21.9	106	70	53	11:03:50.1	20	7	62	196	0.944	0.936	
Stara Zagora	42°25′N	025°38′E	–	09:41:36.5	292	308	62	–			–			12:31:15.6	107	68	51	11:08:31.0	20	2	61	203	0.958	0.954	
Sumen	43°16′N	026°55′E	–	09:43:35.3	290	303	61	–			–			12:31:44.3	109	70	50	11:09:49.9	20	1	60	206	0.991	0.993	
Tutrakan	44°03′N	026°37′E	–	09:42:35.0	289	302	60	11:07:15.7	134	116	11:09:26.0	266	248	12:30:03.2	110	72	50	11:08:20.9	20	2	59	204	1.029	1.000	02m10s
Varna	43°13′N	027°55′E	35	09:45:33.5	290	301	61	–			–			12:33:10.5	110	70	49	11:11:40.8	20	359	59	208	1.000	1.000	
CROATIA																									
Osijek	45°33′N	018°41′E	–	09:27:33.5	289	314	55	–			–			12:16:03.2	108	81	55	10:52:26.1	18	18	60	181	0.983	0.984	
Rijeka	45°20′N	014°27′E	–	09:20:19.0	290	321	53	–			–			12:09:15.4	105	85	57	10:44:40.5	18	25	60	169	0.948	0.941	
Slavonski Brod	45°10′N	018°01′E	–	09:26:25.7	289	317	55	–			–			12:15:28.6	107	81	55	10:51:29.3	18	19	60	179	0.967	0.966	
Split	43°31′N	016°27′E	–	09:23:54.8	292	324	55	–			–			12:14:47.7	104	77	58	10:49:41.3	18	22	62	175	0.911	0.895	
Zadar	44°07′N	015°14′E	–	09:21:41.2	292	324	54	–			–			12:11:57.6	104	80	58	10:46:53.7	18	24	61	171	0.919	0.905	
Zagreb	45°48′N	015°58′E	–	09:22:52.8	289	318	53	–			–			12:11:18.4	106	84	56	10:47:14.3	18	23	59	173	0.970	0.969	
CZECH REPUBLIC																									
Brno	49°12′N	016°37′E	–	09:24:29.3	283	308	51	–			–			12:08:03.6	111	92	53	10:46:24.6	197	201	56	175	0.962	0.959	
Ceské Budejovice	48°59′N	014°28′E	–	09:21:04.3	284	311	50	–			–			12:05:06.2	110	94	54	10:42:54.6	197	204	56	169	0.981	0.982	
Hradec Králové	50°12′N	015°50′E	–	09:23:36.8	282	306	50	–			–			12:05:37.0	112	95	53	10:44:36.8	197	202	55	173	0.941	0.932	
Liberec	50°46′N	015°03′E	–	09:22:40.6	281	306	49	–			–			12:03:46.1	113	97	53	10:43:06.3	197	203	54	171	0.930	0.919	
Most	50°32′N	013°39′E	–	09:20:30.0	281	308	48	–			–			12:02:01.6	112	98	53	10:40:57.6	197	205	54	168	0.944	0.936	
Olomouc	49°36′N	017°16′E	–	09:25:37.7	282	306	51	–			–			12:08:28.2	112	93	53	10:47:15.8	197	200	56	176	0.948	0.941	
Ostrava	49°50′N	018°17′E	–	09:27:17.9	282	304	51	–			–			12:09:35.1	113	92	52	10:48:47.3	197	198	56	179	0.935	0.926	
Plzen	49°45′N	013°23′E	–	09:19:44.3	283	310	49	–			–			12:02:32.9	111	96	54	10:40:48.2	197	205	55	167	0.966	0.964	
Praha (Prague)	50°05′N	014°26′E	202	09:21:27.0	282	308	49	–			–			12:03:43.4	111	96	53	10:42:23.5	197	204	55	169	0.952	0.946	
Ústí nad Labem	50°40′N	014°02′E	–	09:21:07.6	281	307	49	–			–			12:02:25.6	112	98	53	10:41:31.5	197	204	54	169	0.938	0.929	
Zlin	49°13′N	017°41′E	–	09:26:11.1	283	306	52	–			–			12:09:35.4	112	91	53	10:48:10.5	197	199	56	177	0.955	0.951	
DENMARK																									
Ålborg	57°03′N	009°56′E	–	09:20:38.7	272	294	42	–			–			11:49:34.8	119	114	48	10:34:30.9	195	205	47	162	0.791	0.742	
Århus	56°09′N	010°13′E	49	09:20:05.1	273	296	43	–			–			11:50:55.0	118	112	49	10:34:54.0	195	206	48	162	0.813	0.769	
København (Cope…	55°40′N	012°35′E	13	09:22:32.8	273	295	44	–			–			11:54:23.4	118	109	49	10:38:04.8	196	204	49	166	0.815	0.773	
Odense	55°24′N	010°23′E	–	09:19:34.9	274	298	43	–			–			11:51:57.3	117	111	50	10:35:09.6	195	206	49	162	0.831	0.793	
ESTONIA																									
Tallinn	59°25′N	024°45′E	–	09:40:52.1	266	274	45	–			–			12:01:25.3	127	111	43	10:51:39.8	197	192	46	189	0.669	0.591	
FINLAND																									
Espoo (Esbo)	60°13′N	024°40′E	–	09:41:13.4	265	273	44	–			–			11:59:57.3	128	112	42	10:51:03.9	196	192	45	188	0.652	0.570	
Helsinki	60°10′N	024°58′E	9	09:41:34.3	265	273	44	–			–			12:00:17.5	128	112	42	10:51:25.6	196	192	45	189	0.652	0.570	
Lahti	60°58′N	025°40′E	–	09:42:54.4	264	270	44	–			–			11:59:26.8	129	114	41	10:51:40.3	196	191	44	190	0.631	0.545	
Oulu	65°01′N	025°28′E	–	09:45:20.0	258	263	40	–			–			11:52:12.1	133	121	38	10:49:05.2	196	192	40	188	0.550	0.450	
Tampere	61°30′N	023°45′E	–	09:40:55.5	263	271	43	–			–			11:56:59.9	129	115	41	10:49:19.8	196	193	44	186	0.629	0.543	
Turku (Åbo)	60°27′N	022°17′E	–	09:38:25.7	265	275	44	–			–			11:57:32.4	127	114	43	10:48:18.4	196	194	45	184	0.659	0.578	
Vaanta (Vanda)	60°16′N	025°03′E	–	09:41:44.0	265	272	44	–			–			12:00:11.1	128	112	42	10:51:27.2	196	192	45	189	0.649	0.567	

TABLE 15
LOCAL CIRCUMSTANCES FOR EUROPE - ENGLAND
TOTAL SOLAR ECLIPSE OF 1999 AUGUST 11

Location Name	Latitude	Longitude	Elev.	First Contact U.T.	P	V	Alt	Second Contact U.T.	P	V	Third Contact U.T.	P	V	Fourth Contact U.T.	P	V	Alt	Maximum Eclipse U.T.	P	V	Alt	Azm	Eclip. Mag.	Eclip. Obs.	Umbral Durat.
			m	h m s	°	°	°	h m s	°	°	h m s	°	°	h m s	°	°	°	h m s	°	°	°	°			
ENGLAND																									
Birmingham	52°30'N	001°50'W	163	09:02:52.1	278	313	37	—			—			11:37:00.1	109	118	52	10:17:59.1	194	219	46	138	0.943	0.935	
Blackburn	53°45'N	002°29'W	—	09:03:40.3	276	310	37	—			—			11:35:35.5	111	120	51	10:17:45.6	194	218	45	138	0.910	0.892	
Blackpool	53°50'N	003°03'W	—	09:03:13.9	276	310	36	—			—			11:34:45.1	111	121	51	10:17:05.3	193	218	45	137	0.908	0.890	
Bodmin	50°29'N	004°43'W	—	08:57:50.5	281	319	36	10:12:22.1	53	83	10:13:40.7	333	3	11:32:59.9	105	119	53	10:13:01.4	193	223	46	131	1.027	1.000	01m19s
Bolton	53°35'N	002°26'W	—	09:03:31.2	277	310	37	—			—			11:35:43.5	110	120	51	10:17:44.3	194	218	45	138	0.914	0.898	
Bournemouth	50°43'N	001°54'W	—	09:00:55.9	281	318	38	—			—			11:37:29.0	106	116	54	10:17:04.1	194	221	48	136	0.993	0.995	
Bradford	53°48'N	001°45'W	—	09:04:26.5	276	309	37	—			—			11:36:37.4	111	119	51	10:18:43.3	194	217	46	139	0.907	0.889	
Brighton	50°50'N	000°08'W	—	09:02:56.7	281	317	39	—			—			11:40:16.9	107	114	54	10:19:39.1	194	219	49	139	0.986	0.988	
Bristol	51°27'N	002°35'W	—	09:00:58.6	280	316	37	—			—			11:36:11.4	107	117	53	10:16:27.7	194	221	47	136	0.973	0.972	
Camborne	50°12'N	005°19'W	—	08:56:58.0	282	320	35	10:11:01.7	96	127	10:13:02.9	290	320	11:32:03.4	105	119	53	10:12:02.2	193	224	46	130	1.026	1.000	02m01s
Chesterfield	53°15'N	001°25'W	—	09:04:07.9	277	311	37	—			—			11:37:19.9	110	118	52	10:18:53.5	194	218	46	139	0.922	0.908	
Coventry	52°25'N	001°30'W	—	09:03:07.0	279	313	38	—			—			11:37:32.2	109	117	52	10:18:23.8	194	219	47	138	0.945	0.937	
Derby	52°55'N	001°29'W	—	09:03:41.2	278	312	38	—			—			11:37:22.1	110	118	52	10:18:39.0	194	218	46	139	0.931	0.919	
Dudley	52°30'N	002°05'W	—	09:02:37.1	278	313	37	—						11:36:37.6	109	118	52	10:17:39.2	194	219	46	137	0.943	0.935	
Exeter	50°43'N	003°31'W	—	08:59:16.2	281	319	37				—			11:34:53.[illegible]	106	118	53	10:14:48.1	193	222	47	133	0.995	0.997	
Falmouth	50°08'N	005°04'W	—	08:57:08.3	282	321	36	10:11:[illegible]	104	134	10:13:21.7	283	313	11:32:28.0	105	119	54	10:12:20.4	13	44	46	130	1.027	1.000	02m02s
Gloucester	51°53'N	002°14'W		09:01:47.6	279	315	37	—			—			11:36:36.[illegible]	108	118	53	10:17:09.1	194	220	47	137	0.961	0.957	
Helston	50°05'N	005°16'W	—	08:56:53.6	282	321	36	10:11:[illegible]	111	142	10:13:[illegible]	276	306	11:32:08.7	104	119	54	10:12:02.[illegible]	13	44	46	130	1.026	1.000	02m[illegible]s
High Wycombe	51°38'N	000°46'W	—	09:03:3[illegible]	280	315	38							11:38:5[illegible]	108	116	53	10:19:03.[illegible]	194	219	48	139	0.965	0.962	
Huddersfield	53°39'N	001°47'W	—	09:04:13.9	277	310	37	—						11:36:38.2	111	119	51	10:18:36.4	194	218	46	139	0.911	0.894	
Hugh Town	49°55'N	006°17'W	—	08:55:45.2	282	321	35	10:09:42.5	137	169	10:11:23.5	249	281	11:30:29.9	104	120	53	10:10:32.9	13	45	45	128	1.026	1.000	01m41s
Ipswich	52°04'N	001°10'E	—	09:05:34.8	279	313	40	—			—			11:41:44.9	109	114	53	10:21:56.5	194	217	48	143	0.949	0.942	
Kingsbridge	50°17'N	003°46'W	—	08:58:34.6	282	320	37	10:13:17.6	78	108	10:15:09.6	308	338	11:34:33.8	105	118	54	10:14:13.6	193	223	47	133	1.027	1.000	01m52s
Kingston upon H...	53°45'N	000°20'W	—	09:05:47.8	277	309	38	—			—			11:38:40.7	111	118	51	10:20:32.1	194	217	46	141	0.906	0.888	
Leeds	53°50'N	001°35'W	—	09:04:38.7	276	309	37				—			11:36:50.9	111	119	51	10:18:57.1	194	217	46	139	0.906	0.888	
Leicester	52°38'N	001°05'W		09:03:46.8	278	312	38	—			—			11:38:04.7	109	117	52	10:19:03.3	194	218	47	139	0.938	0.928	
Liverpool	53°25'N	002°55'W	60	09:02:51.4	277	311	36	—			—			11:35:05.0	110	120	51	10:17:01.9	193	219	45	137	0.919	0.905	
London	51°30'N	000°10'W	45	09:03:33.5	280	315	39	—			—			11:39:57.2	108	115	53	10:19:51.2	194	219	48	140	0.968	0.966	
Luton	51°53'N	000°25'W	—	09:03:40.7	279	314	39	—			—			11:39:24.2	108	116	53	10:19:39.2	194	219	48	140	0.957	0.953	
Manchester	53°30'N	002°15'W	—	09:03:36.0	277	310	37	—			—			11:36:01.3	110	119	51	10:17:55.8	194	218	46	138	0.916	0.901	
Mansfield	53°09'N	001°11'W	—	09:04:15.2	277	311	38				—			11:37:42.9	110	118	52	10:19:09.2	194	218	46	139	0.924	0.911	
Middlesbrough	54°35'N	001°14'W		09:05:53.2	275	307	37	—			—			11:37:00.7	112	120	50	10:19:45.3	194	216	45	141	0.885	0.861	
Newcastle upon ...	54°59'N	001°35'W	—	09:06:02.5	275	306	37	—						11:36:20.8	113	120	50	10:19:30.8	194	216	45	140	0.875	0.848	
Newquay	50°25'N	005°05'W	—	08:57:25.0	281	320	36	10:11:40.[illegible]	67	97	10:13:18.1	320	350	11:32:24.9	105	119	53	10:12:29.0	193	223	46	131	1.026	1.000	01m38s
Newton Abbot	50°32'N	003°36'W	—	08:58:59.9	281	319	37	10:14:20.5	27	56	10:14:50.5	359	28	11:34:47.3	106	118	54	10:14:35.4	193	222	47	133	1.027	1.000	00m30s
Northampton	52°14'N	000°54'W	—	09:03:32.2	279	313	38	—			—			11:38:31.1	109	116	53	10:19:0[illegible]	194	219	47	139	0.949	0.942	
Norwich	52°38'N	001°18'E		09:06:18.1	276	[illegible]	[illegible]	—			—			11:41:38.9	110	115	53	10:22:18.9	194	216	48	143	0.933	0.922	
Nottingham	52°58'N	001°10'W		09:04:03.8	278	311	38				—			11:37:49.[illegible]	110	118	52	10:19:05.6	194	218	47	139	0.929	0.917	
Oldbury	52°30'N	002°0[illegible]'W		09:02:42.1	278	313	37	—			—			11:36:45.1	109	118	52	10:17:45.8	194	219	46	138	0.943	0.935	
Oxford	51°46'N	001°15'W	—	09:02:40.8	280	316	38	—			—			11:38:09.7	108	116	53	10:18:26.8	194	219	47	138	0.962	0.959	
Penzance	50°07'N	005°33'W	—	08:56:39.3	282	321	35	10:10:39.4	108	139	10:12:4[illegible]	279	309	11:31:41.0	104	120	53	10:11:40.1	13	44	46	130	1.026	1.000	02m[illegible]s
Plymouth	50°23'N	004°10'W	27	08:58:16.7	282	320	36	10:12:54.2	66	96	10:14:32.[illegible]	320	350	11:33:53.9	105	118	54	10:13:43.4	193	223	46	132	1.027	1.000	01m39s
Poole	50°43'N	001°59'W	—	09:00:50.6	281	318	38	—			—			11:37:21.0	106	116	54	10:16:57.1	194	221	48	136	0.993	0.995	
Portsmouth	50°48'N	001°05'W		09:01:52.8	281	317	38	—			—			11:38:46.1	107	115	54	10:18:16.2	194	220	48	138	0.989	0.991	
Preston	53°46'N	002°42'W	—	09:03:29.0	276	310	36	—			—			11:35:16.6	111	120	51	10:17:29.6	193	218	45	137	0.910	0.892	
Reading	51°28'N	000°59'W	—	09:02:39.1	280	316	38	—			—			11:38:41.3	108	116	53	10:18:41.4	194	219	48	138	0.970	0.969	
Redruth	50°13'N	005°14'W	—	08:57:03.8	282	320	36	10:11:09.4	94	125	10:13:10.0	292	323	11:32:11.4	105	119	53	10:12:09.6	193	224	46	130	1.026	1.000	02m01s
Saint Austell	50°20'N	004°48'W	—	08:57:36.3	282	320	36	10:11:54.4	77	107	10:13:44.4	309	339	11:32:52.8	105	119	53	10:12:49.3	193	223	46	131	1.027	1.000	01m50s
Saltash	50°24'N	004°12'W	—	08:58:15.8	282	320	36	10:12:53.5	64	94	10:14:28.9	322	352	11:33:50.5	105	118	54	10:13:41.2	193	223	46	132	1.027	1.000	01m35s
Sheffield	53°23'N	001°30'W	—	09:04:12.1	277	310	37	—			—			11:37:09.3	110	119	51	10:18:50.7	194	218	46	139	0.918	0.903	
Southampton	50°55'N	001°25'W	—	09:01:38.4	281	317	38	—			—			11:38:11.8	107	115	54	10:17:50.6	194	220	48	137	0.986	0.988	
Southend-on-Sea...	51°33'N	000°43'E	—	09:04:34.4	280	315	39	—			—			11:41:18.7	108	114	54	10:21:07.8	194	218	49	141	0.964	0.961	
Stoke-on-Trent	53°00'N	002°10'W	—	09:03:06.0	278	312	37	—						11:36:19.5	110	119	52	10:17:47.3	194	219	46	138	0.930	0.918	
Sunderland	54°55'N	001°23'W	—	09:06:09.1	275	306	37				—			11:36:39.1	113	120	50	10:19:43.7	194	216	45	141	0.876	0.850	
Swindon	51°34'N	001°47'W	—	09:01:55.0	280	316	38	—			—			11:37:24.2	108	117	53	10:17:37.2	194	220	47	137	0.969	0.967	
Torquay (Torbay...	50°28'N	003°30'W	—	08:59:01.9	281	319	37	10:14:08.0	46	75	10:15:15.4	340	9	11:34:57.8	105	117	54	10:14:41.7	193	222	47	133	1.027	1.000	01m07s
Truro	50°16'N	005°03'W	—	08:57:17.5	282	320	36	10:11:27.8	87	118	10:13:25.2	299	330	11:32:29.0	105	119	53	10:12:26.5	193	224	46	130	1.026	1.000	01m57s
Walsall	52°35'N	001°58'W		09:02:49.7	278	313	37	—			—			11:36:46.4	109	118	52	10:17:50.9	194	219	46	138	0.941	0.932	
Watford	51°40'N	000°25'W	—	09:03:27.4	280	315	39	—						11:39:29.6	108	115	53	10:19:34.0	194	219	48	140	0.963	0.960	
West Bromwich	52°31'N	001°56'W	—	09:02:47.3	278	313	37	—			—			11:36:50.8	109	118	52	10:17:51.7	194	219	46	138	0.943	0.934	
Wolverhampton	52°36'N	002°08'W	—	09:02:40.8	278	313	37	—			—			11:36:31.0	109	118	52	10:17:38.1	194	219	46	137	0.941	0.932	
York	53°58'N	001°05'W	—	09:05:17.7	276	309	37	—			—			11:37:30.3	111	119	51	10:19:39.5	194	217	46	140	0.902	0.882	

TABLE 16
LOCAL CIRCUMSTANCES FOR EUROPE - FRANCE I
TOTAL SOLAR ECLIPSE OF 1999 AUGUST 11

Location Name	Latitude	Longitude	Elev.	First Contact U.T.	P	V	Alt	Second Contact U.T.	P	V	Third Contact U.T.	P	V	Fourth Contact U.T.	P	V	Alt	Maximum Eclipse U.T.	P	V	Alt	Azm	Eclip. Mag.	Eclip. Obs.	Umbral Durat.
			m	h m s	°	°	°	h m s	°	°	h m s	°	°	h m s	°	°	°	h m s	°	°	°	°			
FRANCE																									
Abbeville	50°06′N	001°50′E	–	09:04:31.3	282	318	41	10:21:45.2	47	72	10:22:55.4	342	6	11:43:48.1	106	110	55	10:22:20.3	194	219	50	142	1.027	1.000	01m10s
Albert	50°00′N	002°39′E	–	09:05:24.6	283	318	41	10:22:54.2	52	75	10:24:13.3	337	0	11:45:11.8	107	109	55	10:23:33.7	195	218	51	144	1.027	1.000	01m19s
Aix en Provence	43°32′N	005°26′E	–	09:05:49.9	293	335	46	–			–			11:53:29.3	98	94	62	10:27:53.6	16	42	58	144	0.848	0.815	
Amiens	49°54′N	002°18′E	–	09:04:54.5	283	318	41	10:22:04.4	73	97	10:23:55.8	316	339	11:44:40.3	106	110	55	10:23:00.1	195	218	51	143	1.027	1.000	01m51s
Angers	47°28′N	000°33′W	–	08:59:39.6	287	327	40	–			–			11:40:38.7	102	110	57	10:17:49.7	14	44	51	136	0.944	0.937	
Angoulême	45°39′N	000°09′E	–	08:59:22.3	290	332	41	–			–			11:42:19.8	99	107	59	10:18:26.7	14	45	52	135	0.893	0.872	
Annecy	45°54′N	006°07′E	–	09:07:29.4	289	328	46	–			–			11:53:37.3	102	97	59	10:29:00.8	16	38	56	148	0.920	0.906	
Avignon	43°57′N	004°49′E	–	09:05:00.8	293	334	46	–			–			11:52:03.7	98	96	61	10:26:41.2	15	42	57	143	0.858	0.827	
Bayonne	43°29′N	001°29′W	–	08:56:19.1	293	339	40	–			–			11:39:12.3	95	107	61	10:14:58.2	14	50	53	130	0.825	0.785	
Beauvais	49°26′N	002°05′E	–	09:04:16.4	283	320	41	10:21:36.2	134	158	10:23:30.6	255	280	11:44:32.5	106	109	56	10:22:33.4	15	39	51	142	1.027	1.000	01m54s
Besançon	47°15′N	006°02′E	–	09:07:57.9	287	324	45	–			–			11:52:38.2	104	100	58	10:28:48.9	15	37	55	149	0.958	0.954	
Béthune	50°32′N	002°38′E	–	09:05:50.1	282	317	41	–			–			11:44:52.7	107	110	55	10:23:39.4	195	217	50	144	0.988	0.990	
Bolbec	49°34′N	000°29′E	–	09:02:30.0	283	320	40	10:19:12.2	135	161	10:21:03.4	254	280	11:41:47.6	105	111	55	10:20:07.8	14	40	50	139	1.027	1.000	01m51s
Bordeaux	44°50′N	000°34′W	48	08:58:03.4	291	334	41	–			–			11:41:02.4	98	107	60	10:16:59.0	14	47	52	133	0.867	0.839	
Boulogne Bilian…	48°50′N	002°15′E	–	09:04:01.0	284	322	42	–			–			11:45:06.6	105	108	56	10:22:40.9	15	40	51	142	0.991	0.993	
Brest	48°24′N	004°29′W	–	08:56:04.0	285	325	36	–			–			11:33:33.0	102	117	55	10:12:11.8	13	46	47	130	0.964	0.962	
Caen	49°11′N	000°21′W	–	09:01:13.4	284	322	39	–			–			11:40:31.2	104	112	56	10:18:43.9	14	41	50	138	0.994	0.996	
Calais	50°57′N	001°50′E	–	09:05:16.0	281	316	40	–			–			11:43:22.3	108	112	54	10:22:34.0	194	218	50	143	0.978	0.979	
Cambrai	50°10′N	003°14′E	–	09:06:15.0	282	317	42	–			–			11:46:03.6	107	109	55	10:24:29.6	195	217	51	145	0.996	0.998	
Cannes	43°33′N	007°01′E	–	09:08:13.4	293	334	48	–			–			11:56:42.0	99	91	61	10:30:58.4	16	39	58	148	0.856	0.824	
Châlons sur Mar…	48°57′N	004°22′E	–	09:06:45.2	284	320	43	10:25:49.6	181	203	10:26:22.4	209	231	11:48:40.1	106	105	56	10:26:06.0	15	37	52	146	1.028	1.000	00m33s
Chambéry	45°34′N	005°56′E	–	09:07:06.3	290	329	46	–			–			11:53:28.1	101	97	60	10:28:42.0	16	39	56	147	0.910	0.893	
Charleville Méz…	49°46′N	004°43′E	–	09:07:46.8	283	317	43	10:26:03.1	54	75	10:27:26.7	336	357	11:48:44.4	107	106	56	10:26:44.9	195	216	52	148	1.028	1.000	01m24s
Chauny	49°37′N	003°13′E	–	09:05:47.7	283	319	42	10:23:16.7	97	121	10:25:27.2	292	315	11:46:20.7	106	108	56	10:24:21.9	195	218	51	145	1.027	1.000	02m10s
Cherbourg	49°39′N	001°39′W	–	09:00:11.2	283	321	38	10:16:10.5	145	174	10:17:45.3	242	270	11:38:11.4	105	114	55	10:16:57.8	14	42	49	136	1.027	1.000	01m35s
Clermont Ferrand	45°47′N	003°05′E	–	09:03:14.4	289	330	43	–			–			11:47:55.7	100	102	60	10:23:34.8	15	42	54	141	0.905	0.887	
Compiègne	49°25′N	002°50′E	–	09:05:10.2	284	320	42	10:22:42.3	126	150	10:24:44.8	263	287	11:45:48.7	106	108	56	10:23:43.5	15	38	51	144	1.027	1.000	02m02s
Creil	49°16′N	002°29′E	–	09:04:37.7	284	320	42	10:22:25.0	153	177	10:23:52.5	236	260	11:45:18.0	105	108	56	10:23:08.7	15	39	51	143	1.027	1.000	01m27s
Crépy en Valois	49°14′N	002°54′E	–	09:05:06.8	284	320	42	10:23:02.4	151	175	10:24:33.1	238	262	11:46:01.3	106	108	56	10:23:47.7	15	38	51	144	1.028	1.000	01m31s
Dieppe	49°56′N	001°05′E	–	09:03:30.3	283	319	40	10:20:09.0	83	109	10:22:09.5	305	330	11:42:38.5	106	111	55	10:21:09.2	194	219	50	141	1.027	1.000	02m01s
Dijon	47°19′N	005°01′E	–	09:06:36.6	287	324	44	–			–			11:50:46.4	104	102	58	10:27:03.7	15	38	54	147	0.956	0.952	
Douai	50°22′N	003°04′E	–	09:06:12.8	282	317	42	–			–			11:45:40.4	107	109	55	10:24:16.7	195	217	51	145	0.991	0.993	
Dunkerque	51°03′N	002°22′E	–	09:05:58.5	281	315	41	–			–			11:44:09.9	108	111	54	10:23:22.9	194	217	50	144	0.974	0.974	
Elbeuf	49°17′N	001°00′E	–	09:02:52.0	284	321	40	–			–			11:42:46.7	105	110	56	10:20:50.1	14	40	50	140	1.000	1.000	
Fécamp	49°45′N	000°22′E	–	09:02:31.3	283	320	40	10:18:57.1	113	139	10:21:04.5	276	302	11:41:31.6	105	112	55	10:20:00.7	14	40	50	139	1.027	1.000	02m07s
Forbach	49°11′N	006°54′E	–	09:10:14.1	284	318	45	10:29:07.7	96	115	10:31:20.8	295	313	11:52:46.0	107	103	56	10:30:14.2	196	214	53	152	1.028	1.000	02m13s
Forges les Eaux	49°37′N	001°33′E	–	09:03:47.1	283	320	41	10:20:43.1	117	142	10:22:50.2	272	297	11:43:33.7	106	110	56	10:21:46.6	14	39	50	141	1.027	1.000	02m07s
Grenoble	45°10′N	005°43′E	–	09:06:39.5	291	330	46	–			–			11:53:16.3	101	97	60	10:28:19.4	16	40	56	146	0.897	0.877	
Hagondange	49°15′N	006°10′E	–	09:09:17.9	284	318	44	10:27:55.6	100	120	10:30:09.3	291	310	11:51:29.9	107	104	56	10:29:02.4	195	215	53	150	1.028	1.000	02m14s
Haguenau	48°49′N	007°47′E	–	09:11:13.2	284	318	46	10:30:38.2	124	142	10:32:47.6	267	285	11:54:31.6	107	101	56	10:31:42.9	16	33	54	154	1.028	1.000	02m09s
Hayange	49°20′N	006°03′E	–	09:09:12.0	284	318	44	10:27:46.0	92	112	10:29:56.4	299	318	11:51:14.6	107	104	56	10:28:51.2	195	215	53	150	1.028	1.000	02m10s
Hirson	49°55′N	004°05′E	–	09:07:05.8	283	317	42	10:25:22.5	36	58	10:26:10.0	354	15	11:47:36.2	107	107	55	10:25:46.2	195	217	52	147	1.028	1.000	00m48s
La Rochette	45°28′N	006°07′E	–	09:07:20.1	290	329	46	–			–			11:53:52.4	101	97	60	10:29:02.5	16	39	56	148	0.908	0.890	
Le Havre	49°30′N	000°08′E	–	09:02:02.5	283	321	40	10:18:48.9	149	176	10:20:20.2	239	266	11:41:13.8	105	112	55	10:19:34.5	14	41	50	139	1.027	1.000	01m31s
Le Mans	48°00′N	000°12′E	–	09:00:55.9	286	325	40	–			–			11:41:50.9	103	110	57	10:19:11.9	14	42	51	137	0.961	0.958	
Lens	50°26′N	002°50′E	–	09:05:59.3	282	317	41	–			–			11:45:15.5	107	110	55	10:23:56.2	195	217	50	145	0.990	0.992	
Lille	50°38′N	003°04′E	43	09:06:26.1	282	316	41	–			–			11:45:31.1	108	110	55	10:24:20.0	195	217	50	145	0.984	0.985	
Limoges	45°50′N	001°16′E	–	09:00:53.0	289	331	42	–			–			11:44:26.1	100	105	59	10:20:24.6	15	44	53	137	0.901	0.882	
Longwy	49°31′N	005°46′E	–	09:08:57.1	283	318	44	10:27:27.5	73	93	10:29:20.8	317	337	11:50:38.6	107	105	56	10:28:24.1	195	215	53	150	1.028	1.000	01m53s
Lorient	47°45′N	003°22′W	–	08:56:41.7	286	327	37	–			–			11:35:32.7	101	115	56	10:13:32.2	13	46	49	131	0.947	0.940	
Lyon	45°45′N	004°51′E	286	09:05:38.3	290	329	45	–			–			11:51:18.7	101	99	60	10:26:43.7	15	40	55	145	0.911	0.894	
Marseille	43°18′N	005°24′E	75	09:05:44.6	294	336	46	–			–			11:53:31.0	98	94	62	10:27:50.8	16	42	58	144	0.841	0.806	
Maubeuge	50°17′N	003°58′E	–	09:07:14.4	282	316	42	–			–			11:47:11.0	107	108	55	10:25:38.7	195	216	51	147	0.991	0.993	
Meaux	48°57′N	002°52′E	–	09:04:51.7	284	321	42	–			–			11:46:06.7	105	107	56	10:23:41.3	15	39	52	143	0.996	0.997	
Melun	48°32′N	002°40′E	–	09:04:18.9	285	322	42	–			–			11:45:58.4	104	107	57	10:23:17.3	15	39	52	143	0.983	0.984	
Metz	49°08′N	006°10′E	–	09:09:13.4	284	318	44	10:27:55.6	113	133	10:30:09.0	277	297	11:51:35.0	107	103	56	10:29:02.3	15	35	53	150	1.028	1.000	02m13s
Montbéliard	47°31′N	006°48′E	–	09:09:09.8	287	323	46	–			–			11:53:49.2	105	100	58	10:30:07.9	16	36	55	151	0.969	0.968	
Montpellier	43°36′N	003°53′E	–	09:03:35.0	293	336	45	–			–			11:50:17.1	97	97	62	10:24:54.8	15	44	56	141	0.844	0.809	
Moyeuvre Grande	49°15′N	006°02′E	–	09:09:07.3	284	318	44	10:27:42.4	102	122	10:29:56.4	289	308	11:51:16.6	107	104	56	10:28:49.4	195	215	53	150	1.028	1.000	02m14s
Mulhouse (Mülha…	47°45′N	007°20′E	–	09:10:01.5	286	321	46	–			–			11:54:35.2	105	99	57	10:31:02.2	16	35	55	152	0.978	0.978	

TABLE 16 - continued
LOCAL CIRCUMSTANCES FOR EUROPE - FRANCE II
TOTAL SOLAR ECLIPSE OF 1999 AUGUST 11

Location Name	Latitude	Longitude	Elev.	First Contact U.T.	P	V	Alt	Second Contact U.T.	P	V	Third Contact U.T.	P	V	Fourth Contact U.T.	P	V	Alt	Maximum Eclipse U.T.	P	V	Alt	Azm	Eclip. Mag.	Eclip. Obs.	Umbral Durat.
			m	h m s	°	°	°	h m s	°	°	h m s	°	°	h m s	°	°	°	h m s	°	°	°	°			
FRANCE																									
Nancy	48°41′N	006°12′E	—	09:08:59.1	285	320	45	—			—			11:51:57.6	106	103	57	10:29:05.3	15	35	54	150	1.000	1.000	
Nantes	47°13′N	001°33′W	—	08:58:19.0	287	328	39	—			—			11:38:52.7	101	111	57	10:16:08.4	14	45	50	133	0.935	0.925	
Nice	43°42′N	007°15′E	—	09:08:36.1	293	334	48	—			—			11:57:05.3	99	91	61	10:31:24.2	16	39	58	149	0.861	0.831	
Nîmes	43°50′N	004°21′E	—	09:04:18.6	293	335	45	—			—			11:51:09.6	98	97	61	10:25:48.1	15	43	57	142	0.853	0.820	
Noyon	49°35′N	003°00′E	—	09:05:30.2	283	319	42	10:22:55.5	104	128	10:25:06.9	285	309	11:46:00.2	106	108	56	10:24:01.2	195	218	51	144	1.027	1.000	02m11s
Orléans	47°55′N	001°54′E	—	09:02:56.1	286	324	42	—			—			11:44:54.0	103	107	57	10:21:55.1	15	41	52	141	0.963	0.961	
Paris	48°52′N	002°20′E	50	09:04:08.6	284	321	42	—			—			11:45:14.3	105	108	56	10:22:49.3	15	39	51	142	0.992	0.994	
Pau	43°18′N	000°22′W	—	08:57:40.9	293	339	41	—			—			11:41:31.1	95	105	62	10:16:56.0	14	49	54	132	0.822	0.781	
Perpignan	42°41′N	002°53′E	—	09:01:57.9	295	339	44	—			—			11:48:25.8	95	98	63	10:22:57.7	15	46	56	138	0.813	0.770	
Poitiers	46°35′N	000°20′E	—	09:00:08.3	288	329	41	—			—			11:42:28.7	101	108	58	10:19:00.3	14	44	52	136	0.921	0.907	
Reims	49°15′N	004°02′E	—	09:06:32.4	284	319	43	10:24:36.3	131	154	10:26:35.7	259	281	11:47:55.5	106	106	56	10:25:36.0	15	37	52	146	1.028	1.000	01m59s
Rennes	48°05′N	001°41′W	—	08:58:49.1	285	325	39	—			—			11:38:29.3	102	113	57	10:16:16.1	14	44	50	134	0.960	0.956	
Rethel	49°31′N	004°22′E	—	09:07:09.2	283	318	43	10:25:04.7	94	116	10:27:15.1	296	317	11:48:19.3	107	106	56	10:26:09.9	195	217	52	147	1.028	1.000	02m10s
Rouen	49°27′N	001°05′E	—	09:03:06.0	283	320	40	10:20:11.0	144	170	10:21:50.5	244	270	11:42:51.0	105	111	56	10:21:00.7	14	40	50	140	1.027	1.000	01m40s
Saint Avold	49°06′N	006°42′E	—	09:09:54.8	284	318	45	10:28:47.3	109	128	10:31:02.1	282	301	11:52:29.9	107	103	56	10:29:54.6	15	35	53	152	1.028	1.000	02m15s
Saint Étienne	45°26′N	004°24′E	—	09:04:53.3	290	330	45	—			—			11:50:36.1	100	99	60	10:25:54.7	15	41	55	144	0.900	0.880	
Saint Étienne d…	49°23′N	001°06′E	—	09:03:03.9	284	321	40	10:20:20.3	155	181	10:21:42.0	233	259	11:42:54.4	105	110	56	10:21:01.1	14	40	50	140	1.027	1.000	01m22s
Saint Nazaire	47°17′N	002°12′W	—	08:57:37.4	287	328	39	—			—			11:37:41.2	101	112	57	10:15:08.2	14	45	50	132	0.936	0.926	
Saint Quentin	49°51′N	003°17′E	—	09:06:03.5	283	318	42	10:23:39.3	66	89	10:25:22.2	323	346	11:46:19.5	107	108	55	10:24:30.7	195	217	51	145	1.027	1.000	01m43s
Sarcelles	49°00′N	002°23′E	—	09:04:18.2	284	321	42	—			—			11:45:15.6	105	108	56	10:22:55.8	15	39	51	143	0.996	0.997	
Sarrebourg	48°44′N	007°03′E	—	09:10:09.9	285	319	45	10:29:41.1	150	169	10:31:18.8	242	260	11:53:21.7	106	101	56	10:30:29.9	16	34	54	152	1.028	1.000	01m38s
Saverne	48°44′N	007°22′E	—	09:10:35.9	285	319	45	10:30:07.3	143	161	10:31:55.9	249	267	11:53:53.7	107	101	56	10:31:01.6	16	34	54	153	1.028	1.000	01m49s
Sedan	49°42′N	004°57′E	—	09:08:01.7	283	318	43	10:26:19.0	61	82	10:27:54.3	329	350	11:49:10.2	107	106	56	10:27:06.6	195	216	52	148	1.028	1.000	01m35s
Senlis	49°12′N	002°35′E	—	09:04:42.0	284	320	42	10:22:44.0	164	189	10:23:50.6	225	249	11:45:30.1	105	108	56	10:23:17.3	15	39	51	143	1.027	1.000	01m07s
Soissons	49°22′N	003°20′E	—	09:05:44.9	284	319	42	10:23:28.7	126	149	10:25:31.9	264	287	11:46:40.8	106	107	56	10:24:30.3	15	38	52	145	1.028	1.000	02m03s
Strasbourg	48°35′N	007°45′E	142	09:11:02.5	285	319	46	10:30:58.5	158	176	10:32:22.7	234	252	11:54:39.5	106	100	56	10:31:40.6	16	34	54	153	1.028	1.000	01m24s
Thionville	49°22′N	006°10′E	—	09:09:22.5	284	318	44	10:27:59.3	86	106	10:30:05.8	305	324	11:51:24.8	107	104	56	10:29:02.6	195	215	53	151	1.028	1.000	02m07s
Toulon	43°07′N	005°56′E	—	09:06:31.2	294	336	47	—			—			11:54:41.7	98	92	62	10:28:54.6	16	42	58	145	0.838	0.802	
Toulouse	43°36′N	001°26′E	164	09:00:10.0	293	337	43	—			—			11:45:13.9	96	102	62	10:20:18.2	15	47	55	136	0.836	0.799	
Tours	47°23′N	000°41′E	—	09:01:04.8	287	326	41	—			—			11:42:54.2	102	108	58	10:19:48.1	14	43	52	138	0.945	0.937	
Troyes	48°18′N	004°05′E	—	09:05:57.3	285	322	43	—			—			11:48:33.6	105	105	57	10:25:33.9	15	38	53	145	0.981	0.982	
Valence	44°56′N	004°54′E	—	09:05:24.7	291	331	45	—			—			11:51:48.0	100	98	60	10:26:49.3	15	41	56	144	0.887	0.864	
Valenciennes	50°21′N	003°32′E	—	09:06:45.8	282	317	42	—			—			11:46:26.4	107	109	55	10:24:59.3	195	217	51	146	0.990	0.992	
Verdun	49°10′N	005°23′E	—	09:08:12.9	284	319	44	10:26:41.5	121	142	10:28:50.3	269	290	11:50:14.9	106	104	56	10:27:45.9	15	36	53	149	1.028	1.000	02m09s
Villeurbanne	45°46′N	004°53′E	—	09:05:41.7	290	329	45	—			—			11:51:22.0	101	99	60	10:26:47.4	15	40	55	145	0.911	0.895	

TABLE 17
LOCAL CIRCUMSTANCES FOR EUROPE - GERMANY I
TOTAL SOLAR ECLIPSE OF 1999 AUGUST 11

Location Name	Latitude	Longitude	Elev.	First Contact U.T.	P	V	Alt	Second Contact U.T.	P	V	Third Contact U.T.	P	V	Fourth Contact U.T.	P	V	Alt	Maximum Eclipse U.T.	P	V	Alt	Azm	Eclip. Mag.	Eclip. Obs.	Umbral Durat.
			m	h m s	°	°	°	h m s	°	°	h m s	°	°	h m s	°	°	°	h m s	°	°	°	°			
GERMANY																									
Aachen	50°47′N	006°05′E	—	09:10:16.9	281	314	43	—			—			11:50:13.0	109	106	54	10:28:58.0	195	214	52	151	0.970	0.969	
Aalen	48°50′N	010°05′E	—	09:14:28.4	284	316	47	10:34:32.7	79	94	10:36:35.3	313	327	11:58:18.8	108	98	56	10:35:34.0	196	210	55	159	1.028	1.000	02m03s
Augsburg	48°23′N	010°53′E	—	09:15:26.0	285	317	48	10:35:53.3	116	129	10:38:10.1	277	290	12:00:03.5	108	96	56	10:37:01.7	16	30	56	161	1.028	1.000	02m17s
Backnang	48°56′N	009°25′E	—	09:13:34.3	284	316	47	10:33:23.9	81	96	10:35:27.6	311	326	11:57:07.8	108	99	56	10:34:25.7	196	211	55	157	1.028	1.000	02m04s
Baden-Baden	48°46′N	008°14′E	—	09:11:49.0	285	318	46	10:31:22.7	122	139	10:33:33.9	270	287	11:55:19.1	107	100	56	10:32:28.3	16	33	54	155	1.028	1.000	02m11s
Berlin	52°30′N	013°22′E	—	09:21:12.6	278	303	47	—			—			11:59:15.7	114	102	52	10:39:53.9	196	204	52	167	0.893	0.872	
Bielefeld	52°01′N	008°31′E	—	09:14:21.4	279	308	44	—			—			11:52:52.2	112	106	53	10:32:41.1	196	210	52	157	0.927	0.915	
Böblingen	48°41′N	009°01′E	—	09:12:52.4	285	317	47	10:32:40.6	117	134	10:34:55.0	275	290	11:56:41.6	107	99	56	10:33:47.8	16	32	55	156	1.028	1.000	02m14s
Bockum	51°20′N	006°44′E	—	09:11:32.1	280	312	44	—			—			11:50:47.4	110	107	54	10:29:59.4	195	213	51	153	0.953	0.947	
Bonn	50°44′N	007°05′E	—	09:11:31.8	281	313	44	—			—			11:51:49.6	109	105	54	10:30:31.0	195	213	52	153	0.968	0.966	
Bottrop	51°31′N	006°55′E	—	09:11:54.8	280	311	44	—			—			11:50:55.1	110	107	54	10:30:16.4	195	212	51	153	0.947	0.940	
Braunschweig	52°16′N	010°31′E	—	09:17:10.4	279	306	45	—			—			11:55:31.7	113	104	53	10:35:40.1	196	208	52	161	0.912	0.896	
Bremen	53°04′N	008°49′E	16	09:15:35.1	278	305	44	—			—			11:52:17.6	113	108	52	10:33:05.2	195	209	51	158	0.898	0.878	
Bremerhaven	53°33′N	008°34′E	—	09:15:41.0	277	304	44	—			—			11:51:28.2	114	109	52	10:32:43.0	195	209	50	158	0.886	0.863	
Bruchsal	49°07′N	008°35′E	—	09:12:29.9	284	317	46	10:32:03.2	74	91	10:33:58.8	318	334	11:55:36.0	108	100	56	10:33:00.9	196	212	54	156	1.028	1.000	01m56s
Dachau	48°15′N	011°27′E	—	09:16:13.2	285	316	49	10:36:54.5	119	132	10:39:10.1	274	286	12:01:07.3	108	95	56	10:38:02.3	16	29	56	162	1.028	1.000	02m16s
Darmstadt	49°53′N	008°40′E	—	09:13:04.1	283	314	46	—			—			11:55:03.6	109	102	55	10:33:04.2	196	212	54	156	0.985	0.987	
Dessau	51°50′N	012°14′E	—	09:19:12.8	279	306	47	—			—			11:58:27.6	113	102	53	10:38:21.3	196	206	53	165	0.916	0.901	
Dortmund	51°31′N	007°28′E	—	09:12:37.0	280	311	44	—			—			11:51:45.2	111	106	54	10:31:06.5	195	212	52	155	0.945	0.937	
Dresden	51°03′N	013°44′E	—	09:20:53.1	280	306	48	—			—			12:01:31.8	112	99	53	10:40:54.8	197	204	54	168	0.930	0.918	
Duisburg	51°25′N	006°46′E	—	09:11:38.6	280	311	44	—			—			11:50:46.3	110	107	54	10:30:02.6	195	212	51	153	0.950	0.944	
Düren	50°48′N	006°28′E	—	09:10:47.0	281	313	44	—			—			11:50:48.4	109	106	54	10:29:33.6	195	213	52	152	0.968	0.967	
Düsseldorf	51°12′N	006°47′E	—	09:11:29.7	281	312	44	—			—			11:50:58.6	110	106	54	10:30:03.7	195	213	52	153	0.956	0.951	
Erfurt	50°58′N	011°01′E	—	09:16:58.9	281	309	47	—			—			11:57:38.0	111	101	54	10:36:39.4	196	208	53	162	0.945	0.938	
Erlangen	49°36′N	011°01′E	—	09:16:12.2	283	313	47	—			—			11:59:03.3	109	99	55	10:36:57.2	196	209	55	161	0.982	0.983	
Essen	52°43′N	007°57′E	—	09:14:11.9	278	307	44	—			—			11:51:23.4	112	108	52	10:31:49.8	195	210	51	156	0.911	0.894	
Esslingen	48°45′N	009°16′E	—	09:13:15.8	285	317	47	10:33:03.9	105	121	10:35:21.0	287	302	11:57:02.9	107	99	56	10:34:12.4	196	212	55	157	1.028	1.000	02m17s
Ettlingen	48°56′N	008°24′E	—	09:12:08.4	284	317	46	10:31:36.1	100	117	10:33:51.7	292	308	11:55:27.3	107	100	56	10:32:43.9	196	213	54	155	1.028	1.000	02m16s
Frankfurt am Ma…	50°07′N	008°40′E	103	09:13:12.8	282	314	46	—			—			11:54:51.0	109	102	55	10:33:02.8	196	211	53	156	0.979	0.979	
Freiburg im Bre…	47°59′N	007°51′E	—	09:10:52.1	286	320	46	—			—			11:55:18.2	106	99	57	10:31:54.1	16	34	55	153	0.987	0.988	
Freising	48°23′N	011°44′E	—	09:16:41.8	285	316	49	10:37:20.4	98	110	10:39:37.7	295	307	12:01:26.9	108	95	56	10:38:29.1	197	208	56	163	1.028	1.000	02m17s
Fürstenfeldbruck	48°10′N	011°15′E	—	09:15:53.2	285	317	48	10:36:41.1	133	146	10:38:45.1	260	272	12:00:52.5	107	95	56	10:37:43.0	16	29	56	161	1.028	1.000	02m04s
Fürth	49°28′N	010°59′E	—	09:16:05.3	283	313	47	—			—			11:59:08.3	109	98	55	10:36:55.7	196	209	55	161	0.986	0.988	
Gaggenau	48°48′N	008°19′E	—	09:11:57.1	284	318	46	10:31:29.3	116	134	10:33:43.5	275	292	11:55:25.8	107	100	56	10:32:36.4	16	33	54	155	1.028	1.000	02m14s
Geislingen an d…	48°36′N	009°50′E	—	09:14:00.0	285	317	47	10:34:02.9	112	127	10:36:20.0	281	295	11:58:07.3	107	98	56	10:35:11.4	16	31	55	158	1.028	1.000	02m17s
Gelsenkirchen	51°31′N	007°07′E	—	09:12:10.1	280	311	44	—			—			11:51:13.3	110	107	54	10:30:34.6	195	212	51	154	0.946	0.939	
Gera	50°52′N	012°04′E	—	09:18:23.7	281	308	47	—			—			11:59:18.4	111	100	53	10:38:20.2	196	207	54	164	0.943	0.935	
Giessen	50°35′N	008°40′E	—	09:13:31.2	282	312	45	—			—			11:54:25.7	110	103	54	10:33:00.4	196	211	53	157	0.966	0.963	
Göppingen	48°42′N	009°40′E	—	09:13:48.5	285	317	47	10:33:44.8	103	119	10:36:02.0	289	304	11:57:45.2	107	98	56	10:34:53.4	196	211	55	158	1.028	1.000	02m17s
Göttingen	51°32′N	009°55′E	—	09:15:51.0	280	309	46	—			—			11:55:24.7	111	104	53	10:34:51.3	196	209	52	160	0.935	0.924	
Hagen	51°22′N	007°28′E	—	09:12:30.0	280	311	44	—			—			11:51:53.0	110	106	54	10:31:06.5	195	212	52	154	0.949	0.943	
Hamburg	53°33′N	009°59′E	20	09:17:28.2	277	303	44	—			—			11:53:25.5	114	107	51	10:34:44.5	196	208	51	161	0.881	0.856	
Hamm	51°41′N	007°49′E	—	09:13:11.7	280	310	44	—			—			11:52:08.1	111	106	53	10:31:38.3	195	211	52	155	0.939	0.930	
Hannover	52°24′N	009°44′E	—	09:16:14.2	279	306	45	—			—			11:54:15.9	113	106	53	10:34:28.8	196	208	52	160	0.912	0.896	
Heidelberg	49°25′N	008°43′E	—	09:12:51.4	283	316	46	—			—			11:55:33.3	108	101	56	10:33:12.1	196	212	54	156	0.998	0.999	
Heidenheim an d…	48°40′N	010°08′E	—	09:14:27.9	285	316	47	10:34:33.0	98	113	10:36:49.3	294	308	11:58:33.2	108	98	56	10:35:41.1	196	211	55	159	1.028	1.000	02m16s
Heilbronn	49°08′N	009°13′E	—	09:13:23.7	284	316	46	10:33:19.5	57	72	10:34:48.3	335	351	11:56:37.3	108	100	56	10:34:03.9	196	211	54	157	1.028	1.000	01m29s
Herford	52°06′N	008°40′E	—	09:14:37.0	279	308	44	—			—			11:53:00.7	112	106	53	10:32:54.4	196	210	51	157	0.925	0.912	
Herne	51°32′N	007°13′E	—	09:12:18.5	280	311	44	—			—			11:51:21.6	110	106	54	10:30:43.7	195	212	52	154	0.945	0.938	
Ingolstadt	48°46′N	011°27′E	—	09:16:26.2	284	315	48	10:37:11.2	54	67	10:38:35.6	339	351	12:00:35.8	108	96	56	10:37:53.4	196	209	56	162	1.028	1.000	01m24s
Jena	50°56′N	011°35′E	—	09:17:45.2	281	309	47	—			—			11:58:30.9	111	101	54	10:37:33.4	196	207	53	163	0.943	0.935	
Kaiserslautern	49°26′N	007°46′E	—	09:11:33.7	283	316	45	10:31:11.1	40	57	10:32:06.2	352	9	11:53:59.7	108	102	56	10:31:38.7	196	213	54	154	1.028	1.000	00m55s
Karl-Marx-Stadt…	50°50′N	012°55′E	—	09:19:35.4	281	308	48	—			—			12:00:35.9	112	99	53	10:39:41.5	196	206	54	166	0.940	0.931	
Karlsruhe	49°03′N	008°24′E	—	09:12:12.4	284	317	46	10:31:39.1	86	103	10:33:47.2	306	322	11:55:21.4	107	100	56	10:32:43.1	196	212	54	155	1.028	1.000	02m08s
Kassel	51°19′N	009°29′E	—	09:15:07.0	280	310	45	—			—			11:54:58.8	111	104	54	10:34:12.7	196	210	52	159	0.942	0.934	
Kiel	54°20′N	010°08′E	—	09:18:19.2	276	301	44	—			—			11:52:47.6	115	109	51	10:34:53.5	196	207	50	161	0.860	0.829	
Kirchheim unter…	48°39′N	009°27′E	—	09:13:28.4	285	317	47	10:33:23.8	113	129	10:35:40.2	279	294	11:57:26.5	107	98	56	10:34:32.0	16	31	55	157	1.028	1.000	02m16s
Koblenz	50°21′N	007°35′E	—	09:11:54.8	282	314	45	—			—			11:52:56.0	109	104	55	10:31:18.4	196	212	53	154	0.977	0.977	
Köln (Cologne)	50°56′N	006°59′E	—	09:11:32.9	281	313	44	—			—			11:51:30.3	110	106	54	10:30:21.8	195	213	52	153	0.963	0.960	
Krefeld	51°20′N	006°34′E	—	09:11:19.4	280	312	44	—			—			11:50:32.0	110	107	54	10:29:44.1	195	213	51	153	0.953	0.948	

TABLE 17 - continued
LOCAL CIRCUMSTANCES FOR EUROPE - GERMANY II
TOTAL SOLAR ECLIPSE OF 1999 AUGUST 11

Location Name	Latitude	Longitude	Elev.	First Contact U.T.	P	V	Alt	Second Contact U.T.	P	V	Third Contact U.T.	P	V	Fourth Contact U.T.	P	V	Alt	Maximum Eclipse U.T.	P	V	Alt	Azm	Eclip. Mag.	Eclip. Obs.	Umbral Durat.
			m	h m s	°	°	°	h m s	°	°	h m s	°	°	h m s	°	°	°	h m s	°	°	°	°			
GERMANY																									
Landau	49°12'N	008°07'E	—	09:11:54.0	284	317	46	10:31:17.6	72	90	10:33:10.8	319	336	11:54:45.8	108	101	56	10:32:14.2	196	213	54	155	1.028	1.000	01m53s
Landshut	48°33'N	012°09'E	—	09:17:23.2	285	315	49	10:38:14.6	68	79	10:40:02.9	325	336	12:01:57.0	108	95	55	10:39:08.7	197	208	56	164	1.028	1.000	01m48s
Leipzig	51°19'N	012°20'E	—	09:19:32.1	280	307	47	—			—			11:59:11.5	112	101	53	10:38:38.4	196	206	53	165	0.930	0.918	
Leverkusen	51°03'N	006°59'E	—	09:11:38.2	281	312	44	—			—			11:51:24.5	110	106	54	10:30:21.9	195	213	52	153	0.960	0.956	
Lübeck	53°52'N	010°40'E	—	09:18:36.2	276	302	45	—			—			11:54:00.8	115	107	51	10:35:41.2	196	207	50	162	0.870	0.842	
Ludwigsburg	48°53'N	009°11'E	—	09:13:12.6	284	317	47	10:32:56.6	91	107	10:35:09.0	301	316	11:56:47.5	107	99	56	10:34:02.8	196	212	55	157	1.028	1.000	02m12s
Ludwigshafen	49°29'N	008°26'E	—	09:12:30.3	283	316	46	—			—			11:55:02.3	108	101	56	10:32:43.8	196	212	54	156	0.997	0.998	
Magdeburg	52°07'N	011°38'E	—	09:18:34.4	279	306	46	—			—			11:57:17.1	113	103	52	10:37:22.7	196	207	52	164	0.912	0.895	
Mainz	50°01'N	008°16'E	—	09:12:36.4	283	314	45				—			11:54:18.3	109	102	55	10:32:24.8	196	212	53	155	0.983	0.984	
Mannheim	49°29'N	008°29'E	—	09:12:34.4	283	316	46	—			—			11:55:07.1	108	101	55	10:32:48.7	196	212	54	156	0.997	0.998	
Mönchengladbach	51°12'N	006°28'E	—	09:11:05.5	281	312	44	—			—			11:50:29.2	110	107	54	10:29:34.6	195	213	52	152	0.957	0.953	
Mülheim an der …	51°24'N	006°54'E	—	09:11:48.0	280	311	44	—			—			11:50:59.4	110	107	54	10:30:14.7	195	212	52	153	0.950	0.944	
München	48°08'N	011°34'E	530	09:16:20.6	285	317	49	10:37:12.3	130	143	10:39:20.1	263	275	12:01:25.8	107	95	56	10:36:16.2	17	29	56	162	1.028	1.000	02m08s
Münster	51°57'N	007°37'E	—	09:13:08.9	279	309	44				—			11:51:35.8	111	107	53	10:31:20.1	195	211	51	155	0.933	0.922	
Neunkirchen/Saar	49°20'N	007°10'E	—	09:10:41.2	284	317	45	10:29:43.5	73	91	10:31:36.9	316	336	11:53:05.5	107	102	56	10:30:40.2	196	214	53	153	1.028	1.000	01m53s
Neuss	51°12'N	006°41'E		09:11:22.0	281	312	44							11:50:49.3	110	106	54	10:29:54.5	195	213	52	153	0.957	0.952	
Neustadt an der…	49°21'N	008°08'E	—	09:12:00.7	284	316	46	10:31:39.9	47	64	10:32:50.2	344	1	11:54:39.6	108	101	56	10:32:15.0	196	213	54	155	1.028	1.000	01m10s
Neu-Ulm	48°23'N	010°01'E	—	09:14:09.8	285	317	47	10:34:31.5	134	149	10:36:34.2	259	273	11:58:37.5	107	97	56	10:35:32.8	16	31	55	159	1.028	1.000	02m03s
Neuwied	50°26'N	007°27'E	—	09:11:47.0	282	314	45	—			—			11:52:40.0	109	104	55	10:31:05.7	195	213	53	154	0.975	0.975	
Nürnberg	49°27'N	011°04'E	320	09:16:11.8	283	313	48							11:59:17.2	109	98	55	10:37:04.1	196	209	55	161	0.986	0.988	
Oberhausen	51°28'N	006°50'E	—	09:11:46.1	280	311	44	—			—			11:50:49.9	110	107	54	10:30:08.8	195	212	51	153	0.949	0.942	
Offenburg	48°28'N	007°57'E	—	09:11:15.5	285	319	46	10:31:42.5	180	198	10:32:20.5	212	230	11:55:05.4	106	100	57	10:32:01.5	16	34	55	154	1.028	1.000	00m38s
Oldenburg	53°08'N	008°13'E	—	09:14:53.2	278	306	44	—			—			11:51:22.8	113	108	52	10:32:13.1	195	210	50	157	0.899	0.878	
Osnabrück	52°16'N	008°02'E	—	09:13:56.0	279	308	44				—			11:51:55.6	112	107	53	10:31:57.4	195	210	51	156	0.923	0.909	
Passau	48°35'N	013°28'E	—	09:19:23.8	284	313	50	—						12:04:00.7	109	94	55	10:41:22.9	197	206	56	167	0.997	0.999	
Pforzheim	48°54'N	008°42'E	—	09:12:32.5	284	317	46	10:32:06.5	98	115	10:34:21.8	294	310	11:55:58.8	107	100	56	10:33:14.2	196	212	54	156	1.028	1.000	02m15s
Pirmasens	49°12'N	007°36'E	—	09:11:11.6	284	317	45	10:30:21.1	82	100	10:32:25.2	309	327	11:53:54.7	107	102	56	10:31:23.1	196	213	54	154	1.028	1.000	02m04s
Potsdam	52°24'N	013°04'E	—	09:20:43.8	278	303	47	—			—			11:58:58.1	114	102	52	10:39:28.6	196	205	52	167	0.897	0.877	
Rastatt	48°51'N	008°12'E	—	09:11:49.0	284	318	46	10:31:16.7	113	130	10:33:32.1	279	296	11:55:11.6	107	100	56	10:32:24.4	16	33	54	155	1.028	1.000	02m15s
Recklinghausen	51°36'N	007°13'E	—	09:12:21.7	280	311	44	—			—			11:51:18.1	111	107	54	10:30:43.8	195	212	51	154	0.944	0.936	
Regensburg	49°01'N	012°06'E		09:17:30.4	284	314	48	—			—			12:01:22.4	109	96	55	10:38:54.8	197	208	55	164	0.993	0.995	
Remscheid	51°11'N	007°11'E	—	09:11:59.7	281	312	44	—			—			11:51:36.4	110	106	54	10:30:40.4	195	212	52	154	0.955	0.950	
Reutlingen	48°29'N	009°11'E	—	09:13:00.6	285	318	47	10:33:08.6	139	155	10:35:04.5	253	269	11:57:08.8	107	98	56	10:34:06.6	16	32	55	157	1.028	1.000	01m56s
Rosenheim	47°51'N	012°07'E	—	09:17:04.7	286	317	49	10:38:32.9	155	167	10:40:05.6	238	249	12:02:37.6	107	93	56	10:39:19.2	17	28	57	163	1.028	1.000	01m33s
Rostock	54°05'N	012°07'E	—	09:20:39.2	276	300	45	—			—			11:55:41.6	116	107	51	10:37:42.7	196	205	51	165	0.858	0.827	
Rottenburg am N…	48°28'N	008°56'E		09:12:38.7	285	318	47	10:32:49.2	147	163	10:34:33.4	245	261	11:56:44.6	107	99	56	10:33:41.3	16	32	55	156	1.028	1.000	01m44s
Saarbrücken	49°14'N	006°59'E	—	09:10:22.7	284	317	45	10:29:17.6	89	108	10:31:27.0	302	321	11:52:52.0	107	103	56	10:30:22.3	196	214	53	152	1.028	1.000	02m09s
Saarlouis	49°21'N	006°45'E		09:10:08.3	284	317	45	10:28:59.6	78	97	10:30:59.2	313	331	11:52:23.5	107	103	56	10:29:59.4	195	214	53	152	1.028	1.000	02m00s
Salzgitter	52°10'N	010°25'E	—	09:16:58.0	279	306	45	—			—			11:55:29.4	113	104	53	10:35:31.9	196	208	52	161	0.916	0.900	
Sankt Ingbert	49°17'N	007°06'E	—	09:10:34.0	284	317	45	10:29:32.6	81	99	10:31:34.9	310	329	11:53:01.2	107	102	56	10:30:33.7	196	214	53	152	1.028	1.000	02m02s
Sankt Wendel	49°28'N	007°10'E		09:10:46.3	283	317	45	10:30:00.2	52	70	10:31:19.6	339	357	11:52:59.2	107	103	56	10:30:39.3	196	214	52	153	1.028	1.000	01m19s
Schwäbisch Gmünd	48°48'N	009°47'E	—	09:14:01.6	284	316	47	10:33:58.3	90	105	10:36:09.9	303	317	11:57:51.2	108	98	56	10:35:04.1	196	211	55	158	1.028	1.000	02m12s
Schweinfurt	50°03'N	010°14'E	—	09:15:20.3	282	312	47	—			—			11:57:22.5	110	100	55	10:35:34.9	196	209	54	160	0.974	0.973	
Schwerin	53°38'N	011°25'E	—	09:19:23.3	277	302	45	—			—			11:55:16.9	115	106	51	10:36:47.3	196	206	51	164	0.873	0.845	
Siegburg	50°47'N	007°12'E	—	09:11:43.1	281	313	44	—			—			11:51:58.1	109	105	54	10:30:41.9	195	212	52	154	0.966	0.964	
Siegen	50°52'N	008°02'E	—	09:12:52.3	281	312	45	—			—			11:53:11.5	110	104	54	10:31:59.7	196	211	52	155	0.961	0.957	
Sindelfingen	48°42'N	009°00'E	—	09:12:51.5	285	317	47	10:32:38.4	116	132	10:34:53.5	276	292	11:56:39.1	107	99	56	10:33:46.0	16	32	55	156	1.028	1.000	02m15s
Sinsheim	49°15'N	008°53'E	—	09:12:59.5	284	316	46	10:32:58.1	44	60	10:34:01.5	348	4	11:55:58.4	108	100	56	10:33:29.8	196	212	54	156	1.028	1.000	01m03s
Solingen	51°10'N	007°05'E	—	09:11:51.2	281	312	44							11:51:28.0	110	106	54	10:30:31.2	195	212	52	154	0.956	0.951	
Speyer	49°19'N	008°26'E	—	09:12:24.3	284	316	46	10:32:12.5	44	61	10:33:17.2	347	4	11:55:10.9	108	101	56	10:32:44.8	196	212	54	155	1.028	1.000	01m05s
Stuttgart	48°46'N	009°11'E		09:13:09.2	284	317	47	10:32:55.4	105	121	10:35:12.4	287	303	11:56:53.8	107	99	56	10:34:03.9	196	212	55	157	1.028	1.000	02m17s
Trier	49°45'N	006°38'E	—	09:10:14.9	283	316	44	—			—			11:51:53.6	108	104	55	10:29:48.0	195	214	53	152	0.997	0.998	
Tübingen	48°31'N	009°02'E	—	09:12:48.7	285	318	47	10:32:52.4	137	154	10:34:49.7	255	270	11:56:52.0	107	99	56	10:33:51.0	16	32	55	156	1.028	1.000	01m57s
Ulm	48°24'N	010°00'E	—	09:14:08.8	285	317	47	10:34:28.6	132	147	10:36:33.2	261	275	11:58:34.9	107	97	56	10:35:30.9	16	31	55	159	1.028	1.000	02m05s
Wiesbaden	50°05'N	008°14'E		09:12:36.3	282	314	45	—			—			11:54:11.6	109	103	55	10:32:21.3	196	212	53	155	0.981	0.982	
Wilhelmshaven	53°31'N	008°08'E	—	09:15:06.9	277	305	43	—			—			11:50:53.9	114	109	52	10:32:06.0	195	209	50	157	0.889	0.866	
Witten	51°26'N	007°20'E	—	09:12:22.8	280	311	44	—			—			11:51:37.4	110	106	54	10:30:54.3	195	212	52	154	0.948	0.941	
Wuppertal	51°16'N	007°11'E	—	09:12:03.5	281	312	44	—			—			11:51:32.1	110	106	54	10:30:40.5	195	212	52	154	0.953	0.947	
Würzburg	49°48'N	009°56'E	—	09:14:46.5	283	313	47	—			—			11:57:08.9	109	100	55	10:35:08.3	196	210	54	159	0.982	0.983	
Zweibrücken	49°15'N	007°21'E	—	09:10:53.0	284	317	45	10:29:57.2	80	99	10:31:59.5	311	329	11:53:27.6	107	102	56	10:30:58.3	196	214	54	153	1.028	1.000	02m02s
Zwickau	50°44'N	012°29'E	—	09:18:54.9	281	308	48	—			—			12:00:04.4	111	99	53	10:39:02.0	196	206	54	165	0.944	0.937	

TABLE 18
LOCAL CIRCUMSTANCES FOR EUROPE - GREECE, GUERNSEY, HUNGARY, IRELAND & N. IRELAND
TOTAL SOLAR ECLIPSE OF 1999 AUGUST 11

Location Name	Latitude	Longitude	Elev.	First Contact				Second Contact			Third Contact			Fourth Contact				Maximum Eclipse					Eclip.	Eclip.	Umbral
				U.T.	P	V	Alt	U.T.	P	V	U.T.	P	V	U.T.	P	V	Alt	U.T.	P	V	Alt	Azm	Mag.	Obs.	Durat.
			m	h m s	°	°	°	h m s	°	°	h m s	°	°	h m s	°	°	°	h m s	°	°	°	°			
GREECE																									
Athinai (Athens)	37°58'N	023°43'E	107	09:41:25.6	300	324	65	—			—			12:34:29.7	100	56	54	11:10:14.1	21	1	66	204	0.822	0.781	
Iráklion	35°20'N	025°09'E	—	09:47:39.2	304	325	68	—			—			12:40:18.9	98	47	54	11:16:50.5	21	353	67	214	0.766	0.711	
Kallithéa	37°57'N	023°42'E	—	09:41:24.7	300	324	65	—			—			12:34:29.1	100	56	55	11:10:13.3	21	1	66	204	0.821	0.780	
Lárisa	39°38'N	022°25'E	—	09:37:13.7	298	323	63	—			—			12:30:04.9	102	61	55	11:05:29.4	20	6	65	197	0.854	0.822	
Pátrai	38°15'N	021°44'E	—	09:37:04.2	300	328	63	—			—			12:30:34.8	100	58	56	11:05:40.0	20	7	66	197	0.810	0.766	
Peristérion	38°01'N	023°42'E	—	09:41:20.5	300	323	65	—			—			12:34:24.0	101	56	54	11:10:08.2	21	1	66	204	0.823	0.783	
Piraiévs (Pirae…	37°57'N	023°38'E	—	09:41:16.4	300	324	65	—			—			12:34:22.1	100	56	55	11:10:04.9	21	1	66	204	0.820	0.780	
Thessaloníki (S…	40°38'N	022°56'E	24	09:37:29.4	296	319	62	—			—			12:29:38.0	104	64	54	11:05:23.2	20	6	64	198	0.886	0.863	
Vólos	39°21'N	022°56'E	—	09:38:30.4	298	322	63	—			—			12:31:20.6	102	60	55	11:06:53.5	20	5	65	199	0.851	0.819	
GUERNSEY																									
Saint Peter Port	49°27'N	002°32'W	—	08:59:03.3	283	322	38	—			—			11:36:45.3	104	115	55	10:15:34.6	14	43	48	134	0.997	0.999	
Alderney Island	49°43'N	002°12'W	—	08:59:39.3	283	321	38	10:15:20.6	140	169	10:17:01.8	247	275	11:37:15.3	105	115	55	10:16:11.2	14	42	48	135	1.027	1.000	01m41s
HUNGARY																									
Ajka	47°07'N	017°34'E	—	09:25:38.5	286	312	53	10:48:07.8	98	101	10:50:27.7	297	299	12:12:13.9	109	86	54	10:49:17.8	198	200	58	177	1.029	1.000	02m20s
Baja	46°11'N	018°57'E	—	09:27:58.7	287	312	55	10:51:55.7	173	172	10:52:56.6	223	222	12:15:37.5	108	82	54	10:52:26.2	18	17	59	181	1.029	1.000	01m01s
Békéscsaba	46°41'N	021°06'E	—	09:31:41.4	286	307	56	—			—			12:18:09.2	110	82	53	10:55:52.0	198	193	58	187	0.998	0.999	
Budapest	47°30'N	019°05'E	120	09:28:12.9	285	309	54	—			—			12:14:00.8	110	86	53	10:51:41.6	198	197	58	181	0.991	0.993	
Debrecen	47°32'N	021°38'E	—	09:32:34.1	285	304	55	—			—			12:17:38.8	112	84	52	10:56:04.0	198	193	58	188	0.972	0.971	
Dombóvár	46°23'N	018°08'E	—	09:26:34.1	287	313	54	10:50:23.8	178	178	10:51:13.8	218	219	12:14:04.9	108	84	55	10:50:48.8	18	19	59	179	1.029	1.000	00m50s
Dunaújváros	46°58'N	018°57'E	—	09:27:58.1	286	310	54	10:50:51.8	75	74	10:52:50.2	322	320	12:14:33.2	110	85	54	10:51:51.0	198	197	58	181	1.029	1.000	01m58s
Györ	47°42'N	017°38'E	—	09:25:49.0	285	310	53	—			—			12:11:33.8	110	88	54	10:49:02.0	198	200	58	177	0.996	0.998	
Hódmezövásárhely	46°25'N	020°20'E	—	09:30:22.2	287	309	55	10:53:34.7	95	92	10:55:53.3	302	298	12:17:24.5	110	82	53	10:54:44.0	198	195	59	185	1.029	1.000	02m19s
Kalocsa	46°32'N	018°59'E	—	09:28:01.5	287	311	55	10:51:04.6	122	121	10:53:23.2	274	273	12:15:11.9	109	83	54	10:52:13.9	18	17	59	181	1.029	1.000	02m19s
Kaposvár	46°22'N	017°47'E	—	09:25:58.3	287	314	54	—			—			12:13:33.1	108	84	55	10:50:11.6	18	19	59	178	0.998	0.999	
Kecskemét	46°54'N	019°42'E	—	09:29:15.1	286	309	55	10:52:31.0	55	53	10:53:55.8	341	339	12:15:46.5	110	84	53	10:53:13.5	198	196	58	183	1.029	1.000	01m25s
Keszthely	46°46'N	017°15'E	—	09:25:05.1	287	313	53	10:48:03.0	148	150	10:49:52.4	248	250	12:12:11.5	108	85	55	10:48:57.7	18	20	59	177	1.029	1.000	01m49s
Kiskunfélegyháza	46°43'N	019°52'E	—	09:29:32.6	286	309	55	10:52:40.2	75	73	10:54:39.2	321	318	12:16:17.0	110	83	53	10:53:39.8	198	196	59	184	1.029	1.000	01m59s
Kiskunhalas	46°26'N	019°30'E	—	09:28:55.2	287	311	55	10:52:03.5	118	116	10:54:24.3	279	276	12:16:07.6	109	83	54	10:53:14.0	18	16	59	183	1.029	1.000	02m21s
Körmend	47°01'N	016°37'E	—	09:24:03.1	287	314	53	10:46:37.3	135	139	10:48:43.2	260	263	12:10:52.4	108	87	55	10:47:40.3	18	21	58	175	1.029	1.000	02m06s
Köszeg	47°23'N	016°33'E	—	09:23:59.3	286	313	52	10:46:10.6	96	100	10:48:29.2	299	302	12:10:18.3	109	88	55	10:47:19.9	198	201	58	175	1.028	1.000	02m19s
Makó	46°13'N	020°29'E	—	09:30:39.1	287	310	56	10:53:58.7	112	108	10:56:21.3	285	281	12:17:55.2	109	81	53	10:55:10.1	18	14	59	186	1.029	1.000	02m23s
Miskolc	48°06'N	020°47'E	—	09:31:07.1	284	305	54	—			—			12:15:36.1	112	86	52	10:54:09.3	198	194	57	185	0.964	0.961	
Nyíregyháza	47°59'N	021°43'E	—	09:32:42.2	284	303	55	—			—			12:17:04.6	112	85	52	10:55:49.6	198	193	57	188	0.960	0.956	
Paks	46°39'N	018°53'E	—	09:27:51.1	287	311	54	10:50:46.8	112	112	10:53:09.0	284	283	12:14:53.2	109	84	54	10:51:57.9	18	17	59	181	1.029	1.000	02m22s
Pápa	47°19'N	017°28'E	—	09:25:29.7	286	312	53	10:47:57.8	78	81	10:50:00.8	317	319	12:11:48.9	109	87	54	10:48:59.4	198	200	58	177	1.028	1.000	02m03s
Pécs	46°05'N	018°13'E	—	09:26:42.9	288	314	54	—			—			12:14:36.5	108	83	55	10:51:10.8	18	19	59	179	0.993	0.995	
Sárvár	47°15'N	016°57'E	—	09:24:37.8	286	313	53	10:46:56.9	100	104	10:49:17.4	295	298	12:11:06.0	109	87	55	10:48:07.2	198	201	58	176	1.028	1.000	02m20s
Siófok	46°54'N	018°04'E	—	09:26:28.1	286	312	54	10:49:08.7	108	109	10:51:30.9	288	288	12:13:17.5	109	85	54	10:50:19.8	18	19	58	179	1.029	1.000	02m22s
Sopron	47°41'N	016°36'E	—	09:24:07.0	285	312	52	10:46:31.1	56	59	10:47:57.9	339	343	12:10:00.1	109	88	54	10:47:14.5	198	201	58	175	1.028	1.000	01m27s
Szeged	46°15'N	020°09'E	—	09:30:03.8	287	310	55	10:53:22.2	118	115	10:55:43.0	279	275	12:17:22.2	109	82	54	10:54:32.6	18	15	59	185	1.029	1.000	02m21s
Székesfehérvár	47°12'N	018°25'E	—	09:27:04.5	286	311	54	10:49:55.8	61	62	10:51:33.1	334	334	12:13:25.5	110	86	54	10:50:44.5	198	198	58	180	1.028	1.000	01m37s
Szekszárd	46°21'N	018°42'E	—	09:27:32.5	287	312	55	10:51:01.9	154	154	10:52:41.2	242	241	12:15:00.6	109	83	54	10:51:51.6	18	18	59	181	1.029	1.000	01m39s
Szentes	46°39'N	020°16'E	—	09:30:14.4	286	309	55	10:53:30.2	69	66	10:55:20.7	327	324	12:16:58.5	110	83	53	10:54:25.5	198	195	59	185	1.029	1.000	01m50s
Szombathely	47°14'N	016°38'E	—	09:24:06.3	286	313	53	10:46:23.4	110	114	10:48:44.9	285	288	12:10:37.5	109	87	55	10:47:34.2	18	21	58	175	1.028	1.000	02m22s
Várpalota	47°12'N	018°09'E	—	09:26:37.6	286	311	54	10:49:19.5	71	72	10:51:13.0	325	325	12:13:01.1	109	86	54	10:50:16.3	198	199	58	179	1.028	1.000	01m54s
Veszprém	47°06'N	017°55'E	—	09:26:13.6	286	312	53	10:48:48.0	90	92	10:51:03.3	305	306	12:12:47.7	109	86	54	10:49:55.6	198	199	58	178	1.029	1.000	02m15s
Zalaegerszeg	46°51'N	016°51'E	—	09:24:25.4	287	314	53	10:47:18.7	150	153	10:49:04.3	246	248	12:11:27.2	108	86	55	10:48:11.5	18	21	58	176	1.029	1.000	01m46s
IRELAND																									
Cork	51°54'N	008°28'W	17	08:56:03.2	279	316	33	—			—			11:27:06.8	106	124	51	10:09:06.5	193	224	43	127	0.966	0.963	
Dublin (Baile Á…	53°20'N	006°15'W	47	08:59:42.8	277	312	34	—			—			11:30:16.9	109	123	50	10:12:49.4	193	221	43	131	0.925	0.912	
IRELAND, NORTH																									
Belfast, NI	54°35'N	005°55'W	17	09:01:37.0	275	309	34	—			—			11:30:31.8	111	124	49	10:14:03.7	193	219	43	133	0.891	0.868	

TABLE 19
LOCAL CIRCUMSTANCES FOR EUROPE - ITALY
TOTAL SOLAR ECLIPSE OF 1999 AUGUST 11

Location Name	Latitude	Longitude	Elev.	First Contact U.T.	P	V	Alt	Second Contact U.T.	P	V	Third Contact U.T.	P	V	Fourth Contact U.T.	P	V	Alt	Maximum Eclipse U.T.	P	V	Alt	Azm	Eclip. Mag.	Eclip. Obs.	Umbral Durat.
			m	h m s	°	°	°	h m s	°	°	h m s	°	°	h m s	°	°	°	h m s	°	°	°	°			
ITALY																									
Ancona	43°38´N	013°30´E	–	09:18:45.1	293	327	53	–			–			12:09:17.5	102	81	59	10:43:46.3	18	28	61	166	0.894	0.873	
Bari	41°07´N	016°52´E	–	09:25:28.8	296	330	57	–			–			12:18:11.9	100	70	59	10:52:23.5	19	21	64	177	0.847	0.813	
Bergamo	45°41´N	009°43´E	–	09:12:45.9	290	325	49	–			–			12:00:27.5	103	92	59	10:35:39.5	16	33	58	157	0.930	0.919	
Bologna	44°29´N	011°20´E	–	09:15:08.1	291	327	51	–			–			12:04:27.6	102	86	59	10:39:07.3	17	32	60	160	0.905	0.887	
Bolzano (Bozen)	46°31´N	011°22´E	–	09:15:31.4	288	321	50	–			–			12:02:40.6	105	91	58	10:38:26.3	17	30	58	161	0.963	0.960	
Brescia	45°33´N	013°15´E	–	09:18:21.0	289	322	52	–			–			12:06:55.8	105	87	58	10:42:18.5	17	28	59	166	0.946	0.939	
Cagliari	39°13´N	009°07´E	1	09:12:17.1	301	346	52	–			–			12:03:19.4	93	77	65	10:36:40.4	17	40	64	151	0.738	0.677	
Catania	37°30´N	015°06´E	–	09:24:28.4	303	344	59	–			–			12:17:51.6	94	61	63	10:51:31.8	19	26	68	172	0.729	0.666	
Catanzaro	38°54´N	016°26´E	–	09:25:58.1	300	337	59	–			–			12:19:32.2	97	64	61	10:53:21.0	19	22	66	176	0.781	0.730	
Como	45°47´N	009°05´E	–	09:11:49.4	289	326	48	–			–			11:59:13.0	103	93	59	10:34:27.6	16	34	57	155	0.930	0.919	
Cosenza	39°17´N	016°15´E	–	09:25:21.0	300	336	58	–			–			12:18:48.4	97	65	61	10:52:36.5	19	23	66	175	0.790	0.742	
Ferrara	44°50´N	011°35´E	–	09:15:34.1	291	326	51	–			–			12:04:37.0	103	87	59	10:39:27.8	17	31	59	161	0.916	0.901	
Firenze (Floren…	43°46´N	011°15´E	–	09:14:58.3	293	330	51	–			–			12:04:54.6	101	85	60	10:39:15.0	17	32	60	160	0.884	0.860	
Foggia	41°27´N	015°34´E	–	09:22:55.8	296	331	56	–			–			12:15:21.7	100	73	60	10:49:24.0	18	24	64	173	0.847	0.813	
Forlè	44°13´N	012°03´E	–	09:16:17.8	292	327	51	–			–			12:06:02.2	102	85	59	10:40:37.6	17	30	60	162	0.902	0.883	
Genova (Genoa)	44°25´N	008°57´E	97	09:11:19.9	292	330	49	–			–			11:59:58.8	101	90	60	10:34:32.5	16	36	59	154	0.890	0.868	
La Spezia	44°07´N	009°50´E	–	09:12:41.5	292	330	50	–			–			12:01:54.3	101	88	60	10:36:20.4	17	34	59	156	0.886	0.863	
Latina	41°28´N	012°52´E	–	09:18:03.7	297	335	54	–			–			12:10:00.1	99	76	61	10:43:42.2	18	30	63	164	0.828	0.789	
Lecce	40°23´N	018°11´E	–	09:28:22.2	297	330	59	–			–			12:21:29.4	100	67	58	10:55:49.0	19	18	65	182	0.837	0.801	
Livorno (Leghor…	43°33´N	010°19´E	–	09:13:26.5	293	331	50	–			–			12:03:16.3	100	86	61	10:37:29.2	17	34	60	157	0.873	0.846	
Massa	44°01´N	010°09´E	–	09:13:11.5	292	330	50	–			–			12:02:35.5	101	87	60	10:36:59.5	17	34	60	157	0.885	0.862	
Messina	38°11´N	015°34´E	–	09:24:49.4	302	341	59	–			–			12:18:20.4	95	63	62	10:52:01.1	19	25	67	173	0.753	0.695	
Mestre	45°29´N	012°15´E	–	09:16:43.2	290	323	51	–			–			12:05:13.4	104	88	58	10:40:27.7	17	29	59	163	0.939	0.930	
Milano (Milan)	45°28´N	009°12´E	–	09:11:55.3	290	327	48	–			–			11:59:40.6	103	92	59	10:34:45.1	16	34	58	155	0.922	0.908	
Modena	44°40´N	010°55´E	–	09:14:28.7	291	327	50	–			–			12:03:31.4	102	87	59	10:38:14.9	17	32	59	159	0.908	0.891	
Monza	45°35´N	009°16´E	–	09:12:03.1	290	326	48	–			–			11:59:42.6	103	92	59	10:34:50.8	16	34	58	155	0.925	0.913	
Napoli (Naples)	40°51´N	014°17´E	25	09:20:49.5	297	335	56	–			–			12:13:23.5	99	72	61	10:47:06.6	18	27	64	168	0.820	0.779	
Novara	45°28´N	008°38´E	–	09:11:03.5	290	327	48	–			–			11:58:37.3	102	93	59	10:33:41.7	16	35	58	154	0.919	0.905	
Padova	45°25´N	011°53´E	–	09:16:07.2	290	324	51	–			–			12:04:37.7	104	88	58	10:39:48.0	17	30	59	162	0.935	0.924	
Palermo	38°07´N	013°22´E	108	09:20:39.2	302	344	57	–			–			12:13:34.2	94	66	63	10:47:00.9	18	31	67	165	0.734	0.671	
Parma	44°48´N	010°20´E	–	09:13:33.8	291	327	50	–			–			12:02:19.2	102	89	59	10:37:05.0	17	33	59	158	0.908	0.891	
Perugia	43°08´N	012°22´E	–	09:16:52.1	294	330	52	–			–			12:07:35.9	101	81	60	10:41:46.0	17	30	61	163	0.873	0.846	
Pescara	42°28´N	014°13´E	–	09:20:11.0	295	330	54	–			–			12:11:46.2	101	77	60	10:45:54.2	18	27	62	168	0.866	0.837	
Piacenza	45°01´N	009°40´E	–	09:12:32.9	291	327	49	–			–			12:00:53.8	102	90	59	10:35:44.9	17	34	58	156	0.911	0.895	
Pisa	43°43´N	010°23´E	–	09:13:33.2	293	331	50	–			–			12:03:16.5	101	86	60	10:37:33.4	17	34	60	157	0.878	0.852	
Prato	43°53´N	011°06´E	–	09:14:43.6	292	329	51	–			–			12:04:31.5	101	85	60	10:38:54.4	17	32	60	159	0.887	0.863	
Ravenna	44°25´N	012°12´E	–	09:16:32.9	291	327	51	–			–			12:06:08.1	103	85	59	10:40:49.7	17	30	60	163	0.908	0.891	
Reggio di Calab…	38°06´N	015°39´E	–	09:25:03.1	302	341	59	–			–			12:18:35.2	95	62	62	10:52:16.9	19	24	67	174	0.751	0.693	
Reggio nell'Emi…	44°43´N	010°36´E	–	09:13:58.6	291	327	50	–			–			12:02:53.3	102	88	59	10:37:37.4	17	33	59	158	0.908	0.890	
Rimini	44°04´N	012°34´E	–	09:17:09.2	292	327	52	–			–			12:07:08.4	102	84	59	10:41:42.4	17	30	60	164	0.900	0.881	
Roma (Rome)	41°54´N	012°29´E	115	09:17:16.4	296	334	53	–			–			12:08:52.4	99	78	61	10:42:39.1	18	31	63	163	0.838	0.802	
Salerno	40°41´N	014°47´E	–	09:21:48.8	298	334	56	–			–			12:14:32.9	99	71	61	10:48:18.0	18	26	64	170	0.819	0.778	
Sassari	40°43´N	008°34´E	–	09:10:51.0	298	342	50	–			–			12:01:24.7	95	82	64	10:34:53.0	17	40	62	150	0.780	0.729	
Savona	44°17´N	008°30´E	–	09:10:36.8	292	331	49	–			–			11:59:12.1	101	90	60	10:33:42.5	16	36	59	153	0.884	0.860	
Siracusa (Syrac…	37°04´N	015°18´E	–	09:25:15.0	304	345	59	–			–			12:18:37.4	94	59	63	10:52:22.6	19	25	68	172	0.719	0.652	
Taranto	40°28´N	017°15´E	–	09:26:32.3	298	331	58	–			–			12:19:37.1	100	68	59	10:53:45.4	19	20	65	179	0.832	0.794	
Terni	42°34´N	012°37´E	–	09:17:22.3	295	332	53	–			–			12:08:34.6	100	80	60	10:42:34.1	18	30	62	163	0.858	0.827	
Torino (Turin)	45°03´N	007°40´E	–	09:09:29.6	291	329	47	–			–			11:57:05.2	101	93	60	10:31:58.0	16	37	57	151	0.902	0.883	
Torre del Greco	40°47´N	014°22´E	–	09:21:00.3	297	335	56	–			–			12:13:37.2	98	72	61	10:47:20.1	18	27	64	169	0.819	0.778	
Trento	46°04´N	011°08´E	–	09:15:03.1	289	323	50	–			–			12:02:41.0	104	90	58	10:38:09.8	17	31	58	160	0.949	0.942	
Trieste (Triest)	45°40´N	013°46´E	–	09:19:12.5	289	321	52	–			–			12:07:42.7	105	86	57	10:43:13.1	17	27	59	167	0.953	0.947	
Udine	46°03´N	013°14´E	–	09:18:23.2	289	321	51	–			–			12:06:23.0	105	88	57	10:42:02.7	17	27	59	166	0.960	0.957	
Venezia (Venice)	45°27´N	012°21´E	25	09:16:52.6	290	323	51	–			–			12:05:26.0	104	87	58	10:40:39.8	17	29	59	163	0.938	0.929	
Verona	45°27´N	011°00´E	–	09:14:43.1	290	325	50	–			–			12:02:59.9	103	89	59	10:38:07.6	17	31	59	160	0.931	0.919	
Vicenza	45°33´N	011°33´E	–	09:15:36.5	290	324	50	–			–			12:03:54.3	104	89	58	10:39:07.3	17	30	59	161	0.936	0.927	

TABLE 20
LOCAL CIRCUMSTANCES FOR EUROPE - LATVIA, LIECHTENSTEIN, LITHUANIA, LUXEMBOURG, MACEDONIA, MOLDOVA, MONACO & NETHERLANDS
TOTAL SOLAR ECLIPSE OF 1999 AUGUST 11

Location Name	Latitude	Longitude	Elev.	First Contact				Second Contact			Third Contact			Fourth Contact				Maximum Eclipse					Eclip.	Eclip.	Umbral
			m	U.T. h m s	P °	V °	Alt °	U.T. h m s	P °	V °	U.T. h m s	P °	V °	U.T. h m s	P °	V °	Alt °	U.T. h m s	P °	V °	Alt °	Azm °	Mag.	Obs.	Durat.
LATVIA																									
Daugavpils	55°53′N	026°32′E	–	09:41:45.2	271	279	49	–			–			12:09:18.7	124	102	44	10:56:23.4	197	189	49	194	0.737	0.674	
Liepaja	56°31′N	021°01′E	–	09:34:15.4	271	285	47	–			–			12:02:45.4	122	106	46	10:48:52.9	197	195	49	183	0.754	0.696	
Riga	56°57′N	024°06′E	–	09:38:43.5	270	280	47	–			–			12:05:06.8	124	106	45	10:52:31.3	197	192	48	189	0.727	0.662	
LIECHTENSTEIN																									
Vaduz	47°09′N	009°31′E	–	09:12:54.9	287	321	48	–			–			11:58:52.5	105	95	58	10:34:56.4	16	32	56	157	0.971	0.970	
LITHUANIA																									
Kaunas	54°54′N	023°54′E	–	09:37:34.1	273	284	49	–			–			12:08:26.1	122	101	46	10:53:42.5	197	192	50	189	0.775	0.722	
Klaipeda (Memel)	55°43′N	021°07′E	–	09:33:57.1	272	286	48	–			–			12:04:08.7	121	104	47	10:49:27.8	197	195	50	183	0.772	0.719	
Panevezys	55°44′N	024°21′E	–	09:38:32.7	271	282	49	–			–			12:07:27.6	123	103	45	10:53:41.8	197	192	49	190	0.753	0.695	
Siauliai	55°56′N	023°19′E	–	09:37:09.3	271	283	48	–			–			12:06:04.7	122	104	46	10:52:12.8	197	193	49	187	0.755	0.697	
Vilnius	54°41′N	025°19′E	–	09:39:35.3	273	283	50	–			–			12:10:15.8	122	100	46	10:55:46.3	197	190	50	192	0.772	0.718	
LUXEMBOURG																									
Differdange	49°32′N	005°52′E	–	09:09:05.6	283	317	44	10:27:39.9	69	89	10:29:27.8	321	341	11:50:47.8	107	104	56	10:28:33.8	195	215	53	150	1.028	1.000	01m48s
Dudelange	49°28′N	006°05′E	–	09:09:19.9	283	317	44	10:27:57.1	74	94	10:29:52.2	316	335	11:51:12.2	107	104	56	10:28:54.6	195	215	53	150	1.028	1.000	01m55s
Esch-Sur-Alzette	49°30′N	005°59′E	–	09:09:13.4	283	317	44	10:27:49.3	72	92	10:29:40.8	319	338	11:51:00.8	107	104	56	10:28:45.0	195	215	53	150	1.028	1.000	01m51s
Luxembourg	49°36′N	006°09′E	334	09:09:30.4	283	317	44	10:28:21.2	52	72	10:29:41.1	339	358	11:51:12.8	107	104	56	10:29:01.1	195	215	53	151	1.028	1.000	01m20s
MACEDONIA																									
Bitola	41°01′N	021°20′E	–	09:34:04.9	296	322	61	–			–			12:26:25.0	103	66	55	11:01:43.8	20	10	64	192	0.882	0.857	
Gostivar	41°47′N	020°54′E	–	09:32:47.8	294	321	60	–			–			12:24:40.6	104	69	55	11:00:03.7	19	12	63	190	0.899	0.879	
Kumanovo	42°08′N	021°43′E	–	09:34:10.7	294	318	60	–			–			12:25:34.6	105	69	55	11:01:19.3	19	10	63	192	0.915	0.900	
Skopje	41°59′N	021°26′E	240	09:33:42.7	294	319	60	–			–			12:25:18.6	104	69	55	11:00:55.1	19	11	63	191	0.909	0.892	
Tetovo	42°01′N	020°58′E	–	09:32:48.1	294	320	60	–			–			12:24:29.0	104	69	55	10:59:57.5	19	12	63	190	0.905	0.888	
MOLDOVA																									
Bel'cy	47°46′N	027°56′E	–	09:43:37.4	283	292	57	–			–			12:25:23.6	115	82	48	11:06:15.8	199	183	56	203	0.916	0.901	
Bendery	46°48′N	029°29′E	–	09:46:44.0	284	290	58	–			–			12:28:55.4	115	79	47	11:09:52.3	199	180	56	208	0.926	0.914	
Kisin'ov (Kishi…	47°00′N	028°50′E	–	09:45:28.4	283	291	58	–			–			12:27:48.6	115	80	47	11:08:35.3	199	181	56	206	0.927	0.915	
Tiraspol'	46°51′N	029°38′E	–	09:46:59.3	283	290	58	–			–			12:29:00.3	115	79	47	11:10:03.0	200	180	56	208	0.924	0.911	
MONACO																									
Monaco	43°42′N	007°23′E	55	09:08:48.3	293	333	48	–			–			11:57:21.4	99	91	61	10:31:39.8	16	39	59	149	0.862	0.832	
NETHERLANDS																									
Alkmaar	52°37′N	004°44′E	–	09:10:11.2	278	309	42	–			–			11:46:46.3	111	111	53	10:27:09.5	195	213	50	150	0.924	0.911	
Amersfoort	52°09′N	005°24′E	–	09:10:33.6	279	310	42	–			–			11:48:07.0	111	109	53	10:28:03.6	195	213	50	151	0.935	0.925	
Amsterdam	52°22′N	004°54′E	2	09:10:09.2	279	310	42	–			–			11:47:12.3	111	110	53	10:27:21.6	195	213	50	150	0.931	0.919	
Apeldoorn	52°13′N	005°58′E	–	09:11:18.6	279	310	43	–			–			11:48:54.7	111	109	53	10:28:54.0	195	213	50	152	0.931	0.920	
Arnhem	51°59′N	005°55′E	–	09:11:02.8	279	310	43	–			–			11:49:01.4	111	109	53	10:28:48.2	195	213	51	152	0.938	0.928	
Breda	51°35′N	004°46′E	–	09:09:17.9	280	312	42	–			–			11:47:34.5	110	109	54	10:27:03.0	195	214	50	149	0.952	0.947	
Dordrecht	51°49′N	004°40′E	–	09:09:22.8	280	312	42	–			–			11:47:15.3	110	110	54	10:26:56.2	195	214	50	149	0.946	0.939	
Eindhoven	51°26′N	005°28′E	–	09:10:01.7	280	312	43	–			–			11:48:45.8	110	108	54	10:28:04.8	195	214	51	150	0.954	0.949	
Enschede	52°12′N	006°53′E	–	09:12:25.7	279	309	43	–			–			11:50:17.4	111	108	53	10:30:15.0	195	212	51	154	0.929	0.917	
Geleen	50°58′N	005°52′E	–	09:10:08.9	281	313	43	–			–			11:49:44.2	109	107	54	10:28:38.7	195	214	51	151	0.966	0.963	
Groningen	53°13′N	006°33′E	–	09:12:54.7	278	307	43	–			–			11:48:56.0	113	110	52	10:29:49.9	195	211	50	154	0.902	0.883	
Haarlem	52°23′N	004°38′E	–	09:09:51.1	279	310	42	–			–			11:46:47.6	111	111	53	10:26:58.6	195	214	50	149	0.931	0.920	
Haarlemmermeer	52°15′N	004°38′E	–	09:09:43.8	279	311	42	–			–			11:46:53.5	111	110	53	10:26:57.3	195	214	50	149	0.935	0.924	
Heerlen	50°54′N	005°59′E	–	09:10:14.7	281	313	43	–			–			11:49:58.2	109	107	54	10:28:49.2	195	214	52	151	0.967	0.965	
Leiden	52°09′N	004°30′E	–	09:09:28.8	279	311	42	–			–			11:46:45.8	110	110	53	10:26:44.7	195	214	50	149	0.938	0.928	
Maastricht	50°52′N	005°43′E	–	09:09:52.9	281	314	43	–			–			11:49:34.6	109	107	54	10:28:24.5	195	214	51	151	0.969	0.968	
Nijmegen	51°50′N	005°50′E	–	09:10:49.0	280	311	43	–			–			11:49:00.9	110	108	53	10:28:40.0	195	213	51	151	0.942	0.934	
Rotterdam	51°55′N	004°28′E	–	09:09:13.8	280	312	42	–			–			11:46:52.7	110	110	53	10:26:39.5	195	214	50	149	0.944	0.936	
's-Gravenhage (…	52°06′N	004°18′E	–	09:09:11.8	279	311	42	–			–			11:46:29.8	110	111	53	10:26:26.7	195	214	50	149	0.940	0.931	
's-Hertogenbosch	51°41′N	005°19′E	–	09:10:03.2	280	312	43	–			–			11:48:20.7	110	109	54	10:27:52.9	195	214	51	150	0.948	0.941	
Tilburg	51°34′N	005°05′E	–	09:09:40.1	280	312	42	–			–			11:48:04.4	110	109	54	10:27:31.2	195	214	51	150	0.952	0.946	
Utrecht	52°05′N	005°08′E	–	09:10:10.7	279	311	42	–			–			11:47:46.0	110	110	53	10:27:39.6	195	214	50	150	0.938	0.928	
Zaandam	52°26′N	004°49′E	–	09:10:06.9	279	310	42	–			–			11:47:01.8	111	111	53	10:27:15.0	195	214	50	150	0.929	0.917	
Zoetermeer	52°03′N	004°30′E	–	09:09:23.4	279	311	42	–			–			11:46:50.1	110	110	53	10:26:43.7	195	214	50	149	0.941	0.932	

TABLE 21
LOCAL CIRCUMSTANCES FOR EUROPE - NORWAY, POLAND, PORTUGAL
TOTAL SOLAR ECLIPSE OF 1999 AUGUST 11

Location Name	Latitude	Longitude	Elev.	First Contact				Second Contact			Third Contact			Fourth Contact				Maximum Eclipse					Eclip.	Eclip.	Umbral
			m	U.T. h m s	P °	V °	Alt °	U.T. h m s	P °	V °	U.T. h m s	P °	V °	U.T. h m s	P °	V °	Alt °	U.T. h m s	P °	V °	Alt °	Azm °	Mag.	Obs.	Durat.
NORWAY																									
Bergen	60°23'N	005°20'E	43	09:19:41.7	267	289	38	—			—			11:41:01.6	122	122	45	10:29:34.5	194	207	43	155	0.722	0.656	
Oslo	59°55'N	010°45'E	94	09:24:38.1	267	286	41	—			—			11:47:14.0	123	118	45	10:35:29.3	195	203	45	164	0.718	0.651	
Stavanger	58°58'N	005°45'E	—	09:18:17.2	269	292	39	—			—			11:42:46.7	120	120	46	10:29:41.7	194	208	44	155	0.756	0.698	
Trondheim	63°25'N	010°25'E	127	09:28:29.4	262	278	38	—			—			11:42:50.1	127	124	42	10:35:17.5	195	202	41	164	0.637	0.553	
POLAND																									
Bialystok	53°09'N	023°09'E	—	09:35:51.1	276	289	51	—			—			12:10:34.7	119	97	48	10:53:57.7	198	192	52	188	0.822	0.781	
Bielsko-Biala	49°49'N	019°02'E	—	09:28:29.2	282	303	52	—			—			12:10:38.9	113	92	52	10:50:01.0	198	197	56	180	0.931	0.920	
Bydgoszcz	53°08'N	018°00'E	—	09:28:09.6	277	296	49	—			—			12:04:29.4	117	100	50	10:46:31.2	197	199	52	177	0.853	0.820	
Bytom (Beuthen)	50°22'N	018°54'E	—	09:28:25.3	281	302	51	—			—			12:09:40.2	114	93	51	10:49:27.1	197	198	55	180	0.918	0.903	
Chorzow	50°19'N	018°57'E	—	09:28:29.1	281	302	51	—			—			12:09:48.6	114	93	51	10:49:33.8	197	197	55	180	0.919	0.904	
Czestochowa	50°49'N	019°06'E	—	09:28:52.1	280	300	51	—			—			12:09:17.0	114	94	51	10:49:29.3	197	197	55	180	0.905	0.887	
Elblag (Elbing)	54°10'N	019°25'E	—	09:30:43.9	275	292	48	—			—			12:04:40.1	119	102	48	10:48:00.8	197	197	51	180	0.819	0.778	
Gdansk (Danzig)	54°23'N	018°40'E	11	09:29:46.7	275	292	48	—			—			12:03:27.7	119	103	48	10:46:50.9	197	198	51	179	0.818	0.776	
Gdynia	54°32'N	018°33'E	—	09:29:41.9	274	292	48	—			—			12:03:36.0	119	103	48	10:46:36.7	197	198	51	178	0.815	0.772	
Gliwice (Gleiwi...	50°17'N	018°40'E	—	09:28:01.8	281	302	51	—			—			12:09:28.4	114	93	52	10:49:07.9	197	198	55	179	0.921	0.908	
Gorzow Wielkopo...	52°44'N	015°15'E	—	09:23:59.7	278	300	48				—			12:01:30.3	115	101	51	10:42:38.2	197	202	52	171	0.878	0.852	
Grudziadz	53°29'N	018°45'E	—	09:29:25.5	276	294	49	—			—			12:04:54.0	118	101	49	10:47:25.7	197	198	52	179	0.840	0.804	
Kalisz	51°46'N	018°06'E	—	09:27:40.8	279	299	50	—			—			12:06:35.2	115	97	51	10:47:23.0	197	199	54	178	0.887	0.864	
Katowice	50°16'N	019°00'E	—	09:28:33.0	281	302	51	—			—			12:09:57.0	114	93	51	10:49:40.5	197	197	55	180	0.920	0.906	
Kielce	50°52'N	020°37'E	—	09:31:16.0	280	298	52	—			—			12:11:11.4	115	93	50	10:51:49.7	198	195	54	184	0.894	0.873	
Koszalin (Kösli...	54°12'N	016°09'E	—	09:26:09.1	275	296	47	—			—			12:00:41.2	117	104	49	10:43:23.4	196	200	51	173	0.836	0.799	
Kraków	50°03'N	019°58'E	220	09:30:02.0	281	301	52	—			—			12:11:34.3	114	92	51	10:51:21.9	198	196	55	183	0.919	0.905	
Legnica (Liegni...	51°13'N	016°09'E	—	09:24:31.1	280	303	49	—			—			12:04:43.6	114	97	52	10:44:38.5	197	201	54	173	0.912	0.896	
Lódz	51°46'N	019°30'E	—	09:29:48.3	276	297	50	—			—			12:08:23.7	116	96	50	10:49:31.2	197	197	54	181	0.878	0.853	
Lublin	51°15'N	022°35'E	—	09:34:28.9	279	294	52	—			—			12:13:02.6	117	93	49	10:54:34.5	198	193	54	188	0.872	0.844	
Olsztyn (Allens...	53°48'N	020°29'E	—	09:32:06.2	275	291	49	—			—			12:06:28.6	119	100	48	10:49:43.7	197	196	52	182	0.822	0.781	
Opole (Oppeln)	50°41'N	017°55'E	—	09:26:59.3	280	302	51	—			—			12:07:52.9	114	95	52	10:47:42.0	197	199	55	178	0.916	0.900	
Plock	52°33'N	019°43'E	—	09:30:25.9	277	295	50	—			—			12:07:29.2	117	98	49	10:49:22.1	197	196	53	181	0.858	0.826	
Poznan	52°25'N	016°55'E	—	09:26:13.3	278	299	49	—			—			12:04:07.5	116	99	51	10:45:15.7	197	200	53	175	0.877	0.851	
Radom	51°25'N	021°10'E	—	09:32:16.7	279	296	51	—			—			12:11:02.0	116	94	50	10:52:17.5	198	194	54	185	0.877	0.851	
Ruda Slaska	50°18'N	018°51'E	—	09:28:19.4	281	302	51	—			—			12:09:41.9	114	93	51	10:49:24.8	197	198	55	180	0.920	0.906	
Rybnik	50°06'N	018°32'E	—	09:27:46.1	281	303	51	—			—			12:09:33.2	113	93	52	10:49:01.8	197	198	55	179	0.927	0.915	
Rzeszów	50°03'N	022°00'E	—	09:33:19.9	281	298	53	—			—			12:14:14.2	115	90	50	10:54:36.5	198	193	55	187	0.905	0.887	
Slupsk (Stolp)	54°28'N	017°01'E	—	09:27:31.1	275	294	47				—			12:01:22.4	119	104	49	10:44:30.2	197	199	51	175	0.825	0.786	
Sosnowiec	50°18'N	019°08'E		09:28:46.2	281	302	52	—			—			12:10:05.3	114	93	51	10:49:52.0	198	197	55	180	0.916	0.903	
Szczecin (Stett...	53°24'N	014°32'E	—	09:23:24.3	277	299	47	—			—			11:59:41.9	116	103	51	10:41:21.1	196	203	52	170	0.864	0.835	
Tarnów	50°01'N	021°00'E	—	09:31:41.7	281	299	53							12:12:59.5	114	91	51	10:53:02.6	198	194	55	185	0.913	0.897	
Torun	53°02'N	018°35'E	—	09:28:57.7	277	296	49	—						12:05:21.7	117	100	50	10:47:25.5	197	198	52	179	0.852	0.819	
Tychy	50°09'N	018°59'E	—	09:28:29.6	281	302	52	—			—			12:10:05.8	114	93	52	10:49:43.3	198	197	55	180	0.923	0.909	
Walbrzych (Wald...	50°46'N	016°17'E	—	09:24:31.5	281	304	50	—			—			12:05:30.7	113	96	52	10:45:03.9	197	201	54	174	0.923	0.910	
Warszawa (Warsa...	52°15'N	021°00'E	90	09:32:16.3	277	294	51	—			—			12:09:31.3	117	96	49	10:51:28.0	197	195	53	184	0.857	0.826	
Wloclawek	52°39'N	019°02'E	—	09:29:27.1	277	296	50	—			—			12:06:29.3	117	98	50	10:48:17.9	197	197	53	180	0.859	0.828	
Wodzislaw Slaski	50°00'N	018°28'E	—	09:27:38.1	281	303	51	—			—			12:09:36.2	113	93	52	10:48:59.0	197	198	55	179	0.930	0.919	
Wroclaw (Bresla...	51°06'N	017°00'E	147	09:25:44.6	280	302	50	—			—			12:06:03.1	114	96	52	10:46:01.8	197	200	54	175	0.910	0.894	
Zabrze	50°18'N	018°46'E	—	09:28:11.5	281	302	51	—			—			12:09:35.1	114	93	52	10:49:16.8	197	198	55	180	0.920	0.906	
Zielona Góra (G...	51°56'N	015°31'E	—	09:23:56.1	279	302	48	—			—			12:02:54.8	114	99	52	10:43:21.1	197	202	53	172	0.897	0.877	
PORTUGAL																									
Lisboa (Lisbon)	38°43'N	009°08'W	95	08:46:01.9	300	354	34	—			—			11:20:16.8	84	118	61	09:59:19.1	12	61	47	112	0.672	0.595	
Porto	41°11'N	008°36'W	—	08:47:10.8	296	347	34	—			—			11:23:26.5	89	118	59	10:01:40.0	12	57	47	115	0.748	0.688	

TABLE 22
LOCAL CIRCUMSTANCES FOR EUROPE - ROMANIA
TOTAL SOLAR ECLIPSE OF 1999 AUGUST 11

Location Name	Latitude	Longitude	Elev.	First Contact U.T.	P	V	Alt	Second Contact U.T.	P	V	Third Contact U.T.	P	V	Fourth Contact U.T.	P	V	Alt	Maximum Eclipse U.T.	P	V	Alt	Azm	Eclip. Mag.	Eclip. Obs.	Umbral Durat.
			m	h m s	°	°	°	h m s	°	°	h m s	°	°	h m s	°	°	°	h m s	°	°	°	°			
ROMANIA																									
Arad	46°11′N	021°20′E	–	09:32:09.3	287	308	56	10:55:35.5	89	84	10:57:49.9	308	302	12:19:13.9	110	81	53	10:56:42.8	199	193	59	188	1.029	1.000	02m14s
Bacau	46°34′N	026°55′E	–	09:42:07.4	285	296	58	–			–			12:26:14.5	113	79	49	11:05:57.3	199	184	57	202	0.954	0.949	
Baia-Mare	47°40′N	023°35′E	–	09:35:56.8	284	301	56	–			–			12:20:06.2	113	83	51	10:59:14.9	199	190	57	193	0.954	0.949	
Botosani	47°45′N	026°40′E	–	09:41:22.3	283	295	57	–			–			12:23:54.1	115	82	49	11:04:14.8	199	185	56	200	0.927	0.915	
Braila	45°16′N	027°58′E	–	09:44:33.3	287	297	59	–			–			12:29:47.2	112	75	49	11:09:13.8	200	181	58	206	0.977	0.978	
Brasov	45°39′N	025°37′E	–	09:40:02.0	287	301	58	–			–			12:26:05.3	111	77	50	11:04:45.9	199	185	58	200	0.988	0.990	
Bucuresti (Buch…	44°26′N	026°06′E	82	09:41:24.9	288	303	60	11:05:47.7	113	97	11:08:10.0	287	270	12:28:43.6	110	74	50	11:06:58.9	20	3	59	202	1.029	1.000	02m22s
Buzau	45°09′N	026°49′E	–	09:42:27.1	287	299	59	–			–			12:28:30.3	112	75	49	11:07:24.2	200	183	58	203	0.991	0.993	
Calarasi	44°11′N	027°20′E	–	09:43:52.9	288	300	60	11:08:19.0	93	75	11:10:34.9	306	287	12:30:47.8	111	73	49	11:09:27.1	200	181	59	206	1.028	1.000	02m16s
Caransebes	45°25′N	022°13′E	–	09:33:52.7	288	309	57	10:58:01.4	144	137	10:59:58.1	253	245	12:21:40.1	109	78	53	10:58:59.8	19	11	60	191	1.029	1.000	01m57s
Cluj	46°47′N	023°36′E	392	09:36:05.3	285	303	57	–			–			12:21:32.0	112	81	51	11:00:07.1	199	189	58	193	0.976	0.976	
Constanta	44°11′N	028°39′E	4	09:46:24.0	288	297	61	–			–			12:32:30.3	111	72	48	11:11:44.1	200	179	58	209	0.998	0.999	
Craiova	44°19′N	023°48′E	–	09:37:08.2	289	308	59	–			–			12:25:39.3	109	74	52	11:02:58.6	19	7	60	196	0.991	0.993	
Curtea-De-Arges	45°08′N	024°41′E	–	09:38:29.0	288	304	59	11:02:34.3	90	77	11:04:48.7	309	295	12:25:38.8	110	76	51	11:03:41.6	199	186	59	198	1.029	1.000	02m14s
Deva	45°53′N	022°55′E	–	09:35:02.4	287	306	57	10:58:54.3	68	59	11:00:41.5	330	321	12:21:58.8	110	79	52	10:59:48.0	199	190	59	192	1.029	1.000	01m47s
Dragasani	44°40′N	024°16′E	–	09:37:52.5	289	306	59	11:02:39.8	158	146	11:04:14.1	240	227	12:25:47.5	109	75	52	11:03:27.0	19	7	60	197	1.029	1.000	01m34s
Drobeta-Turnu-S…	44°38′N	022°39′E	–	09:34:53.2	289	310	58	–			–			12:23:28.7	109	76	53	11:00:32.9	19	10	60	193	0.989	0.991	
Focsani	45°41′N	027°11′E	–	09:42:55.2	286	298	59	–			–			12:28:04.9	112	77	49	11:07:24.6	199	183	58	204	0.974	0.974	
Galati	45°26′N	028°03′E	–	09:44:38.3	286	296	59	–			–			12:29:36.3	113	76	48	11:09:10.2	200	181	58	206	0.973	0.972	
Giurgiu	43°53′N	025°57′E	–	09:41:23.7	289	304	60	–			–			12:29:24.3	109	72	51	11:07:21.3	20	3	60	202	0.998	0.999	
Hateg	45°37′N	022°57′E	–	09:35:09.7	287	307	57	10:58:57.0	98	89	11:01:17.1	300	291	12:22:26.1	110	78	52	11:00:07.1	199	190	59	193	1.029	1.000	02m20s
Hunedoara	45°45′N	022°54′E	–	09:35:02.3	287	306	57	10:58:48.8	85	76	11:00:59.0	313	304	12:22:09.6	110	79	52	10:59:54.0	199	190	59	192	1.029	1.000	02m10s
Iasi	47°10′N	027°35′E	–	09:43:09.3	284	294	58	–			–			12:26:01.7	114	81	48	11:06:22.6	199	183	56	203	0.934	0.924	
Jimbolia	45°47′N	020°43′E	–	09:31:07.0	288	310	56	10:55:08.3	155	151	10:56:46.4	242	237	12:18:53.5	109	80	53	10:55:57.4	19	14	59	186	1.029	1.000	01m38s
Lipova	46°05′N	021°40′E	–	09:32:45.7	287	308	56	10:56:16.3	89	83	10:58:30.9	308	301	12:19:52.1	110	80	53	10:57:23.7	199	192	59	189	1.029	1.000	02m15s
Lugoj	45°41′N	021°54′E	–	09:33:15.0	288	309	57	10:57:01.7	125	118	10:59:19.5	273	265	12:20:48.3	109	79	53	10:58:10.6	19	12	59	190	1.029	1.000	02m18s
Lupeni	45°22′N	023°13′E	–	09:35:42.7	288	307	58	10:59:39.3	115	106	11:02:01.5	283	273	12:23:12.1	110	77	52	11:00:50.5	19	9	59	194	1.029	1.000	02m22s
Mangalia	43°50′N	028°35′E	–	09:46:28.5	289	298	61	11:11:02.2	81	60	11:13:05.4	319	297	12:33:00.9	111	71	49	11:12:03.9	200	178	59	209	1.028	1.000	02m03s
Medgidia	44°15′N	028°16′E	–	09:45:37.8	288	298	61	–			–			12:31:54.1	111	73	49	11:10:59.5	200	179	59	208	1.000	1.000	
Moreni	45°00′N	025°39′E	–	09:40:19.8	288	302	59	11:04:40.3	67	52	11:06:24.8	332	317	12:27:11.5	111	75	50	11:05:32.6	199	185	59	200	1.029	1.000	01m45s
Oradea	47°03′N	021°57′E	–	09:33:08.8	285	305	56	–			–			12:18:49.5	111	83	52	10:57:01.8	199	192	58	189	0.982	0.983	
Petrila	45°27′N	023°25′E	–	09:36:03.2	288	306	58	10:59:56.3	100	90	11:02:17.3	298	288	12:23:21.7	110	78	52	11:01:06.9	199	189	59	194	1.029	1.000	02m21s
Petrosani	45°25′N	023°22′E	–	09:35:58.3	288	306	58	10:59:52.2	105	95	11:02:14.7	293	282	12:23:20.5	110	78	52	11:01:03.5	199	189	59	194	1.029	1.000	02m23s
Piatra-Neamt	46°56′N	026°22′E	–	09:41:01.3	284	297	57	–			–			12:24:55.5	113	80	49	11:04:38.1	199	185	57	200	0.950	0.944	
Pitesti	44°52′N	024°52′E	–	09:38:55.2	288	304	59	11:03:07.0	111	98	11:05:29.7	287	273	12:26:19.5	110	75	51	11:04:18.4	19	6	59	198	1.029	1.000	02m23s
Ploiesti	44°56′N	026°02′E	–	09:41:04.4	288	302	59	11:05:34.3	57	42	11:07:00.5	342	326	12:27:49.3	111	75	50	11:06:17.5	200	184	59	201	1.028	1.000	01m26s
Resita	45°17′N	021°53′E	–	09:33:18.3	288	310	57	–			–			12:21:22.2	109	78	53	10:58:31.2	19	12	60	190	1.000	1.000	
Rimnicu-Vilcea	45°06′N	024°22′E	–	09:37:54.5	288	305	58	11:01:58.8	104	92	11:04:21.1	294	281	12:25:15.3	110	76	51	11:03:10.0	199	187	59	197	1.029	1.000	02m22s
Satu Mare	47°48′N	022°53′E	–	09:34:43.1	284	302	55	–			–			12:18:57.2	113	84	51	10:57:57.0	199	191	57	191	0.956	0.951	
Sibiu	45°48′N	024°09′E	–	09:37:17.8	287	304	58	–			–			12:23:50.8	111	78	51	11:02:03.2	199	188	59	196	0.997	0.998	
Sinnicolau Mare	46°05′N	020°38′E	–	09:30:55.8	287	310	56	10:54:23.2	121	117	10:56:42.6	276	271	12:18:20.1	109	81	53	10:55:32.9	19	14	59	186	1.029	1.000	02m19s
Slatina	44°26′N	024°22′E	–	09:38:09.1	289	307	59	–			–			12:26:17.8	109	74	52	11:03:52.9	19	6	60	198	0.998	0.999	
Slobozia	44°34′N	027°23′E	–	09:43:47.0	288	299	60	11:08:56.3	26	8	11:09:12.5	13	355	12:30:13.4	111	74	49	11:09:04.4	200	181	59	205	1.028	1.000	00m16s
Suceava	47°39′N	026°19′E	–	09:40:46.1	283	295	57	–			–			12:23:38.6	114	82	49	11:03:47.1	199	185	56	199	0.933	0.922	
Timisoara	45°45′N	021°13′E	–	09:32:00.7	288	310	56	10:55:52.1	140	135	10:57:54.3	257	251	12:19:41.5	109	80	53	10:56:53.3	19	13	59	188	1.029	1.000	02m02s
Tirgoviste	44°56′N	025°27′E	–	09:39:59.0	288	303	59	11:04:11.9	84	69	11:06:19.9	315	300	12:27:01.6	110	75	51	11:05:16.0	199	185	59	200	1.029	1.000	02m08s
Tirgu-Jiu	45°02′N	023°17′E	–	09:35:55.8	288	307	58	11:00:24.5	151	141	11:02:11.0	247	236	12:23:48.5	109	77	52	11:01:17.8	19	9	60	194	1.029	1.000	01m46s
Tirgu Mures	46°33′N	024°33′E	–	09:37:50.3	286	301	57	–			–			12:23:11.6	112	80	51	11:01:58.7	199	188	58	196	0.975	0.974	
Urziceni	44°43′N	026°38′E	–	09:42:17.7	288	301	60	11:06:51.3	58	41	11:08:19.4	341	324	12:28:58.9	111	74	50	11:07:35.4	200	183	59	203	1.028	1.000	01m28s
Vulcan	45°23′N	023°17′E	–	09:35:49.7	288	307	58	10:59:45.1	111	102	11:02:08.0	287	276	12:23:16.4	110	78	52	11:00:56.6	19	9	59	194	1.029	1.000	02m23s

TABLE 23
LOCAL CIRCUMSTANCES FOR EUROPE - SCOTLAND & SPAIN
TOTAL SOLAR ECLIPSE OF 1999 AUGUST 11

Location Name	Latitude	Longitude	Elev.	First Contact U.T.	P	V	Alt	Second Contact U.T.	P	V	Third Contact U.T.	P	V	Fourth Contact U.T.	P	V	Alt	Maximum Eclipse U.T.	P	V	Alt	Azm	Eclip. Mag.	Eclip. Obs.	Umbral Durat.
			m	h m s	°	°	°	h m s	°	°	h m s	°	°	h m s	°	°	°	h m s	°	°	°	°			
SCOTLAND																									
Aberdeen	57°10′N	002°04′W	–	09:08:27.9	271	300	36	–			–			11:34:41.6	116	123	48	10:20:04.6	193	214	43	141	0.818	0.776	
Ayr	55°28′N	004°38′W	–	09:03:53.8	274	306	35	–			–			11:32:02.6	113	124	49	10:16:08.1	193	218	43	136	0.866	0.837	
Dundee	56°28′N	003°00′W	–	09:06:40.8	272	303	35	–			–			11:33:49.4	114	123	48	10:18:37.3	193	216	43	139	0.838	0.801	
Dunfermline	56°04′N	003°29′W	–	09:05:42.6	273	304	35	–			–			11:33:21.4	114	123	48	10:17:50.0	193	216	43	138	0.849	0.815	
Edinburgh	55°57′N	003°13′W	134	09:05:47.3	273	304	35	–			–			11:33:45.0	114	123	49	10:18:04.4	193	216	43	138	0.852	0.818	
Falkirk	56°00′N	003°48′W	–	09:05:20.6	273	304	35	–			–			11:32:58.0	114	124	48	10:17:25.7	193	216	43	138	0.851	0.818	
Glasgow	55°53′N	004°15′W	–	09:04:47.6	273	305	35	–			–			11:32:25.0	113	124	48	10:16:50.3	193	217	43	137	0.855	0.822	
Greenock	55°57′N	004°45′W	–	09:04:27.3	273	305	35	–			–			11:31:44.4	113	124	48	10:16:18.5	193	217	43	136	0.853	0.821	
Kirkcaldy	56°07′N	003°10′W	–	09:06:03.5	273	304	35	–			–			11:33:45.0	114	123	48	10:18:13.7	193	216	43	139	0.847	0.813	
SPAIN																									
Albacete	38°59′N	001°51′W	–	08:55:07.6	301	353	41	–			–			11:37:18.0	87	104	65	10:13:03.7	14	56	55	124	0.687	0.614	
Alcalá de Henar...	40°29′N	003°22′W	–	08:53:08.7	298	348	39	–			–			11:34:26.1	89	108	63	10:10:32.7	14	55	53	123	0.731	0.667	
Alcorcón	40°21′N	003°50′W	–	08:52:32.0	298	349	39	–			–			11:33:19.5	89	109	63	10:09:37.4	13	56	52	122	0.726	0.661	
Alicante	38°21′N	000°29′W	–	08:57:09.6	302	354	42	–			–			11:40:19.7	87	100	66	10:15:45.0	14	56	56	126	0.671	0.594	
Almería	36°50′N	002°27′W	–	08:54:44.5	305	359	41	–			–			11:34:35.4	83	104	67	10:11:20.7	14	60	55	120	0.619	0.532	
Avilés	43°33′N	005°55′W	–	08:51:07.0	292	339	36	–			–			11:30:10.0	94	115	59	10:07:24.1	13	53	49	122	0.821	0.779	
Badajoz	38°53′N	006°58′W	–	08:48:32.5	300	353	36	–			–			11:25:17.4	85	114	62	10:03:13.9	13	60	50	115	0.678	0.602	
Badalona	41°27′N	002°15′E	–	09:00:54.5	297	343	44	–			–			11:47:12.7	93	97	64	10:21:39.1	15	49	57	135	0.774	0.721	
Barcelona	41°23′N	002°11′E	95	09:00:48.3	297	344	44	–			–			11:47:04.1	93	97	64	10:21:30.8	15	49	57	135	0.772	0.718	
Bilbao	43°15′N	002°58′W	–	08:54:23.6	293	340	39	–			–			11:36:06.5	94	110	61	10:12:16.1	14	51	51	127	0.815	0.773	
Burgos	42°21′N	003°42′W	–	08:53:10.2	295	343	38	–			–			11:34:21.8	92	110	61	10:10:37.7	13	53	51	124	0.787	0.737	
Cádiz	36°32′N	006°18′W	–	08:49:35.2	305	1	37	–			–			11:24:43.9	81	113	64	10:03:25.9	13	63	51	113	0.605	0.515	
Cartagena	37°36′N	000°59′W	–	08:56:37.1	304	357	42	–			–			11:38:44.9	85	101	67	10:14:35.3	14	58	56	124	0.646	0.564	
Castellón de la...	39°59′N	000°02′W	–	08:57:36.6	299	349	42	–			–			11:41:54.5	90	100	65	10:16:54.9	14	53	56	128	0.723	0.657	
Córdoba	37°53′N	004°46′W	–	08:51:19.4	303	356	38	–			–			11:29:39.1	84	110	64	10:06:56.7	13	60	52	117	0.649	0.567	
Elche	38°15′N	000°42′W	–	08:56:52.0	302	355	42	–			–			11:39:45.4	86	100	66	10:15:16.9	14	56	56	125	0.667	0.589	
El Ferrol del C...	43°29′N	008°14′W	–	08:48:37.3	292	340	34	–			–			11:25:31.9	93	118	58	10:03:39.5	12	54	47	118	0.817	0.775	
Fuenlabrada	40°17′N	003°48′W	–	08:52:34.0	298	349	39	–			–			11:33:22.1	89	109	63	10:09:39.7	13	56	52	122	0.724	0.659	
Getafe	40°18′N	003°43′W	–	08:52:40.5	298	349	39	–			–			11:33:33.9	89	109	63	10:09:49.4	13	56	52	122	0.725	0.659	
Gijón	43°32′N	005°40′W	–	08:51:23.1	292	339	36	–			–			11:30:39.8	94	114	59	10:07:48.3	13	53	49	122	0.820	0.779	
Granada	37°13′N	003°41′W	–	08:52:54.0	304	358	39	–			–			11:31:47.1	83	107	66	10:08:53.3	14	60	54	118	0.629	0.543	
Hospitalet	41°22′N	002°08′E	–	09:00:44.0	297	344	44	–			–			11:46:57.5	93	97	64	10:21:24.8	15	49	57	135	0.771	0.717	
Huelva	37°16′N	006°57′W	–	08:48:37.3	303	358	36	–			–			11:23:53.7	82	114	63	10:02:30.1	13	62	50	113	0.628	0.541	
Jaén	37°46′N	003°47′W	–	08:52:38.8	303	357	39	–			–			11:31:56.6	84	108	65	10:08:51.0	13	60	53	119	0.646	0.564	
Jerez de la Fro...	36°41′N	006°08′W	–	08:49:45.9	305	0	37	–			–			11:25:17.4	81	112	64	10:03:49.0	13	63	51	114	0.610	0.520	
La Coruña	43°22′N	008°23′W	–	08:48:24.2	292	340	34	–			–			11:25:10.7	93	118	58	10:03:21.1	12	54	47	118	0.814	0.771	
Leganés	40°19′N	003°45′W	–	08:52:38.1	298	349	39	–			–			11:33:29.8	89	109	63	10:09:46.0	13	56	52	122	0.725	0.660	
León	42°36′N	005°34′W	–	08:51:04.1	294	342	37	–			–			11:30:31.9	92	114	60	10:07:30.2	13	54	50	122	0.792	0.744	
Lérida	41°37′N	000°37′E	–	08:58:36.5	296	344	43	–			–			11:43:35.7	93	101	63	10:18:27.1	15	50	55	132	0.774	0.721	
Logroño	42°28′N	002°27′W	–	08:54:45.0	295	342	39	–			–			11:37:02.8	93	108	62	10:12:54.5	14	52	52	127	0.793	0.744	
Madrid	40°24′N	003°41′W	667	08:52:43.1	298	349	39	–			–			11:33:41.0	89	109	63	10:09:54.8	13	56	52	122	0.728	0.663	
Málaga	36°43′N	004°25′W	–	08:52:02.0	305	0	39	–			–			11:29:33.4	82	109	66	10:07:14.7	13	62	53	116	0.612	0.524	
Mataró	41°32′N	002°27′E	–	09:01:12.2	297	343	44	–			–			11:47:38.9	93	97	64	10:22:03.3	15	48	57	136	0.777	0.725	
Móstoles	40°19′N	003°51′W	–	08:52:30.5	298	349	39	–			–			11:33:16.3	89	109	63	10:09:34.8	13	56	52	122	0.725	0.660	
Murcia	37°59′N	001°07′W	–	08:56:19.5	303	355	42	–			–			11:38:36.6	86	101	66	10:14:22.0	14	57	56	124	0.658	0.578	
Orense	42°20′N	007°51′W	–	08:48:26.3	294	343	34	–			–			11:25:42.2	91	117	59	10:03:34.4	12	55	48	118	0.783	0.732	
Oviedo	43°22′N	005°50′W	–	08:51:07.0	293	340	36	–			–			11:30:16.2	93	115	59	10:07:26.6	13	53	49	122	0.815	0.772	
Palma (de Mallo...	39°34′N	002°39′E	–	09:01:35.8	300	349	45	–			–			11:48:08.4	90	93	66	10:22:26.1	15	51	58	134	0.718	0.652	
Pamplona	42°49′N	001°38′W	–	08:55:53.5	294	341	40	–			–			11:38:50.0	94	107	62	10:14:29.9	14	51	53	129	0.805	0.760	
Sabadell	41°33′N	002°06′E	–	09:00:41.9	297	343	44	–			–			11:46:52.4	93	97	64	10:21:21.5	15	49	57	135	0.777	0.724	
Salamanca	40°58′N	005°39′W	–	08:50:25.7	297	347	37	–			–			11:29:36.3	89	113	61	10:06:34.7	13	56	50	120	0.743	0.682	
San Cristóbal d...	28°29′N	016°19′W	–	08:42:38.1	319	24	27	–			–			10:46:11.0	61	124	54	09:40:54.5	10	76	40	94	0.362	0.248	
San Sebastián	43°19′N	001°59′W	–	08:55:37.7	293	339	40	–			–			11:38:09.3	95	108	61	10:14:01.6	14	50	52	129	0.819	0.778	
Santander	43°28′N	003°48′W	–	08:53:29.4	293	339	38	–			–			11:34:26.1	94	111	60	10:10:54.7	13	51	51	125	0.820	0.779	
Sevilla	37°23′N	005°59′W	30	08:49:49.3	303	358	37	–			–			11:26:18.8	83	112	64	10:04:23.5	13	62	51	115	0.632	0.546	
Tarrasa	41°34′N	002°01′E	–	09:00:34.8	297	343	44	–			–			11:46:41.3	93	98	64	10:21:11.7	15	49	57	135	0.777	0.725	
Valencia	39°28′N	000°22′W	24	08:57:09.9	300	351	42	–			–			11:40:59.7	89	101	65	10:16:09.5	14	54	56	127	0.706	0.636	
Valladolid	41°39′N	004°43′W	–	08:51:43.4	296	345	38	–			–			11:31:56.8	91	112	61	10:08:32.2	13	55	51	122	0.764	0.709	
Vigo	42°14′N	008°43′W	–	08:47:29.1	294	343	34	–			–			11:23:52.9	91	118	59	10:02:07.2	12	56	47	116	0.780	0.728	
Vitoria	42°51′N	002°40′W	–	08:54:36.5	294	341	39	–			–			11:36:39.7	94	109	61	10:12:39.1	14	51	52	127	0.804	0.758	
Zaragoza	41°38′N	000°53′W	–	08:56:34.2	296	344	41	–			–			11:40:17.5	92	104	63	10:15:34.3	14	52	54	129	0.771	0.717	

TABLE 24

LOCAL CIRCUMSTANCES FOR EUROPE - SLOVAKIA, SLOVENIA, SWEDEN, SWITZERLAND, WALES & YUGOSLAVIA

TOTAL SOLAR ECLIPSE OF 1999 AUGUST 11

Location Name	Latitude	Longitude	Elev.	First Contact				Second Contact			Third Contact			Fourth Contact				Maximum Eclipse					Eclip.	Eclip.	Umbral
				U.T.	P	V	Alt	U.T.	P	V	U.T.	P	V	U.T.	P	V	Alt	U.T.	P	V	Alt	Azm	Mag.	Obs.	Durat.
			m	h m s	°	°	°	h m s	°	°	h m s	°	°	h m s	°	°	°	h m s	°	°	°	°			
SLOVAKIA																									
Bratislava	48°09′N	017°07′E	–	09:25:02.4	285	310	52	–			–			12:10:11.2	110	89	54	10:47:51.5	198	200	57	176	0.987	0.989	
Kosice	48°43′N	021°15′E	–	09:31:56.5	283	302	54	–			–			12:15:19.0	113	88	51	10:54:26.8	198	194	56	186	0.944	0.937	
SLOVENIA																									
Ljubljana	46°03′N	014°31′E	–	09:20:28.8	288	319	52	–			–			12:08:34.9	106	86	57	10:44:24.8	17	25	59	169	0.968	0.966	
Maribor	46°33′N	015°39′E	–	09:22:24.1	287	316	52	–			–			12:09:53.4	107	86	56	10:46:12.6	18	23	59	172	0.989	0.991	
SWEDEN																									
Borås	57°43′N	012°55′E	–	09:24:48.3	270	290	43	–			–			11:52:14.7	121	113	47	10:38:11.8	196	203	47	167	0.763	0.707	
Göteborg (Gothe…	57°43′N	011°58′E	17	09:23:40.9	270	291	43	–			–			11:51:10.4	120	114	47	10:37:01.2	195	203	47	166	0.767	0.712	
Helsingborg	56°03′N	012°42′E	–	09:23:01.4	273	294	44	–			–			11:54:03.7	118	110	49	10:38:10.3	196	203	49	167	0.805	0.760	
Jönköping	57°47′N	014°11′E	–	09:26:23.3	270	289	43	–			–			11:53:34.0	121	112	47	10:39:45.2	196	201	47	170	0.756	0.699	
Linköping	58°25′N	015°37′E	–	09:28:41.4	269	286	44	–			–			11:54:15.4	122	113	46	10:41:22.0	196	200	47	172	0.735	0.672	
Malmö	55°36′N	013°00′E	–	09:23:01.0	273	295	45	–			–			11:54:58.9	118	109	49	10:38:39.0	196	203	49	167	0.815	0.773	
Norrköping	58°36′N	016°11′E	–	09:29:32.2	269	285	44	–			–			11:54:35.5	123	113	46	10:42:00.0	196	199	47	173	0.728	0.664	
Örebro	59°17′N	015°13′E	–	09:29:00.0	268	284	43	–			–			11:52:39.1	123	115	45	10:40:41.3	196	200	46	172	0.716	0.649	
Stockholm	59°20′N	018°03′E	45	09:32:25.3	267	281	44	–			–			11:55:24.3	124	113	45	10:43:58.9	196	198	46	177	0.703	0.633	
Uppsala	59°52′N	017°38′E	–	09:32:22.5	267	281	43	–			–			11:54:12.8	125	114	44	10:43:19.6	196	198	45	176	0.693	0.620	
Västerås	59°37′N	016°33′E	–	09:30:52.4	267	282	43	–			–			11:53:31.3	124	115	45	10:42:09.4	196	199	46	174	0.703	0.633	
SWITZERLAND																									
Basel (Bâle)	47°33′N	007°35′E	–	09:10:16.8	287	322	46	–			–			11:55:10.4	105	98	58	10:31:29.2	16	35	55	152	0.973	0.973	
Bern (Berne)	46°57′N	007°26′E	572	09:09:47.4	288	324	46	–			–			11:55:20.9	104	98	58	10:31:16.8	16	36	56	152	0.956	0.951	
Genève (Geneva)	46°12′N	006°09′E	405	09:07:39.0	289	327	46	–			–			11:53:30.3	102	98	59	10:29:03.3	16	38	56	148	0.929	0.917	
Lausanne	46°31′N	006°38′E	–	09:08:28.3	288	326	46	–			–			11:54:11.8	103	98	59	10:29:54.2	16	37	56	150	0.940	0.931	
Luzern	47°03′N	008°18′E	–	09:11:05.4	287	323	47	–			–			11:56:49.1	104	96	58	10:32:48.2	16	34	56	154	0.962	0.959	
Sankt Gallen	47°25′N	009°23′E	–	09:12:49.3	287	321	48	–			–			11:58:24.9	106	96	57	10:34:38.9	16	32	56	157	0.978	0.978	
Winterthur	47°30′N	008°43′E	–	09:11:52.9	287	321	47	–			–			11:57:11.4	105	97	57	10:33:27.9	16	33	56	155	0.977	0.977	
Zürich	47°23′N	008°32′E	493	09:11:33.6	287	322	47	–			–			11:56:57.8	105	97	57	10:33:09.6	16	34	56	155	0.973	0.972	
WALES																									
Cardiff	51°29′N	003°13′W	62	09:00:22.5	280	316	37	–			–			11:35:11.5	107	118	53	10:15:36.9	193	221	46	135	0.973	0.972	
Newport	51°35′N	003°00′W	–	09:00:41.9	280	316	37	–			–			11:35:30.2	107	118	53	10:15:57.6	193	221	46	135	0.970	0.969	
Port Talbot	51°36′N	003°47′W	–	08:59:56.6	280	316	36	–			–			11:34:17.0	107	119	53	10:14:54.8	193	221	46	134	0.971	0.969	
Swansea	51°38′N	003°57′W	–	08:59:49.1	280	316	36	–			–			11:34:01.1	107	119	53	10:14:42.5	193	221	46	134	0.970	0.968	
YUGOSLAVIA																									
Beograd (Belgra…	44°50′N	020°30′E	138	09:30:54.1	289	313	57	–			–			12:19:54.0	108	78	54	10:56:23.5	19	14	60	186	0.977	0.977	
Kikinda	45°50′N	020°28′E	–	09:30:40.0	288	311	56	10:54:42.0	159	155	10:56:13.4	238	234	12:18:26.4	109	80	54	10:55:27.7	19	14	59	186	1.029	1.000	01m31s
Kragujevac	44°01′N	020°55′E	–	09:31:53.0	291	315	58	–			–			12:21:41.5	107	75	54	10:57:55.0	19	13	61	188	0.959	0.955	
Nis	43°19′N	021°54′E	–	09:33:58.2	292	314	59	–			–			12:24:13.9	106	72	54	11:00:27.9	19	10	62	192	0.948	0.942	
Novi Sad	45°15′N	019°50′E	–	09:29:37.6	289	313	56	–			–			12:18:16.7	108	79	54	10:54:47.9	19	15	60	184	0.983	0.984	
Pristina	42°39′N	021°10′E	–	09:32:52.5	293	318	59	–			–			12:23:58.2	105	71	55	10:59:43.1	19	12	62	190	0.924	0.911	
Subotica	46°06′N	019°39′E	–	09:29:12.1	287	311	55	10:52:55.1	153	151	10:54:36.8	244	241	12:16:49.2	109	82	54	10:53:46.0	18	16	59	183	1.029	1.000	01m42s

TABLE 25
LOCAL CIRCUMSTANCES FOR EUROPE - UKRAINE
TOTAL SOLAR ECLIPSE OF 1999 AUGUST 11

Location Name	Latitude	Longitude	Elev.	First Contact				Second Contact			Third Contact			Fourth Contact				Maximum Eclipse					Eclip.	Eclip.	Umbral
			m	U.T. h m s	P °	V °	Alt °	U.T. h m s	P °	V °	U.T. h m s	P °	V °	U.T. h m s	P °	V °	Alt °	U.T. h m s	P °	V °	Alt °	Azm °	Mag.	Obs.	Durat.
UKRAINE																									
Aleksandrija	48°40′N	033°07′E	-	09:52:34.8	279	279	57	-			-			12:29:16.9	119	83	44	11:13:04.2	199	177	53	213	0.851	0.818	
Belaja Cerkov'	49°49′N	030°07′E	-	09:47:00.6	279	284	55	-			-			12:24:04.6	119	87	45	11:07:16.9	199	182	53	206	0.850	0.816	
Berd'ansk	46°45′N	036°49′E	-	10:00:13.3	281	273	58	-			-			12:36:28.8	119	79	42	11:21:00.8	200	171	52	223	0.861	0.830	
Cerkassy	49°26′N	032°04′E	-	09:50:28.8	279	281	56	-			-			12:26:46.5	119	85	44	11:10:35.0	199	179	53	210	0.843	0.807	
Cernigov	51°30′N	031°18′E	-	09:48:45.1	276	279	54	-			-			12:22:03.2	121	91	44	11:07:02.9	199	182	51	206	0.802	0.756	
Cernovcy	48°18′N	025°56′E	-	09:39:59.1	282	295	56	-			-			12:22:03.7	115	84	49	11:02:29.2	199	186	56	198	0.920	0.905	
Char'kov (Khark…	50°00′N	036°15′E	-	09:57:32.3	276	271	55	-			-			12:29:27.2	122	87	41	11:15:40.2	199	175	50	218	0.794	0.746	
Cherson	46°38′N	032°35′E	-	09:52:30.3	283	283	59	-			-			12:32:38.0	116	78	45	11:14:56.2	200	176	55	215	0.903	0.884	
Chmel'nickij	49°25′N	027°00′E	-	09:41:42.3	280	291	55	-			-			12:21:23.5	117	87	48	11:03:01.1	199	186	55	199	0.884	0.860	
Dneprodzerzinsk	48°30′N	034°37′E	-	09:55:18.7	279	276	57	-			-			12:31:01.5	120	83	43	11:15:26.5	200	175	52	216	0.842	0.806	
Dnepropetrovsk	48°27′N	034°59′E	79	09:55:59.2	279	276	57	-			-			12:31:27.6	120	83	43	11:16:01.7	200	175	52	217	0.840	0.803	
Doneck	48°00′N	037°48′E	-	10:01:14.4	278	270	57	-			-			12:34:48.3	121	82	41	11:20:33.6	200	171	51	223	0.824	0.784	
Gorlovka	48°18′N	038°03′E	-	10:01:30.7	278	269	57	-			-			12:34:23.8	121	82	40	11:20:27.8	200	171	50	223	0.815	0.773	
Ivano-Frankovsk	48°55′N	024°43′E	-	09:37:49.3	282	296	55	-			-			12:19:30.9	115	86	49	10:59:55.4	199	189	56	194	0.914	0.898	
Jenakijevo	48°14′N	038°13′E	-	10:01:50.8	278	269	57	-			-			12:34:40.0	121	82	40	11:20:47.0	200	171	50	223	0.815	0.772	
Jevpatorija	45°12′N	033°22′E	-	09:54:46.8	285	283	60	-			-			12:36:08.8	115	75	44	11:18:06.1	200	173	55	218	0.929	0.917	
Kamenec-Podol's…	48°41′N	026°36′E	-	09:41:05.8	282	293	56	-			-			12:22:12.4	116	85	48	11:03:09.2	199	186	55	199	0.905	0.887	
Kerc'	45°22′N	036°27′E	-	10:00:28.4	283	276	60	-			-			12:38:52.3	117	75	42	11:22:31.0	200	170	53	224	0.895	0.874	
Kijev (Kiev)	50°26′N	030°31′E	-	09:47:35.2	278	282	55	-			-			12:23:19.5	120	88	45	11:07:09.5	199	182	53	206	0.832	0.794	
Kirovograd	48°30′N	032°18′E	-	09:51:11.5	280	281	57	-			-			12:28:47.8	118	83	44	11:12:05.3	199	178	53	212	0.862	0.832	
Kommunarsk	48°30′N	038°47′E	-	10:02:41.6	277	268	56	-			-			12:34:34.1	122	83	40	11:21:08.7	200	171	50	224	0.804	0.759	
Konstantinovka	48°32′N	037°43′E	-	10:00:47.5	278	270	56	-			-			12:33:39.0	121	83	41	11:19:40.8	200	172	50	222	0.813	0.770	
Kramatorsk	48°43′N	037°32′E	-	10:00:22.2	278	270	56	-			-			12:33:07.6	121	83	41	11:19:10.3	200	172	50	222	0.811	0.767	
Krasnoarmejsk	48°17′N	037°11′E	-	09:59:59.0	278	271	57	-			-			12:33:43.1	121	82	41	11:19:18.9	200	172	51	221	0.823	0.783	
Krasnodon	48°17′N	039°48′E	-	10:04:36.8	277	266	56	-			-			12:35:47.6	122	83	39	11:22:47.6	200	170	49	226	0.800	0.753	
Krasnyj Luc	48°08′N	038°56′E	-	10:03:10.6	278	268	56	-			-			12:35:26.4	121	82	40	11:21:53.0	200	170	50	225	0.811	0.767	
Kremencug	49°04′N	033°25′E	-	09:52:57.8	279	278	56	-			-			12:28:47.5	120	84	44	11:12:59.2	199	177	52	213	0.839	0.803	
Krivoj Rog	47°55′N	033°21′E	-	09:53:18.1	280	280	57	-			-			12:30:57.6	118	82	44	11:14:23.3	200	176	53	215	0.866	0.837	
Lisicansk	48°55′N	038°26′E	-	10:01:50.6	277	268	56	-			-			12:33:26.2	122	84	40	11:20:04.8	200	172	50	223	0.799	0.752	
Luck	50°44′N	025°20′E	-	09:38:52.6	279	291	53	-			-			12:17:10.2	117	91	48	10:59:11.2	198	189	54	195	0.865	0.836	
L'vov	49°50′N	024°00′E	298	09:36:36.7	281	295	54	-			-			12:17:06.1	116	89	49	10:57:56.6	198	190	55	192	0.897	0.876	
Makejevka	48°02′N	037°58′E	-	10:01:31.1	278	270	57	-			-			12:34:52.4	121	82	41	11:20:44.3	200	171	51	223	0.822	0.781	
Melitopol'	46°50′N	035°22′E	-	09:57:30.9	281	277	58	-			-			12:34:59.9	118	79	43	11:18:48.9	200	173	53	220	0.872	0.845	
Nikolajev	46°58′N	032°00′E	-	09:51:16.4	282	284	58	-			-			12:31:23.7	116	79	45	11:13:36.3	200	177	55	213	0.900	0.880	
Nikopol'	47°35′N	034°25′E	-	09:55:22.9	281	278	58	-			-			12:32:38.2	118	81	43	11:16:23.7	200	175	53	217	0.864	0.835	
Odessa	46°28′N	030°44′E	65	09:49:10.1	284	288	59	-			-			12:30:56.8	115	78	46	11:12:16.5	200	178	56	211	0.923	0.910	
Pavlograd	48°32′N	035°53′E	-	09:57:32.8	279	274	57	-			-			12:32:05.9	120	83	42	11:17:10.5	200	174	51	219	0.830	0.791	
Poltava	49°35′N	034°34′E	-	09:54:47.2	278	275	56	-			-			12:28:50.1	121	86	43	11:13:56.0	199	176	51	215	0.818	0.776	
Rovno	50°37′N	026°15′E	-	09:40:23.5	279	290	54	-			-			12:18:25.4	118	90	47	11:00:40.7	198	187	54	197	0.861	0.831	
Sevastopol'	44°36′N	033°32′E	-	09:55:29.5	285	284	61	-			-			12:37:26.6	115	73	44	11:19:12.3	200	172	55	219	0.941	0.933	
Severodoneck	48°58′N	038°27′E	-	10:01:50.7	277	268	56	-			-			12:33:20.8	122	84	40	11:20:01.8	200	172	50	223	0.797	0.750	
Simferopol'	44°57′N	034°06′E	-	09:56:19.9	285	282	60	-			-			12:37:22.7	115	74	44	11:19:35.4	200	172	55	220	0.927	0.915	
Slav'ansk	48°52′N	037°37′E	-	10:00:26.3	277	270	56	-			-			12:32:53.3	121	84	41	11:19:04.2	200	173	50	222	0.807	0.762	
Stachanov	48°34′N	038°40′E	-	10:02:26.9	277	268	56	-			-			12:34:20.4	122	83	40	11:20:53.7	200	171	50	224	0.804	0.758	
Sumy	50°55′N	034°45′E	-	09:54:39.6	275	273	54	-			-			12:26:19.8	122	89	42	11:12:27.4	199	178	50	214	0.787	0.737	
Sverdlovsk	48°05′N	039°40′E	272	10:04:30.3	278	266	56	-			-			12:36:06.5	122	82	39	11:22:55.2	200	170	49	226	0.805	0.760	
Ternopol'	49°34′N	025°36′E	-	09:39:18.3	281	293	55	-			-			12:19:29.4	116	88	48	11:00:41.8	199	188	55	196	0.891	0.869	
Torez	48°01′N	038°37′E	-	10:02:41.1	278	268	57	-			-			12:35:25.8	121	82	40	11:21:38.1	200	171	50	224	0.816	0.774	
Uzgorod	48°37′N	022°18′E	-	09:33:42.3	283	301	54	-			-			12:16:53.2	113	87	51	10:56:15.9	198	192	56	189	0.939	0.931	
Vinnica	49°14′N	028°29′E	-	09:44:17.3	280	288	56	-			-			12:23:23.8	117	86	47	11:05:29.4	199	184	54	203	0.877	0.851	
Vorosilovgrad (…	48°34′N	039°20′E	-	10:03:37.2	277	266	56	-			-			12:34:51.2	122	83	39	11:21:46.0	200	171	49	225	0.798	0.751	
Zaporozje	47°50′N	035°10′E	-	09:56:36.5	280	276	57	-			-			12:32:50.8	119	81	43	11:17:07.9	200	174	52	218	0.852	0.819	
Zdanov	47°06′N	037°33′E		10:01:20.2	280	271	58	-			-			12:36:24.9	120	80	41	11:21:31.8	200	171	51	224	0.846	0.811	
Zitomir	50°16′N	028°40′E	-	09:44:28.0	279	286	55	-			-			12:21:42.8	118	88	46	11:04:38.5	199	184	53	202	0.851	0.818	

TABLE 26
LOCAL CIRCUMSTANCES FOR RUSSIA
TOTAL SOLAR ECLIPSE OF 1999 AUGUST 11

Location Name	Latitude	Longitude	Elev.	First Contact U.T.	P	V	Alt	Second Contact U.T.	P	V	Third Contact U.T.	P	V	Fourth Contact U.T.	P	V	Alt	Maximum Eclipse U.T.	P	V	Alt	Azm	Eclip. Mag.	Eclip. Obs.	Umbral Durat.
			m	h m s	°	°	°	h m s	°	°	h m s	°	°	h m s	°	°	°	h m s	°	°	°	°			
Archangel'sk	64°34′N	040°32′E	–	10:01:43.7	255	250	40	–			–			12:01:22.7	138	118	34	11:02:21.0	197	183	38	211	0.488	0.380	
Astrachan'	46°21′N	048°03′E	–	10:20:30.5	276	250	54	–			–			12:45:01.1	124	80	33	11:35:46.8	200	161	44	241	0.761	0.704	
Barnaul	53°22′N	083°45′E	–	10:56:16.8	248	211	26	–			–			12:30:31.7	146	108	12	11:44:38.6	197	159	19	270	0.371	0.257	
Belgorod	50°36′N	036°35′E	–	09:57:52.4	275	270	55	–			–			12:28:30.2	123	88	41	11:15:18.1	199	175	50	218	0.779	0.727	
Br'ansk	53°15′N	034°22′E	–	09:53:31.3	272	271	52	–			–			12:21:17.9	124	94	42	11:09:03.8	199	180	49	211	0.740	0.678	
Bratsk	56°05′N	101°48′E	–	11:02:22.2	238	202	15	–			–			12:17:12.4	154	121	5	11:40:26.7	196	161	10	283	0.253	0.148	
Ceboksary	56°09′N	047°15′E	–	10:12:13.0	263	248	47	–			–			12:22:55.7	133	101	34	11:19:18.0	198	173	41	228	0.587	0.493	
Cel'abinsk	55°10′N	061°24′E	–	10:31:58.8	258	230	41	–			–			12:28:55.1	139	103	25	11:32:11.6	198	165	33	247	0.498	0.391	
Cerepovec	59°08′N	037°54′E	–	09:58:20.0	263	258	46	–			–			12:11:33.8	132	107	38	11:06:08.3	198	181	43	211	0.598	0.506	
Dzerzinsk	56°15′N	043°24′E	–	10:06:37.4	265	253	48	–			–			12:20:53.4	132	101	36	11:15:24.6	198	175	43	223	0.613	0.524	
Gor'kij (Gorky)	56°20′N	044°00′E	–	10:07:27.8	264	252	48	–			–			12:21:00.7	132	101	35	11:15:53.6	198	175	42	223	0.608	0.517	
Groznyj	43°20′N	045°42′E	–	10:19:26.1	281	254	58	–			–			12:50:01.6	120	72	34	11:38:10.4	201	159	47	242	0.845	0.810	
Irkutsk	52°16′N	104°20′E	467	11:07:12.6	240	201	12	–			–			12:23:42.2	151	116	1	11:46:11.4	196	158	6	287	0.280	0.171	
Ivanovo	57°00′N	040°59′E	–	10:02:54.5	265	256	48	–			–			12:17:56.1	131	103	37	11:11:55.8	198	178	43	218	0.617	0.528	
Izevsk	56°51′N	053°14′E	–	10:20:04.1	260	240	44	–			–			12:23:25.4	136	104	30	11:23:24.7	198	170	37	236	0.532	0.430	
Jaroslavl'	57°37′N	039°52′E	–	10:01:11.7	264	257	47	–			–			12:15:57.6	131	104	37	11:09:59.1	198	179	43	216	0.613	0.524	
Kalinin	56°52′N	035°55′E	–	09:55:35.4	267	264	48	–			–			12:15:02.6	129	103	39	11:06:39.2	198	181	45	210	0.655	0.573	
Kaliningrad	55°55′N	037°49′E	–	09:58:28.3	267	262	49	–			–			12:18:19.1	129	100	39	11:09:54.9	198	179	45	214	0.660	0.580	
Kaluga	54°31′N	036°16′E	–	09:56:19.5	270	266	51	–			–			12:20:10.3	127	97	40	11:09:51.5	198	179	47	213	0.699	0.628	
Kazan'	55°49′N	049°08′E	–	10:15:04.4	263	245	46	–			–			12:24:26.8	134	101	33	11:21:32.5	198	171	40	231	0.579	0.483	
Kemerovo	55°20′N	086°05′E	–	10:55:34.8	245	209	25	–			–			12:25:24.6	148	113	12	11:41:33.4	197	161	18	271	0.336	0.223	
Kirov	58°38′N	049°42′E	–	10:14:20.5	259	243	44	–			–			12:18:12.4	136	107	32	11:17:44.6	198	174	38	229	0.528	0.426	
Krasnodar	45°02′N	039°00′E	–	10:05:29.7	282	270	59	–			–			12:41:46.3	118	75	40	11:26:39.9	200	167	51	229	0.877	0.851	
Krasnojarsk	56°01′N	092°50′E	152	10:58:45.7	242	206	20	–			–			12:21:25.3	151	117	9	11:40:56.8	196	161	15	276	0.294	0.184	
Kujbysev	53°12′N	050°09′E	–	10:18:13.6	266	245	48	–			–			12:30:47.6	132	96	32	11:26:36.6	199	168	41	236	0.616	0.527	
Kurgan	55°26′N	065°18′E	–	10:36:23.3	256	226	38	–			–			12:28:32.9	140	104	23	11:34:06.0	198	164	31	251	0.466	0.356	
Kursk	51°42′N	036°12′E	–	09:56:51.3	274	270	54	–			–			12:25:56.3	124	91	41	11:13:20.7	199	177	49	216	0.759	0.701	
Lipeck	52°37′N	039°35′E	–	10:02:04.9	271	262	52	–			–			12:26:31.0	126	93	39	11:16:17.2	199	175	47	220	0.712	0.644	
Machackala	42°58′N	047°30′E	–	10:23:03.5	281	251	57	–			–			12:51:48.5	121	72	33	11:40:55.3	201	157	45	245	0.834	0.796	
Magnitogorsk	53°27′N	059°04′E	–	10:30:35.3	261	233	43	–			–			12:32:36.2	136	98	27	11:33:34.0	199	164	35	246	0.541	0.440	
Moskva (Moscow)	55°45′N	037°35′E	154	09:58:08.9	268	263	49	–			–			12:18:30.9	128	100	39	11:09:51.6	198	179	46	214	0.665	0.586	
Murmansk	68°58′N	033°05′E	–	09:55:16.3	251	251	36	–			–			11:49:12.2	139	127	33	10:52:37.2	195	189	36	198	0.449	0.338	
Nabereznyje Cel…	55°42′N	052°19′E	–	10:19:36.2	262	241	45	–			–			12:25:48.8	135	101	31	11:24:30.4	198	169	38	236	0.557	0.458	
Niznij Tagil	57°55′N	059°57′E	–	10:27:46.2	256	232	40	–			–			12:22:18.0	140	108	27	11:26:32.0	198	168	33	243	0.469	0.360	
Novokuzneck	53°45′N	087°06′E	–	10:58:07.9	246	209	24	–			–			12:28:34.2	147	110	11	11:44:28.0	197	159	17	273	0.348	0.234	
Novomoskovsk	54°05′N	038°13′E	–	09:59:26.9	270	263	51	–			–			12:22:28.1	127	96	39	11:12:42.4	199	177	47	217	0.694	0.621	
Novosibirsk	55°02′N	082°55′E	–	10:53:41.3	247	211	27	–			–			12:27:02.9	147	111	13	11:41:32.1	197	161	20	268	0.357	0.243	
Omsk	55°00′N	073°24′E	85	10:45:30.4	252	218	33	–			–			12:29:07.0	143	107	19	11:38:45.6	198	162	26	260	0.417	0.304	
Ordzonikidze	43°03′N	044°40′E	–	10:17:51.7	282	257	58	–			–			12:49:57.3	119	71	35	11:37:23.1	201	159	48	241	0.862	0.832	
Orenburg	51°54′N	055°06′E	–	10:26:31.0	265	238	46	–			–			12:35:20.4	133	94	29	11:33:08.9	199	164	38	243	0.597	0.505	
Or'ol	52°59′N	036°05′E	–	09:56:19.6	272	268	52	–			–			12:23:12.4	125	94	41	11:11:32.9	199	178	48	214	0.732	0.669	
Penza	53°13′N	045°00′E	–	10:10:23.0	268	253	50	–			–			12:28:26.8	129	95	35	11:21:28.4	199	171	44	228	0.657	0.576	
Perm'	58°00′N	056°15′E	–	10:23:09.7	257	236	42	–			–			12:21:28.7	139	107	29	11:23:50.9	198	170	35	238	0.493	0.386	
Prokopjevsk	53°53′N	086°45′E	–	10:57:44.8	246	209	24	–			–			12:28:23.9	147	110	11	11:44:11.5	197	159	17	272	0.348	0.234	
R'azan'	54°38′N	039°44′E	–	10:01:39.1	268	260	50	–			–			12:22:18.1	128	97	38	11:13:43.1	199	176	46	218	0.671	0.594	
Rostov-Na-Donu	47°14′N	039°42′E	–	10:05:07.6	279	267	57	–			–			12:37:53.0	121	80	39	11:24:14.6	200	169	50	227	0.823	0.782	
Saint Petersburg	59°55′N	030°15′E	5	09:48:14.3	264	267	45	–			–			12:04:52.9	130	110	40	10:57:21.5	197	188	44	198	0.628	0.542	
Saransk	54°11′N	045°11′E	–	10:10:09.6	267	252	49	–			–			12:26:23.5	130	97	35	11:20:12.8	199	172	43	227	0.638	0.553	
Saratov	51°34′N	046°02′E	60	10:12:59.5	270	252	51	–			–			12:32:37.9	128	91	35	11:25:06.0	199	169	44	232	0.679	0.603	
Smolensk	54°47′N	032°03′E	–	09:49:48.3	271	272	51	–			–			12:16:18.9	125	98	42	11:04:24.2	198	183	48	205	0.724	0.659	
Soci	43°35′N	039°45′E	–	10:08:09.9	284	268	60	–			–			12:45:15.8	117	71	39	11:29:59.8	201	164	51	233	0.902	0.882	
Stavropol'	45°02′N	041°59′E	–	10:11:00.3	281	262	58	–			–			12:44:05.5	120	75	38	11:30:41.3	201	164	49	234	0.847	0.813	
Taganrog	47°12′N	038°56′E	–	10:03:46.3	279	268	57	–			–			12:37:21.1	120	80	40	11:23:16.3	200	169	50	226	0.831	0.792	
Tambov	52°43′N	041°25′E	–	10:04:59.1	270	259	52	–			–			12:27:29.8	127	93	38	11:18:16.7	199	173	46	223	0.695	0.623	
Toljatti (Togli…	53°31′N	049°26′E	–	10:16:56.5	266	246	48	–			–			12:29:47.8	132	96	32	11:25:25.8	199	169	41	234	0.616	0.527	
Tomsk	56°30′N	084°58′E	–	10:53:25.0	245	210	25	–			–			12:23:09.1	149	114	13	11:39:18.8	197	162	19	269	0.330	0.217	
Tula	54°12′N	037°37′E	–	09:58:29.0	270	264	51	–			–			12:21:48.2	127	96	39	11:11:51.2	199	178	47	215	0.696	0.624	
T'umen'	57°09′N	065°32′E	–	10:34:56.6	254	226	37	–			–			12:24:31.9	142	108	23	11:31:12.6	198	166	30	250	0.442	0.330	
Ufa	54°44′N	055°56′E	174	10:25:16.5	261	237	44	–			–			12:29:01.2	136	100	29	11:29:01.7	199	167	37	241	0.545	0.444	
Uljanovsk	54°20′N	048°24′E	–	10:14:53.3	265	247	48	–			–			12:27:31.6	132	98	33	11:23:09.7	199	170	41	232	0.610	0.520	
Vladimir	56°10′N	040°25′E	–	10:02:16.7	266	258	49	–			–			12:19:25.0	130	101	37	11:12:25.9	198	177	44	218	0.637	0.552	
Volgograd (Stal…	48°44′N	044°25′E	–	10:12:17.8	274	256	54	–			–			12:37:55.7	125	84	36	11:27:46.1	200	167	46	233	0.748	0.688	
Vologda	59°12′N	039°55′E	–	10:01:03.8	262	256	46	–			–			12:12:34.5	133	107	37	11:08:03.9	198	180	42	214	0.584	0.489	
Volzskij	48°50′N	044°44′E	–	10:12:45.5	274	255	54	–			–			12:37:53.9	125	85	36	11:27:58.5	200	166	46	233	0.743	0.682	
Voronez	51°40′N	039°10′E	–	10:01:46.6	273	264	53	–			–			12:28:14.5	125	91	39	11:17:06.3	199	174	48	221	0.735	0.672	

TABLE 27
LOCAL CIRCUMSTANCES FOR TURKEY
TOTAL SOLAR ECLIPSE OF 1999 AUGUST 11

Location Name	Latitude	Longitude	Elev.	First Contact U.T.	P	V	Alt	Second Contact U.T.	P	V	Third Contact U.T.	P	V	Fourth Contact U.T.	P	V	Alt	Maximum Eclipse U.T.	P	V	Alt	Azm	Eclip. Mag.	Eclip. Obs.	Umbral Durat.
			m	h m s	°	°	°	h m s	°	°	h m s	°	°	h m s	°	°	°	h m s	°	°	°	°			
TURKEY																									
Adana	37°01′N	035°18′E	25	10:06:45.8	296	284	68	–			–			12:52:47.3	107	53	43	11:33:45.0	22	339	57	236	0.925	0.912	
Adapazari	40°46′N	030°24′E	–	09:52:23.2	293	298	64	–			–			12:40:31.0	108	63	48	11:19:24.9	21	351	60	218	0.963	0.960	
Adiyaman	37°46′N	038°17′E	–	10:11:50.6	293	274	66	–			–			12:54:44.2	110	56	40	11:37:20.8	22	337	54	239	0.977	0.977	
Amasya	40°39′N	035°51′E	–	10:03:28.8	290	281	64	11:27:43.5	95	58	11:29:53.4	307	269	12:47:08.1	112	63	43	11:28:48.6	201	164	56	230	1.028	1.000	02m10s
Ankara	39°56′N	032°52′E	861	09:58:12.9	293	291	65	–			–			12:45:01.4	109	61	45	11:24:58.4	21	346	58	225	0.969	0.967	
Antakya (Antioc…	36°14′N	036°07′E	–	10:09:33.3	297	281	68	–			–			12:55:00.5	106	52	42	11:36:25.0	22	336	57	238	0.915	0.900	
Antalya	36°53′N	030°42′E	–	09:57:20.2	299	302	68	–			–			12:47:04.7	104	52	48	11:25:44.5	22	345	62	226	0.868	0.840	
Aydin	37°51′N	027°51′E	–	09:50:09.5	299	311	67	–			–			12:41:26.8	103	55	50	11:18:48.7	21	352	63	217	0.862	0.832	
Balikesir	39°39′N	027°53′E	–	09:48:19.1	296	307	65	–			–			12:38:48.9	105	60	50	11:16:20.7	21	354	62	214	0.909	0.892	
Bartin	41°38′N	032°21′E	–	09:55:31.0	291	291	64	11:20:42.8	155	124	11:22:23.0	247	215	12:41:28.2	111	65	46	11:21:33.0	21	350	58	221	1.028	1.000	01m40s
Batman	37°52′N	041°07′E	–	10:17:21.5	292	265	64	11:40:21.4	119	71	11:42:28.4	285	237	12:57:11.5	111	58	37	11:41:25.0	22	334	52	244	1.027	1.000	02m07s
Bismil	37°51′N	040°40′E	–	10:16:29.6	292	266	65	11:39:52.9	144	97	11:41:42.1	260	213	12:56:49.7	111	57	38	11:40:47.7	22	335	52	243	1.027	1.000	01m49s
Bursa	40°11′N	029°04′E	–	09:50:13.5	294	303	65	–			–			12:39:40.2	107	61	49	11:17:49.3	21	353	61	216	0.934	0.924	
Çetinkaya	39°15′N	037°38′E	–	10:08:38.7	292	276	65	11:33:03.5	154	112	11:34:41.4	249	207	12:51:28.5	111	60	41	11:33:52.6	22	340	55	236	1.028	1.000	01m38s
Cizre	37°20′N	042°12′E	–	10:20:13.2	292	262	64	11:42:49.3	117	68	11:44:54.8	287	238	12:59:03.3	112	57	36	11:43:52.2	22	333	51	246	1.027	1.000	02m06s
Çorum	40°33′N	034°58′E	–	10:01:48.9	291	284	65	11:26:36.7	147	110	11:28:28.0	256	219	12:46:22.0	111	63	43	11:27:32.4	21	345	56	229	1.028	1.000	01m51s
Denizli	37°46′N	029°06′E	–	09:52:52.3	298	307	67	–			–			12:43:26.8	104	54	49	11:21:21.8	21	349	62	220	0.873	0.847	
Divrigi	39°23′N	038°07′E	–	10:09:27.0	291	274	65	11:33:18.1	115	73	11:35:30.2	288	246	12:51:42.8	111	60	41	11:34:24.3	22	339	54	236	1.027	1.000	02m12s
Diyarbakir	37°55′N	040°14′E	–	10:15:32.7	292	268	65	11:39:22.5	163	117	11:40:42.8	240	193	12:56:19.4	111	57	38	11:40:02.8	22	335	53	242	1.027	1.000	01m20s
Edirne	41°40′N	026°34′E	–	09:43:57.3	293	307	63	–			–			12:33:47.0	107	66	51	11:11:11.2	20	359	61	207	0.947	0.941	
Elâzig	38°41′N	039°14′E	–	10:12:32.9	292	271	65	11:36:13.8	131	86	11:38:17.6	273	228	12:54:02.2	111	59	39	11:37:15.9	22	337	53	239	1.027	1.000	02m04s
Ergani	38°17′N	039°46′E	–	10:14:07.7	292	269	65	11:37:49.8	145	100	11:39:38.0	258	212	12:55:14.7	111	58	39	11:38:44.1	22	336	53	241	1.027	1.000	01m48s
Erzurum	39°55′N	041°17′E	1951	10:15:02.7	289	265	63	–			–			12:53:35.5	114	63	38	11:38:11.3	201	157	51	241	0.966	0.964	
Eskisehir	39°46′N	030°32′E	–	09:53:37.7	294	299	65	–			–			12:42:19.9	107	60	48	11:21:06.0	21	350	60	220	0.939	0.931	
Gaziantep	37°05′N	037°22′E	–	10:10:54.9	295	277	67	–			–			12:54:58.1	108	54	41	11:37:02.5	22	336	55	239	0.950	0.944	
Gebze	40°48′N	029°25′E	–	09:50:22.1	293	301	64	–			–			12:39:09.4	108	63	48	11:17:35.9	21	353	60	216	0.954	0.949	
Gümüshaciköy	40°53′N	035°14′E	–	10:02:00.1	290	283	64	11:26:20.4	100	64	11:28:33.8	302	266	12:46:03.7	111	64	43	11:27:27.2	201	165	56	229	1.028	1.000	02m13s
Iskenderun	36°37′N	036°07′E	–	10:09:00.3	297	281	68	–			–			12:54:22.9	107	53	42	11:35:46.7	22	337	57	238	0.925	0.912	
Iskilip	40°45′N	034°29′E	–	10:00:38.2	291	285	64	11:25:31.9	149	113	11:27:20.6	253	218	12:45:28.7	111	63	44	11:26:26.4	21	346	57	227	1.028	1.000	01m49s
Istanbul	41°01′N	028°58′E	18	09:49:16.3	293	302	64	–			–			12:38:11.7	108	64	49	11:16:28.8	21	354	60	214	0.954	0.950	
Izmir	38°25′N	027°09′E	28	09:48:04.0	298	312	66	–			–			12:39:32.7	103	56	51	11:16:38.0	21	354	63	214	0.869	0.841	
Izmit (Kocaeli)	40°46′N	029°55′E	–	09:51:24.5	293	299	64	–			–			12:39:52.8	108	63	48	11:18:32.9	21	352	60	217	0.958	0.954	
Kastamonu	41°22′N	033°47′E	–	09:58:37.5	290	287	64	11:23:12.0	116	83	11:25:29.1	286	252	12:43:36.0	111	65	44	11:24:20.7	21	347	57	225	1.028	1.000	02m17s
Kayseri	38°43′N	035°30′E	–	10:04:58.6	294	283	66	–			–			12:50:09.4	109	58	43	11:31:19.3	22	341	57	233	0.968	0.966	
Kirikkale	39°50′N	033°31′E	–	09:59:38.9	293	289	65	–			–			12:45:57.9	109	60	45	11:26:14.2	21	345	58	227	0.973	0.972	
Konya	37°52′N	032°31′E	–	09:59:52.7	296	294	67	–			–			12:47:59.9	106	55	46	11:27:32.8	22	344	60	228	0.914	0.898	
Kurtalan	37°57′N	041°42′E	–	10:18:23.4	291	263	64	11:41:12.3	80	33	11:43:00.2	323	276	12:57:32.2	112	58	37	11:42:06.4	202	154	51	244	1.027	1.000	01m48s
Maden	38°23′N	039°40′E	–	10:13:47.9	292	270	65	11:37:27.4	140	95	11:39:22.4	264	218	12:54:58.5	111	58	39	11:38:25.0	22	336	53	241	1.027	1.000	01m55s
Malatya	38°21′N	038°19′E	–	10:11:08.6	293	274	65	–			–			12:53:44.9	110	58	40	11:36:25.4	22	338	54	239	0.991	0.993	
Manisa	38°36′N	027°26′E	–	09:48:27.5	298	311	66	–			–			12:39:42.8	104	57	51	11:16:56.0	21	354	63	214	0.877	0.851	
Maras	37°36′N	036°55′E	–	10:09:17.7	295	278	67	–			–			12:53:36.8	109	55	41	11:35:27.5	22	338	56	237	0.957	0.953	
Mersin	36°48′N	034°38′E	–	10:05:40.8	297	287	68	–			–			12:52:21.1	106	53	44	11:32:58.0	22	339	58	235	0.912	0.895	
Merzifon	40°53′N	035°29′E	–	10:02:30.1	290	282	64	11:26:48.7	89	52	11:28:53.8	314	277	12:46:19.7	112	64	43	11:27:51.4	201	165	56	229	1.028	1.000	02m05s
Midyat	37°25′N	041°23′E	–	10:18:30.2	292	264	64	11:41:45.8	153	104	11:43:22.4	251	203	12:58:13.4	111	57	37	11:42:34.2	22	334	51	245	1.027	1.000	01m37s
Osmancik	40°59′N	034°49′E	–	10:01:04.0	290	284	64	11:25:29.1	109	73	11:27:45.5	294	258	12:45:25.8	111	64	44	11:26:37.4	201	165	56	228	1.028	1.000	02m16s
Palu	38°42′N	039°57′E	–	10:13:57.0	291	269	64	11:37:16.3	94	49	11:39:19.8	309	264	12:54:39.9	112	59	39	11:38:18.2	202	157	53	241	1.027	1.000	02m04s
Safranbolu	41°15′N	032°45′E	–	09:56:40.1	291	290	64	–			–			12:42:36.7	110	64	45	11:22:47.5	21	349	58	223	0.999	1.000	
Samsun	41°17′N	036°20′E	40	10:03:46.9	289	279	64	–			–			12:46:29.2	113	65	42	11:28:34.6	201	164	55	230	0.990	0.992	
Siirt	37°56′N	041°57′E	–	10:18:54.0	291	263	64	11:41:44.8	66	19	11:43:13.5	337	289	12:57:46.4	112	58	37	11:42:29.3	202	154	51	245	1.027	1.000	01m29s
Silvan (Miyafar…	38°08′N	041°01′E	–	10:16:48.1	291	265	64	11:39:47.0	97	51	11:41:50.9	306	259	12:56:37.6	112	58	38	11:40:49.1	202	155	52	243	1.027	1.000	02m04s
Sivas	39°45′N	037°02′E	–	10:06:50.9	291	278	65	11:31:03.6	130	90	11:33:10.8	273	232	12:49:58.5	111	61	42	11:32:07.4	21	341	55	234	1.028	1.000	02m07s
Suluova (Suluca)	40°47′N	035°42′E	–	10:03:02.3	290	281	64	11:27:19.0	89	52	11:29:23.9	314	277	12:46:44.3	112	63	43	11:28:21.6	201	164	56	230	1.028	1.000	02m05s
Tarsus	36°55′N	034°53′E	–	10:06:02.3	297	286	68	–			–			12:52:27.7	106	53	43	11:33:12.3	22	339	58	235	0.917	0.903	
Tasköprü	41°30′N	034°14′E	–	09:59:23.7	290	285	64	11:23:52.7	83	49	11:25:52.9	320	285	12:43:52.0	112	65	44	11:24:52.9	201	167	57	226	1.028	1.000	02m00s
Tokat	40°19′N	036°34′E	–	10:05:16.4	291	279	64	11:29:23.8	95	57	11:31:33.0	307	268	12:48:28.9	112	62	42	11:30:28.6	201	162	55	232	1.028	1.000	02m09s
Tosya	41°01′N	034°02′E	–	09:59:27.9	291	286	64	11:24:17.5	141	107	11:26:16.9	261	226	12:44:30.1	111	64	44	11:25:17.3	21	347	57	226	1.028	1.000	01m59s
Trabzon	41°00′N	039°43′E	–	10:10:44.1	288	269	63	–			–			12:50:12.7	114	65	39	11:34:08.1	201	160	52	237	0.960	0.956	
Tunceli	39°07′N	039°32′E	–	10:12:35.8	291	270	64	11:36:10.1	71	27	11:37:48.2	333	289	12:53:31.9	112	60	39	11:36:59.3	202	158	53	239	1.027	1.000	01m38s
Turhal	40°24′N	036°06′E	–	10:04:14.9	291	280	64	11:28:28.7	109	71	11:30:43.5	294	256	12:47:50.8	111	62	42	11:29:36.2	201	163	56	231	1.028	1.000	02m15s
Urfa	37°08′N	038°46′E	–	10:13:41.1	294	272	66	–			–			12:56:18.7	109	55	40	11:39:09.6	22	335	54	241	0.968	0.966	
Van	38°28′N	043°20′E	–	10:20:51.4	289	259	62	–			–			12:57:53.4	113	60	36	11:43:28.3	202	154	50	246	0.976	0.976	
Yildizeli	39°52′N	036°38′E	–	10:05:54.7	291	279	65	11:30:17.5	137	98	11:32:18.6	266	226	12:49:21.4	111	61	42	11:31:18.2	21	342	55	233	1.028	1.000	02m01s
Zara	39°55′N	037°46′E	–	10:08:07.3	290	275	64	11:32:04.4	78	37	11:33:54.5	325	284	12:50:24.5	112	62	41	11:32:59.6	201	161	54	235	1.028	1.000	01m50s
Zile	40°18′N	035°54′E	–	10:03:57.4	291	281	65	11:28:21.4	128	90	11:30:31.4	275	237	12:47:49.0	111	62	43	11:29:26.6	21	343	56	231	1.028	1.000	02m10s

TABLE 28
LOCAL CIRCUMSTANCES FOR THE MIDDLE EAST
TOTAL SOLAR ECLIPSE OF 1999 AUGUST 11

Location Name	Latitude	Longitude	Elev.	First Contact U.T.	P	V	Alt	Second Contact U.T.	P	V	Third Contact U.T.	P	V	Fourth Contact U.T.	P	V	Alt	Maximum Eclipse U.T.	P	V	Alt	Azm	Eclip. Mag.	Eclip. Obs.	Umbral Durat.
			m	h m s	°	°	°	h m s	°	°	h m s	°	°	h m s	°	°	°	h m s	°	°	°	°			
ISRAEL																									
Bat Yam	32°01′N	034°45′E	–	10:13:37.9	305	285	72	–			–			12:59:40.7	100	40	43	11:41:10.2	23	330	59	244	0.794	0.747	
Be'er Sheva' (B…	31°14′N	034°47′E	–	10:15:10.9	306	284	73	–			–			13:00:46.9	99	38	43	11:42:34.2	23	329	59	246	0.775	0.722	
Bene Beraq	32°05′N	034°50′E	–	10:13:41.0	305	285	72	–			–			12:59:41.7	100	40	43	11:41:12.3	23	330	59	244	0.797	0.750	
Ghazzah (Gaza)	31°30′N	034°28′E	–	10:14:00.4	306	286	73	–			–			12:59:59.9	99	38	43	11:41:32.6	23	330	59	245	0.778	0.726	
Hefa (Haifa)	32°49′N	034°59′E	7	10:12:40.5	303	285	71	–			–			12:58:51.7	101	42	43	11:40:13.3	23	331	59	243	0.817	0.776	
Holon	32°01′N	034°46′E	–	10:13:40.0	305	285	72	–			–			12:59:42.0	100	40	43	11:41:12.0	23	330	59	244	0.795	0.747	
Netanya	32°20′N	034°51′E	–	10:13:15.7	304	285	72	–			–			12:59:22.2	101	41	43	11:40:48.5	23	331	59	244	0.804	0.758	
Petah Tiqwa	32°05′N	034°53′E	–	10:13:47.3	305	285	72	–			–			12:59:45.7	100	40	43	11:41:17.7	23	330	59	244	0.798	0.751	
Ramat Gan	32°05′N	034°49′E	–	10:13:38.9	305	285	72	–			–			12:59:40.4	100	40	43	11:41:10.5	23	330	59	244	0.797	0.750	
Rishon leZiyyon	31°58′N	034°48′E	–	10:13:49.8	305	285	72	–			–			12:59:48.7	100	40	43	11:41:20.7	23	330	59	244	0.794	0.746	
Tel Aviv-Yafo	32°04′N	034°46′E	10	10:13:34.5	305	285	72	–			–			12:59:37.9	100	40	43	11:41:06.9	23	330	59	244	0.796	0.748	
Yerushalayim (A…	31°46′N	035°14′E	809	10:15:06.4	305	283	72	–			–			13:00:39.9	100	39	43	11:42:27.7	23	329	58	245	0.794	0.747	
JORDAN																									
'Amman	31°57′N	035°56′E	776	10:16:13.5	304	280	72	–			–			13:01:18.2	101	40	42	11:43:22.2	23	329	58	246	0.808	0.763	
Az-Zarqa'	32°05′N	036°06′E	–	10:16:19.6	304	279	71	–			–			13:01:18.8	101	41	42	11:43:25.4	23	329	58	246	0.813	0.770	
Irbid	32°33′N	035°51′E	–	10:14:57.6	303	281	71	–			–			13:00:20.3	102	42	42	11:42:11.5	23	330	58	245	0.822	0.781	
KUWAIT																									
Al-Jahrah	29°20′N	047°40′E	–	10:43:51.4	300	243	61	–			–			13:16:08.9	105	41	29	12:04:53.7	23	319	44	263	0.894	0.872	
Al-Kuwayt	29°20′N	047°59′E	5	10:44:24.6	300	243	61	–			–			13:16:21.2	105	41	28	12:05:15.8	23	319	44	263	0.898	0.878	
LEBANON																									
Bayrut (Beirut)	33°53′N	035°30′E	–	10:11:55.5	301	283	70	–			–			12:57:57.1	103	45	43	11:39:18.9	23	333	58	242	0.850	0.817	
Sayda (Sidon)	33°33′N	035°22′E	–	10:12:12.4	302	284	71	–			–			12:58:16.9	103	44	43	11:39:38.9	23	332	58	242	0.840	0.805	
Tarabulus (Trip…	34°26′N	035°51′E	–	10:11:45.3	300	282	70	–			–			12:57:32.5	104	47	42	11:38:59.2	22	333	57	241	0.868	0.840	
OMAN																									
Masqat (Muscat)	23°37′N	058°35′E	–	11:11:00.5	300	228	47	–			–			13:28:26.3	104	35	16	12:24:07.0	22	310	30	275	0.892	0.868	
SAUDI ARABIA																									
Ad-Dammam	26°26′N	050°07′E	–	10:53:36.8	302	238	58	–			–			13:21:38.7	103	35	25	12:12:31.0	23	314	41	268	0.857	0.824	
Al-Hufuf	25°22′N	049°34′E	–	10:54:53.9	304	238	58	–			–			13:22:40.7	101	32	25	12:13:42.7	23	313	41	269	0.824	0.783	
Al-Madinah (Med…	24°28′N	039°36′E	–	10:39:47.7	314	254	71	–			–			13:14:51.4	93	23	36	12:02:24.0	24	314	52	263	0.669	0.591	
Ar-Riyad (Riyad…	24°38′N	046°43′E	591	10:51:50.8	308	241	62	–			–			13:21:37.3	99	29	28	12:11:46.0	23	313	44	268	0.770	0.715	
At-Ta'if	21°16′N	040°24′E	–	10:49:31.0	319	248	69	–			–			13:18:38.1	89	15	34	12:09:02.7	24	309	50	269	0.599	0.507	
Jiddah	21°30′N	039°12′E	6	10:46:51.1	319	250	70	–			–			13:16:55.6	89	14	36	12:06:50.6	24	309	52	268	0.587	0.494	
Makkah (Mecca)	21°27′N	039°49′E	–	10:48:02.2	319	249	70	–			–			13:17:45.9	89	15	35	12:07:52.1	24	309	51	268	0.595	0.503	
SYRIA																									
Al-Ladhiqiyah (…	35°31′N	035°47′E	–	10:09:55.5	298	282	69	–			–			12:55:46.4	105	50	42	11:37:03.1	22	335	57	239	0.894	0.873	
Al-Qamishli	37°02′N	041°14′E	–	10:18:44.9	293	264	65	–			–			12:58:46.3	111	56	37	11:43:00.8	22	333	52	245	0.994	0.996	
Ar-Raqqah	35°56′N	039°01′E	–	10:15:55.5	296	271	67	–			–			12:58:35.7	108	52	39	11:41:35.1	22	333	54	244	0.942	0.934	
Dayr az-Zawr	35°20′N	040°09′E	–	10:19:07.2	296	267	67	–			–			13:00:41.9	108	51	38	11:44:20.6	22	331	53	246	0.942	0.933	
Dimashq (Damasc…	33°30′N	036°18′E	720	10:14:14.1	301	280	70	–			–			12:59:30.1	103	44	42	11:41:20.5	23	331	57	244	0.851	0.818	
Halab (Aleppo)	36°12′N	037°10′E	390	10:11:45.8	297	277	68	–			–			12:56:14.0	107	52	41	11:38:12.2	22	335	56	240	0.927	0.915	
Hamah	35°08′N	036°45′E	–	10:12:30.7	298	279	69	–			–			12:57:29.7	106	49	41	11:39:19.0	22	334	56	242	0.896	0.875	
Hims (Homs)	34°44′N	036°43′E	–	10:13:04.3	299	279	69	–			–			12:58:05.3	105	48	41	11:39:56.3	22	333	57	242	0.886	0.863	
UNITED ARAB EMIRATES																									
Abu Zaby	24°28′N	054°22′E	–	11:03:52.3	302	232	52	–			–			13:26:08.8	103	34	20	12:19:40.7	22	312	35	272	0.862	0.831	
Al-Ayn	24°13′N	055°46′E	–	11:06:16.3	301	230	51	–			–			13:26:57.0	103	34	19	12:21:11.5	22	311	33	273	0.873	0.845	
Ash-Shariqah	25°22′N	055°23′E	–	11:03:33.9	300	231	51	–			–			13:25:23.3	105	37	19	12:19:04.8	22	313	34	272	0.896	0.874	
Dubayy	25°18′N	055°18′E	–	11:03:34.5	300	231	51	–			–			13:25:26.8	105	36	20	12:19:07.1	22	313	34	272	0.893	0.871	
WEST BANK																									
Al-Quds (Jerusa…	31°46′N	035°14′E	809	10:15:06.4	305	283	72	–			–			13:00:39.9	100	39	43	11:42:27.7	23	329	58	245	0.794	0.747	
YEMEN																									
'Adan (Aden)	12°45′N	045°12′E	–	11:20:52.1	329	239	57	–			–			13:27:55.2	79	356	26	12:28:24.0	24	298	41	279	0.444	0.333	
Al-Hudaydah	14°48′N	042°57′E	–	11:12:04.0	328	240	61	–			–			13:24:56.9	81	359	30	12:22:48.1	24	300	44	277	0.466	0.357	
San'A'	15°23′N	044°12′E	–	11:11:47.9	325	239	60	–			–			13:26:18.2	83	2	28	12:23:27.2	24	301	43	277	0.500	0.394	
Ta'Izz	13°38′N	044°04′E	–	11:16:57.4	329	239	59	–			–			13:26:30.4	80	357	28	12:25:51.7	24	299	42	278	0.451	0.341	

TABLE 29
LOCAL CIRCUMSTANCES FOR IRAN
TOTAL SOLAR ECLIPSE OF 1999 AUGUST 11

Location Name	Latitude	Longitude	Elev.	First Contact				Second Contact			Third Contact			Fourth Contact				Maximum Eclipse					Eclip.	Eclip.	Umbral
				U.T.	P	V	Alt	U.T.	P	V	U.T.	P	V	U.T.	P	V	Alt	U.T.	P	V	Alt	Azm	Mag.	Obs.	Durat.
			m	h m s	°	°	°	h m s	°	°	h m s	°	°	h m s	°	°	°	h m s	°	°	°	°			
IRAN																									
Ahvaz	31°19'N	048°42'E	–	10:42:02.4	296	243	60	–			–			13:13:47.3	108	46	28	12:02:39.2	22	321	43	261	0.953	0.947	
Aligudarz	33°24'N	049°41'E	–	10:40:13.0	293	242	59	11:59:11.7	117	58	12:01:02.7	288	229	13:10:56.7	111	51	28	12:00:07.3	22	324	43	260	1.025	1.000	01m51s
Amol	36°23'N	052°20'E	–	10:40:07.1	287	239	55	–			–			13:06:55.5	116	59	27	11:57:40.7	202	146	41	258	0.918	0.903	
Arak	34°05'N	049°41'E	–	10:39:06.4	292	243	58	–			–			13:09:47.0	112	53	28	11:58:55.7	202	144	43	259	0.998	0.999	
Ardabil	38°15'N	048°18'E	–	10:30:24.0	287	248	58	–			–			13:01:31.6	116	62	31	11:50:03.0	202	151	45	252	0.924	0.910	
Asadabad	34°47'N	048°07'E	–	10:35:12.2	292	246	60	11:55:28.1	51	354	11:56:22.6	354	298	13:07:42.4	112	54	30	11:55:55.5	202	146	45	256	1.025	1.000	00m54s
Babol	36°34'N	052°42'E	–	10:40:27.8	286	239	55	–			–			13:06:44.0	117	60	27	11:57:43.2	201	147	40	259	0.910	0.892	
Bakhtaran (Kerm…	34°19'N	047°04'E	1320	10:34:02.6	293	248	61	11:55:03.8	175	119	11:55:57.0	229	173	13:07:53.1	111	52	31	11:55:30.5	22	326	46	256	1.026	1.000	00m53s
Bam	29°06'N	058°21'E	–	11:01:03.0	292	229	49	12:14:36.4	89	24	12:16:02.0	314	249	13:20:58.5	111	47	18	12:15:19.3	202	137	33	270	1.022	1.000	01m26s
Bandar 'Abbas	27°11'N	056°17'E	–	11:01:29.4	296	231	51	–			–			13:23:14.5	108	41	19	12:16:52.8	22	315	34	271	0.949	0.941	
Borujerd	33°54'N	048°46'E	–	10:37:46.9	292	244	59	11:57:16.6	115	58	11:59:09.8	289	231	13:09:35.9	111	52	29	11:58:13.4	22	324	44	258	1.025	1.000	01m53s
Bushehr	28°59'N	050°50'E	–	10:49:53.1	298	238	58	–			–			13:18:29.9	107	42	25	12:08:58.0	22	318	41	266	0.925	0.912	
Deh Kord	33°49'N	048°53'E	–	10:38:07.5	292	244	59	11:57:33.6	118	60	11:59:26.2	287	229	13:09:48.3	111	52	29	11:58:30.1	22	324	44	258	1.025	1.000	01m53s
Dezful	32°23'N	048°24'E	–	10:39:39.4	295	244	60	–			–			13:11:54.5	109	48	29	12:00:27.3	22	323	44	260	0.974	0.972	
Dow Rud	33°28'N	049°04'E	–	10:39:01.3	293	244	59	11:58:31.7	145	87	12:00:06.5	259	201	13:10:29.9	111	51	29	11:59:19.2	22	324	44	259	1.025	1.000	01m35s
Esfahan (Isfaha…	32°40'N	051°38'E	1597	10:44:46.6	292	239	57	12:02:41.9	82	22	12:04:15.0	322	262	13:13:08.5	112	51	26	12:03:28.6	202	142	41	262	1.024	1.000	01m33s
Eslamshahr	35°40'N	051°10'E	–	10:39:12.5	288	241	57	–			–			13:07:43.5	115	57	28	11:57:44.6	202	146	42	258	0.947	0.939	
Golpayegan	33°27'N	050°18'E	–	10:41:12.4	292	241	58	11:59:58.0	75	17	12:01:26.3	329	270	13:11:10.7	112	52	28	12:00:42.3	202	143	42	260	1.025	1.000	01m28s
Gorgan	36°50'N	054°29'E	–	10:43:00.4	285	236	53	–			–			13:06:52.7	118	62	25	11:58:57.2	201	146	39	260	0.885	0.860	
Hamadan	34°48'N	048°30'E	–	10:35:52.0	291	246	60	–			–			13:07:53.9	112	54	30	11:56:20.4	202	146	44	257	0.997	0.998	
Harsin	34°16'N	047°35'E	–	10:35:03.6	293	247	61	11:55:22.6	143	87	11:57:01.6	261	204	13:08:16.7	111	52	30	11:56:12.2	22	325	45	257	1.026	1.000	01m39s
Homayunshahr	32°41'N	051°31'E	–	10:44:32.6	292	239	57	12:02:29.2	89	29	12:04:08.1	315	255	13:13:02.7	112	51	26	12:03:18.8	202	142	41	262	1.025	1.000	01m39s
Iranshahr	27°13'N	060°41'E	–	11:07:17.0	293	226	46	–			–			13:24:06.4	110	44	15	12:19:54.0	21	314	30	273	0.999	1.000	
Kangavar	34°30'N	047°58'E	–	10:35:22.9	292	246	60	11:55:18.9	98	41	11:57:10.4	306	250	13:08:06.5	112	53	30	11:56:14.8	202	146	45	257	1.025	1.000	01m51s
Karaj	35°48'N	050°59'E	–	10:38:41.5	288	241	57	–			–			13:07:24.1	115	57	28	11:57:19.0	202	146	42	258	0.946	0.938	
Kashan	33°59'N	051°29'E	–	10:42:22.3	290	240	57	–			–			13:10:49.8	113	54	27	12:01:01.0	202	144	41	261	0.979	0.979	
Kerman	30°17'N	057°05'E	–	10:57:16.6	291	230	50	–			–			13:18:52.3	112	49	20	12:12:25.1	202	138	34	268	0.995	0.997	
Khomeyn	33°38'N	050°04'E	–	10:40:30.2	292	242	58	11:59:28.4	68	9	12:00:47.3	337	278	13:10:44.9	112	52	28	12:00:08.0	202	144	42	260	1.025	1.000	01m19s
Khorramabad	33°30'N	048°20'E	–	10:37:39.8	293	245	60	–			–			13:10:01.5	111	51	29	11:58:25.6	22	324	44	258	0.998	0.999	
Khorramshahr	30°25'N	048°11'E	–	10:42:44.9	298	243	61	–			–			13:14:52.0	107	43	29	12:03:37.4	23	320	44	262	0.926	0.913	
Khvoy	38°33'N	044°58'E	–	10:23:50.9	288	255	61	–			–			12:58:54.5	115	61	34	11:45:28.3	202	153	48	248	0.955	0.950	
Malayer	34°17'N	048°50'E	–	10:37:17.0	292	245	59	11:56:55.9	69	12	11:58:19.1	335	278	13:08:58.8	112	53	29	11:57:37.6	202	145	44	258	1.025	1.000	01m23s
Maragheh	37°23'N	046°13'E	–	10:27:48.4	289	252	61	–			–			13:01:53.8	114	59	33	11:49:04.7	202	150	47	251	0.967	0.964	
Marivan	35°31'N	046°10'E	–	10:30:29.0	292	251	62	11:51:21.4	92	38	11:53:12.7	312	258	13:05:12.7	112	54	32	11:52:17.2	202	148	47	253	1.026	1.000	01m51s
Mashhad	36°18'N	059°36'E	–	10:51:36.3	282	229	48	–			–			13:09:05.1	120	63	21	12:04:06.9	201	144	34	265	0.841	0.804	
Mashiz	29°56'N	056°37'E	–	10:57:11.1	292	231	51	12:11:51.4	96	32	12:13:24.1	308	244	13:19:17.7	111	48	20	12:12:37.9	202	138	35	268	1.023	1.000	01m33s
Masjed Soleyman	31°58'N	049°18'E	–	10:41:57.4	295	242	59	–			–			13:13:05.8	109	48	28	12:02:11.8	22	322	43	261	0.975	0.974	
Nahavand	34°12'N	048°22'E	–	10:36:35.0	292	245	60	11:56:17.3	127	50	11:58:11.0	298	240	13:08:51.5	112	52	30	11:57:14.3	202	145	44	257	1.025	1.000	01m54s
Najafabad	32°37'N	051°21'E	–	10:44:22.4	292	239	57	12:02:21.7	106	46	12:04:08.9	298	238	13:13:04.7	111	50	26	12:03:15.5	202	142	41	262	1.025	1.000	01m47s
Neyshabur	36°12'N	058°50'E	–	10:50:38.2	282	230	49	–			–			13:09:08.5	119	63	21	12:03:42.8	201	144	34	264	0.851	0.817	
Orumiyeh (Reza'…	37°33'N	045°04'E	–	10:25:25.0	290	255	62	–			–			13:00:49.5	113	58	34	11:47:20.3	202	151	48	249	0.976	0.976	
Qa'emshahr	36°28'N	052°53'E	–	10:40:55.0	286	239	55	–			–			13:06:59.3	117	60	26	11:58:04.6	201	146	40	259	0.910	0.892	
Qazvin	36°16'N	050°00'E	–	10:36:16.8	288	243	58	–			–			13:06:05.2	115	58	29	11:55:26.6	202	147	43	256	0.947	0.940	
Qom	34°39'N	050°54'E	–	10:40:19.2	290	241	57	–			–			13:09:24.3	114	55	27	11:59:14.7	202	145	42	259	0.972	0.970	
Rafsanjan	30°24'N	056°01'E	–	10:55:30.8	292	232	52	12:10:41.7	77	13	12:12:01.9	327	263	13:18:24.2	111	48	21	12:11:21.9	202	138	35	268	1.023	1.000	01m20s
Rasht	37°16'N	049°36'E	–	10:34:06.6	287	245	58	–			–			13:04:03.1	116	60	30	11:53:14.8	202	149	43	255	0.930	0.918	
Sabzevar	36°13'N	057°42'E	–	10:48:55.7	283	232	50	–			–			13:08:53.2	119	62	22	12:02:47.9	201	145	35	263	0.862	0.832	
Sanandaj	35°19'N	047°00'E	–	10:32:20.1	292	249	61	11:53:02.1	62	7	11:54:17.0	342	287	13:06:06.0	112	54	31	11:53:39.7	202	147	46	255	1.026	1.000	01m15s
Sari	36°34'N	053°04'E	–	10:41:04.5	286	238	55	–			–			13:06:52.5	117	60	26	11:58:04.8	201	146	40	259	0.906	0.887	
Shahreza	32°01'N	051°52'E	–	10:46:14.5	293	238	56	12:04:02.3	141	80	12:05:35.9	263	202	13:14:17.9	111	49	26	12:04:49.2	22	321	40	263	1.024	1.000	01m34s
Shahr Kord	32°19'N	050°50'E	–	10:44:00.0	293	240	58	12:02:53.8	180	119	12:03:35.0	224	164	13:13:19.6	111	49	27	12:03:14.5	22	322	41	262	1.025	1.000	00m41s
Shiraz	29°37'N	052°33'E	–	10:51:31.0	296	236	56	–			–			13:18:21.9	109	44	24	12:09:35.8	22	318	39	266	0.960	0.956	
Sonqor	34°47'N	047°36'E	–	10:34:16.0	292	247	60	11:54:25.8	89	33	11:56:12.0	316	260	13:07:23.9	112	53	31	11:55:19.0	202	146	45	256	1.026	1.000	01m46s
Tabriz	38°05'N	046°18'E	–	10:26:58.6	288	252	60	–			–			13:00:39.4	115	60	33	11:47:57.5	202	151	47	250	0.950	0.944	
Tehran	35°40'N	051°26'E	1200	10:39:40.4	288	241	56	–			–			13:07:51.3	115	57	27	11:58:01.8	202	146	41	258	0.943	0.935	
Tuysarkan	34°33'N	048°27'E	–	10:36:10.4	292	245	60	11:56:07.8	60	4	11:57:18.4	344	287	13:08:18.2	112	53	30	11:56:43.2	202	145	44	257	1.025	1.000	01m11s
Yazd	31°53'N	054°25'E	–	10:50:35.2	291	234	54	–			–			13:15:31.3	112	51	23	12:07:28.5	202	140	38	265	0.991	0.993	
Zahedan	29°30'N	060°52'E	–	11:03:45.4	290	226	46	–			–			13:20:46.3	113	49	16	12:16:22.8	201	137	30	271	0.970	0.967	
Zanjan	36°40'N	048°29'E	–	10:32:59.5	289	247	59	–			–			13:04:33.8	114	58	30	11:53:02.5	202	148	44	254	0.956	0.951	

TABLE 30
LOCAL CIRCUMSTANCES FOR IRAQ
TOTAL SOLAR ECLIPSE OF 1999 AUGUST 11

Location Name	Latitude	Longitude	Elev.	First Contact U.T.	P	V	Alt	Second Contact U.T.	P	V	Third Contact U.T.	P	V	Fourth Contact U.T.	P	V	Alt	Maximum Eclipse U.T.	P	V	Alt	Azm	Eclip. Mag.	Eclip. Obs.	Umbral Durat.
			m	h m s	°	°	°	h m s	°	°	h m s	°	°	h m s	°	°	°	h m s	°	°	°	°			
IRAQ																									
Al-'Amarah	31°50'N	047°09'E	–	10:38:22.2	297	246	62	–			–			13:12:00.7	108	46	30	11:59:56.7	22	322	45	259	0.946	0.938	
Al-Basrah (Basr...	30°30'N	047°47'E	–	10:41:53.5	298	244	61	–			–			13:14:29.4	107	43	29	12:03:00.9	23	320	44	262	0.923	0.909	
Al-Hillah	32°29'N	044°25'E	–	10:32:10.2	298	253	64	–			–			13:09:01.5	107	46	33	11:55:21.3	23	324	48	256	0.927	0.915	
Al-Mawsil (Mosu...	36°20'N	043°08'E	223	10:23:29.0	293	259	64	11:46:38.8	188	137	11:47:08.4	216	164	13:01:34.7	111	55	35	11:46:53.7	22	331	50	249	1.027	1.000	00m30s
An-Najaf	31°59'N	044°20'E	–	10:32:53.6	298	252	64	–			–			13:09:44.8	107	45	33	11:56:07.2	23	324	48	256	0.915	0.899	
An-Nasiriyah	31°02'N	046°16'E	–	10:38:11.7	298	247	63	–			–			13:12:39.5	107	44	31	12:00:15.7	23	321	46	260	0.917	0.901	
Ar-Ramadi	33°25'N	043°17'E	–	10:28:24.0	297	256	65	–			–			13:06:37.4	108	48	34	11:52:11.5	22	326	50	253	0.935	0.925	
As-Sulaymaniyah	35°33'N	045°26'E	–	10:29:03.6	292	253	62	11:50:20.1	127	74	11:52:16.0	277	223	13:04:39.3	111	54	33	11:51:18.2	22	328	48	253	1.026	1.000	01m56s
Baghdad	33°21'N	044°25'E	34	10:30:41.0	296	253	64	–			–			13:07:37.9	108	48	33	11:53:50.2	22	326	48	255	0.947	0.940	
Ba'Qubah	33°45'N	044°38'E	–	10:30:25.5	296	253	64	–			–			13:07:08.3	109	49	33	11:53:25.3	22	326	48	254	0.959	0.955	
Dahuk	36°52'N	043°00'E	–	10:22:26.6	292	259	64	11:44:45.7	122	72	11:46:48.2	281	231	13:00:31.8	111	56	36	11:45:47.1	22	332	50	248	1.027	1.000	02m03s
Halabjah	35°10'N	045°59'E	–	10:30:41.1	293	251	62	11:51:46.2	137	83	11:53:34.0	267	212	13:05:42.0	111	53	32	11:52:40.3	22	327	47	254	1.026	1.000	01m48s
Irbil	36°11'N	044°01'E	–	10:25:24.5	292	256	63	11:47:25.9	139	87	11:49:15.6	265	213	13:02:31.3	111	55	34	11:48:20.9	22	330	49	250	1.026	1.000	01m50s
Karbala'	32°36'N	044°02'E	–	10:31:14.3	298	254	65	–			–			13:08:32.3	107	46	33	11:54:38.4	23	325	49	255	0.925	0.912	
Kirkuk	35°28'N	044°28'E	–	10:27:21.5	293	255	63	–			–			13:04:06.4	111	53	34	11:50:11.8	22	329	48	252	0.996	0.998	
Kuysanjaq	36°05'N	044°38'E	–	10:26:44.1	292	255	63	11:48:19.4	116	63	11:50:20.7	288	235	13:03:09.1	112	55	34	11:49:20.2	22	329	48	251	1.026	1.000	02m01s
Zakhu	37°08'N	042°41'E	–	10:21:26.7	292	260	64	11:43:50.5	112	63	11:45:55.7	292	242	12:59:48.2	112	56	36	11:44:53.3	22	332	50	247	1.027	1.000	02m05s

TABLE 31
LOCAL CIRCUMSTANCES FOR PAKISTAN
TOTAL SOLAR ECLIPSE OF 1999 AUGUST 11

Location Name	Latitude	Longitude	Elev.	First Contact U.T.	P	V	Alt	Second Contact U.T.	P	V	Third Contact U.T.	P	V	Fourth Contact U.T.	P	V	Alt	Maximum Eclipse U.T.	P	V	Alt	Azm	Eclip. Mag.	Eclip. Obs.	Umbral Durat.
			m	h m s	°	°	°	h m s	°	°	h m s	°	°	h m s	°	°	°	h m s	°	°	°	°			
PAKISTAN																									
Chamburi Kalat	26°09'N	064°43'E	–	11:13:43.3	292	223	41	12:23:14.3	65	357	12:24:09.8	337	269	13:25:49.2	110	45	11	12:23:42.2	201	133	25	276	1.020	1.000	00m55s
Faisalabad (Lya...	31°25'N	073°05'E	–	11:14:13.3	279	217	33	–			–			13:16:56.0	120	61	7	12:18:43.3	200	138	19	277	0.807	0.761	
Gharo	24°44'N	067°35'E	–	11:18:50.0	291	221	37	12:26:22.8	84	15	12:27:28.6	317	248	13:27:32.7	110	43	8	12:26:55.8	201	132	22	278	1.019	1.000	01m06s
Gujranwala	32°26'N	074°33'E	–	11:14:08.5	277	216	32	–			–			13:14:55.1	122	64	7	12:17:31.5	200	139	18	277	0.774	0.718	
Hab Nadi Chowki	25°01'N	066°53'E	–	11:17:43.6	291	221	38	12:25:41.6	88	19	12:26:50.8	313	244	13:27:14.2	110	44	9	12:26:16.3	201	132	22	277	1.020	1.000	01m09s
Hirok Sami	26°02'N	063°25'E	–	11:12:29.1	293	224	42	12:22:55.8	173	105	12:23:34.2	229	161	13:25:57.5	110	44	12	12:23:15.1	21	313	26	275	1.021	1.000	00m38s
Hoshab	26°01'N	063°56'E	–	11:13:05.1	292	223	42	12:22:54.7	134	66	12:24:09.0	268	200	13:25:59.7	110	44	12	12:23:31.9	21	313	26	275	1.021	1.000	01m14s
Hyderabad	25°22'N	068°22'E	–	11:18:35.2	290	220	36	–			–			13:26:38.4	111	45	8	12:26:17.3	201	132	21	278	0.983	0.983	
Islamabad	33°42'N	073°10'E	–	11:11:15.4	276	216	34	–			–			13:13:07.7	123	66	8	12:15:13.1	200	140	20	275	0.761	0.702	
Jati	24°21'N	068°16'E	–	11:20:03.1	291	220	36	12:27:06.9	96	27	12:28:17.0	305	235	13:27:57.6	110	43	7	12:27:42.0	201	131	21	278	1.019	1.000	01m10s
Jungshahi	24°51'N	067°46'E	–	11:18:49.4	291	221	37	12:26:34.8	44	335	12:27:04.2	357	288	13:27:22.6	110	44	8	12:26:49.6	201	132	21	278	1.019	1.000	00m29s
Kandrach	25°29'N	065°29'E	–	11:15:35.6	292	222	40	12:24:23.0	110	42	12:25:40.6	291	223	13:26:41.9	110	44	10	12:25:01.9	201	132	24	276	1.020	1.000	01m18s
Karachi	24°52'N	067°03'E	4	11:18:07.3	292	221	38	12:25:56.4	99	30	12:27:09.3	303	234	13:27:25.1	110	43	9	12:26:32.9	201	132	22	277	1.020	1.000	01m13s
Keti Bandar	24°08'N	067°27'E	–	11:19:38.7	292	221	37	–			–			13:28:18.9	109	42	8	12:27:44.2	21	311	21	278	0.999	1.000	
Kotri Allahrakh...	24°24'N	067°50'E	–	11:19:34.9	292	221	37	12:26:52.9	116	46	12:28:05.6	286	216	13:27:56.6	110	43	8	12:27:29.3	21	311	21	278	1.019	1.000	01m13s
Lahore	31°35'N	074°18'E	–	11:15:02.8	278	216	32	–			–			13:16:22.1	121	62	6	12:18:45.6	200	138	18	277	0.793	0.743	
Mirpur Batoro	24°44'N	068°16'E	–	11:19:27.8	291	220	36	–			–			13:27:28.4	110	44	7	12:27:09.5	201	132	21	278	0.998	0.999	
Mirpur Sakro	24°33'N	067°37'E	–	11:19:09.0	292	221	37	12:26:35.4	108	39	12:27:48.7	293	224	13:27:46.5	110	43	8	12:27:12.1	201	131	21	278	1.019	1.000	01m13s
Multan	30°11'N	071°29'E	122	11:14:28.1	282	218	34	–			–			13:19:13.5	118	57	8	12:20:09.3	200	137	20	276	0.848	0.812	
Naka Kharari	25°15'N	066°44'E	–	11:17:13.0	291	221	38	12:25:28.5	60	352	12:26:16.5	341	273	13:26:56.7	110	44	9	12:25:52.6	201	132	23	277	1.020	1.000	00m48s
Nirwano	26°22'N	062°43'E	–	11:11:08.5	293	224	43	12:22:02.0	171	103	12:22:43.7	231	163	13:25:28.4	110	44	13	12:22:23.0	21	313	27	274	1.021	1.000	00m42s
Peshawar	34°01'N	071°33'E	–	11:09:18.1	277	218	35	–			–			13:12:54.8	123	65	10	12:14:13.6	200	141	22	274	0.769	0.713	
Quetta	30°12'N	067°00'E	–	11:09:58.0	285	221	39	–			–			13:19:48.1	116	55	11	12:18:31.2	201	137	24	274	0.890	0.866	
Rahim ki Bazar	24°19'N	069°09'E	–	11:20:53.0	291	220	35	–			–			13:27:53.4	110	43	7	12:28:01.0	200	131	20	278	0.999	1.000	
Rawalpindi	33°36'N	073°04'E	511	11:11:18.2	276	217	34	–			–			13:13:19.5	123	65	8	12:15:21.1	200	140	20	275	0.763	0.706	
Sargodha	32°05'N	072°40'E	–	11:12:57.2	278	217	34	–			–			13:15:56.7	121	62	8	12:17:35.4	200	139	20	276	0.797	0.748	
Shahbandar	24°10'N	067°54'E	–	11:20:00.3	292	221	36	12:27:20.7	146	76	12:28:19.9	255	186	13:28:13.8	109	42	8	12:27:50.4	21	311	21	278	1.019	1.000	00m59s
Shahbaz Kalat	26°42'N	063°58'E	–	11:12:01.0	291	223	42	–			–			13:25:03.7	111	45	12	12:22:30.9	201	134	26	275	0.998	0.999	
Sialkot	32°30'N	074°31'E	–	11:14:01.6	277	216	32	–			–			13:14:49.1	122	64	7	12:17:25.0	200	139	19	277	0.773	0.717	
Sonmiani	25°26'N	066°36'E	–	11:16:47.9	291	221	38	–			–			13:26:42.7	110	44	9	12:25:33.5	201	132	23	277	0.999	1.000	
Sujawal	24°36'N	068°05'E	–	11:19:30.1	291	221	36	12:26:49.7	70	1	12:27:44.7	331	262	13:27:39.8	110	43	8	12:27:17.3	201	131	21	278	1.019	1.000	00m55s
Tatta	24°45'N	067°55'E	–	11:19:07.0	291	221	37	12:26:41.1	54	345	12:27:21.1	347	278	13:27:29.4	110	44	8	12:27:01.2	201	132	21	278	1.019	1.000	00m40s

TABLE 32
LOCAL CIRCUMSTANCES FOR INDIA - I
TOTAL SOLAR ECLIPSE OF 1999 AUGUST 11

Location Name	Latitude	Longitude	Elev.	First Contact U.T.				Second Contact U.T.			Third Contact U.T.			Fourth Contact U.T.				Maximum Eclipse U.T.					Eclip. Mag.	Eclip. Obs.	Umbral Durat.
			m	h m s	P °	V °	Alt °	h m s	P °	V °	h m s	P °	V °	h m s	P °	V °	Alt °	h m s	P °	V °	Alt °	Azm °			
INDIA																									
Agra	27°11′N	078°01′E	–	11:23:29.5	281	214	27	–			–			13:21:50.2	117	55	1	12:25:37.0	199	134	13	281	0.854	0.819	
Ahmadabad	23°02′N	072°37′E	55	11:25:30.9	290	218	31	–			–			13:28:48.6	110	43	3	12:30:33.3	200	130	16	280	0.995	0.997	
Ajmer	26°27′N	074°38′E	–	11:22:10.9	284	216	30	–			–			13:23:55.6	115	51	3	12:26:15.4	200	133	16	280	0.899	0.877	
Akola	20°44′N	077°00′E	–	11:31:23.5	291	217	25	12:33:23.4	122	51	12:34:18.7	277	205	–				12:33:51.1	19	308	11	282	1.016	1.000	00m55s
Akot	21°11′N	077°04′E	–	11:30:49.1	290	216	25	–			–			–				12:33:18.6	199	129	11	282	0.999	0.999	
Aligarh	27°53′N	078°05′E	–	11:22:38.4	280	214	27	–			–			13:20:49.6	118	57	2	12:24:39.8	199	135	13	281	0.838	0.799	
Allahabad	25°27′N	081°51′E	–	11:27:42.8	281	213	22	–			–			–				12:27:55.1	199	133	9	283	0.861	0.828	
Alleppey	09°29′N	076°19′E	–	11:47:18.8	308	222	20	–			–			–				12:45:36.4	20	298	6	285	0.711	0.640	
Ambala	30°21′N	076°50′E	–	11:18:38.3	278	214	29	–			–			13:17:35.0	121	61	4	12:21:02.5	199	137	16	279	0.797	0.747	
Amravati	20°56′N	077°45′E	–	11:31:29.7	290	216	24	–			–			–				12:33:37.0	199	128	10	283	0.999	0.999	
Amritsar	31°35′N	074°53′E	–	11:15:31.9	278	215	31	–			–			13:16:12.9	121	62	6	12:18:53.1	199	138	18	277	0.788	0.736	
Anand	22°34′N	072°56′E	–	11:26:24.5	291	218	30	12:30:48.3	71	0	12:31:37.6	329	259	13:29:16.9	109	42	3	12:31:13.0	200	130	15	281	1.018	1.000	00m49s
Anjar	23°08′N	070°01′E	–	11:23:23.9	292	220	34	12:29:41.6	181	111	12:30:03.8	219	149	13:29:12.5	109	41	5	12:29:52.8	20	310	18	279	1.018	1.000	00m22s
Asansol	23°41′N	086°59′E	–	11:31:35.5	281	211	16	–			–			–				12:29:24.6	198	132	4	285	0.863	0.829	
Aurangabad	19°53′N	075°20′E	–	11:31:41.9	293	218	26	–			–			–				12:34:49.1	20	307	12	282	0.971	0.968	
Badnera	20°52′N	077°44′E	–	11:31:34.5	290	216	24	–			–			–				12:33:41.8	199	128	10	283	1.016	1.000	
Balapur	20°40′N	076°46′E	–	11:31:21.9	291	217	25	12:33:34.4	151	80	12:34:16.9	248	176	–				12:33:55.7	19	308	11	282	1.016	1.000	00m43s
Ballalpur	19°50′N	079°22′E	–	11:33:39.9	290	216	22	12:34:28.1	138	66	12:35:14.6	260	189	–				12:34:51.5	19	307	8	283	1.015	1.000	00m46s
Bangalore	12°59′N	077°35′E	895	11:42:24.8	301	220	21	–			–			–				12:42:16.1	19	301	6	284	0.816	0.771	
Bareilly	28°21′N	079°25′E	–	11:22:52.6	279	213	26	–			–			13:19:39.9	119	58	1	12:24:05.9	199	135	12	281	0.817	0.773	
Baroda	22°18′N	073°12′E	–	11:26:58.4	291	218	30	12:31:05.1	98	28	12:32:07.0	301	231	13:29:31.4	109	42	2	12:31:36.2	200	129	15	281	1.017	1.000	01m02s
Behala	22°31′N	088°19′E	–	11:33:10.4	282	211	15	–			–			–				12:30:26.7	198	131	2	286	0.880	0.851	
Belgaum	15°52′N	074°31′E	–	11:37:06.0	299	220	25	–			–			–				12:39:23.3	20	303	11	283	0.864	0.832	
Bhatpara	22°52′N	088°24′E	–	11:32:48.7	281	211	15	–			–			–				12:30:02.0	198	131	2	286	0.871	0.840	
Bhaunagar	21°46′N	072°09′E	–	11:27:02.0	292	219	31	–			–			13:30:22.4	108	40	3	12:32:07.8	20	309	16	281	0.988	0.989	
Bhilai	21°13′N	081°26′E	–	11:32:42.9	287	214	21	–			–			–				12:33:06.0	199	129	7	284	0.962	0.955	
Bhopal	23°16′N	077°24′E	–	11:28:13.0	287	216	26	–			–			–				12:30:44.1	199	130	12	282	0.947	0.937	
Bhuj	23°16′N	069°40′E	–	11:22:54.8	292	220	34	12:29:28.2	186	116	12:29:44.7	214	144	13:29:06.3	109	41	6	12:29:36.5	20	310	19	279	1.019	1.000	00m17s
Bhusawal	21°03′N	075°46′E	–	11:30:18.4	291	217	27	12:33:01.8	146	75	12:33:49.1	253	182	–				12:33:25.6	20	308	12	282	1.017	1.000	00m47s
Bikaner	28°01′N	073°18′E	–	11:19:01.3	283	217	32	–			–			13:22:04.6	116	54	5	12:23:48.3	200	135	18	278	0.877	0.849	
Bobbili	18°34′N	083°22′E	–	11:36:34.7	290	215	18	12:35:26.6	136	64	12:36:08.5	261	189	–				12:35:47.6	19	307	4	285	1.014	1.000	00m42s
Bombay	18°58′N	072°50′E	8	11:31:37.1	296	219	29	–			–			13:33:06.5	104	34	1	12:35:43.0	20	306	14	282	0.927	0.912	
Borsad	22°25′N	072°54′E	–	11:26:36.2	291	218	30	12:30:53.1	99	28	12:31:55.6	301	231	13:29:27.7	109	42	3	12:31:24.4	200	129	15	281	1.018	1.000	01m02s
Burhanpur	21°18′N	076°14′E	–	11:30:13.1	290	217	26	12:32:45.7	71	0	12:33:31.2	328	257	–				12:33:08.5	200	129	12	282	1.016	1.000	00m46s
Calcutta	22°32′N	088°22′E	6	11:33:09.9	282	211	15	–			–			–				12:30:24.9	198	131	2	286	0.879	0.851	
Calicut	11°15′N	075°46′E	–	11:44:28.8	305	221	22	–			–			–				12:44:02.9	20	299	7	284	0.754	0.694	
Cambay	22°18′N	072°37′E	–	11:26:34.6	291	218	30	12:31:01.6	134	64	12:32:00.0	266	195	13:29:39.8	109	41	3	12:31:30.9	20	309	16	281	1.018	1.000	00m58s
Chandigarh	30°44′N	076°55′E	–	11:18:12.5	277	214	29	–			–			13:16:57.7	121	62	4	12:20:29.7	199	137	16	279	0.788	0.736	
Chandrapur	19°57′N	079°18′E	–	11:33:29.1	290	216	22	12:34:17.4	120	48	12:35:09.7	278	207	–				12:34:43.6	19	308	8	283	1.015	1.000	00m52s
Chhota Udepur	22°19′N	074°01′E	–	11:27:29.2	290	218	29	–			–			13:29:17.5	109	42	2	12:31:41.2	200	129	14	281	0.999	1.000	
Chopda	21°15′N	075°18′E	–	11:29:46.0	291	217	27	12:32:43.4	140	69	12:33:34.6	259	188	13:30:06.7	108	41	0	12:33:09.1	20	308	13	282	1.017	1.000	00m51s
Cochin	09°58′N	076°14′E	–	11:46:33.2	307	222	20	–			–			–				12:45:11.2	20	298	6	285	0.724	0.656	
Coimbatore	11°00′N	076°58′E	–	11:45:09.0	305	221	20	–			–			–				12:44:11.1	19	299	6	284	0.758	0.698	
Cuttack	20°30′N	085°50′E	–	11:34:53.6	286	213	16	–			–			–				12:33:14.0	198	129	3	285	0.945	0.934	
Dabhoi	22°11′N	073°26′E	–	11:27:17.8	291	218	29	12:31:15.9	102	32	12:32:18.2	297	227	13:29:35.8	109	41	2	12:31:47.1	200	129	15	281	1.017	1.000	01m02s
Darwha	20°19′N	077°46′E	–	11:32:19.6	291	216	24	12:34:00.3	151	80	12:34:41.6	247	176	–				12:34:21.0	19	308	10	283	1.016	1.000	00m41s
Daryapur	20°56′N	077°20′E	–	11:31:17.4	290	216	25	12:33:17.4	63	352	12:33:56.3	336	265	–				12:33:36.9	199	128	11	283	1.016	1.000	00m39s
Dediapada	21°38′N	073°35′E	–	11:28:11.0	292	218	29	–			–			13:30:10.5	108	40	2	12:32:30.4	20	309	14	281	0.998	0.999	
Dehra Dun	30°19′N	078°02′E	–	11:19:31.7	277	214	28	–			–			13:17:14.2	121	62	3	12:21:14.0	199	137	14	280	0.787	0.735	
Delhi	28°40′N	077°13′E	–	11:21:04.1	280	214	28	–			–			13:19:59.6	119	58	3	12:23:29.7	199	135	14	280	0.829	0.787	
Dhanbad	23°48′N	086°27′E	–	11:31:19.1	281	212	17	–			–			–				12:29:22.6	198	132	4	285	0.864	0.831	
Dhandhuka	22°22′N	071°59′E	–	11:26:02.1	292	219	31	12:31:04.7	174	103	12:31:33.6	226	156	13:29:43.8	109	41	3	12:31:19.3	20	309	16	280	1.018	1.000	00m29s
Dharangaon	21°01′N	075°16′E	–	11:30:04.4	291	217	27	–			–			13:30:22.4	108	40	0	12:33:26.2	20	308	12	282	0.998	0.999	
Dholka	22°43′N	072°28′E	–	11:25:52.0	291	218	31	12:30:29.5	78	7	12:31:23.9	323	252	13:29:13.0	109	42	3	12:30:56.8	200	130	16	280	1.018	1.000	00m54s
Dhrangadhra	22°59′N	071°28′E	–	11:24:44.9	291	219	32	12:29:51.8	100	29	12:30:57.0	301	230	13:29:07.2	109	42	4	12:30:24.5	200	130	17	280	1.018	1.000	01m05s
Dondaicha	21°20′N	074°34′E	–	11:29:13.2	291	218	28	–			–			13:30:14.2	108	40	1	12:32:59.4	20	308	13	282	1.000	1.000	
Durgapur	23°29′N	087°20′E	–	11:31:54.1	281	211	16	–			–			–				12:29:34.2	198	131	3	285	0.865	0.832	
Erode	11°21′N	077°44′E	–	11:44:48.1	304	221	20	–			–			–				12:43:47.0	19	300	6	285	0.774	0.718	
Ghaziabad	28°40′N	077°26′E	–	11:21:12.9	280	214	28	–			–			13:19:55.2	119	58	2	12:23:31.0	199	135	14	280	0.827	0.785	

TABLE 32 - continued
LOCAL CIRCUMSTANCES FOR INDIA - II
TOTAL SOLAR ECLIPSE OF 1999 AUGUST 11

Location Name	Latitude	Longitude	Elev.	First Contact				Second Contact			Third Contact			Fourth Contact				Maximum Eclipse					Eclip.	Eclip.	Umbral
				U.T.	P	V	Alt	U.T.	P	V	U.T.	P	V	U.T.	P	V	Alt	U.T.	P	V	Alt	Azm	Mag.	Obs.	Durat.
			m	h m s	°	°	°	h m s	°	°	h m s	°	°	h m s	°	°	°	h m s	°	°	°	°			
INDIA																									
Gorakhpur	26°45′N	083°22′E	–	11:26:50.8	279	212	21	–			–			–				12:26:09.4	199	134	8	283	0.821	0.777	
Guntur	16°18′N	080°27′E	–	11:38:39.1	295	217	19	–			–			–				12:38:36.9	19	304	6	284	0.925	0.909	
Gwalior	26°13′N	078°10′E	–	11:24:49.2	282	214	26	–			–			13:23:06.3	116	53	1	12:26:55.8	199	133	12	281	0.874	0.844	
Hinganghat	20°34′N	078°50′E	–	11:32:28.9	290	216	23	–			–			–				12:34:01.8	199	128	9	283	0.998	0.999	
Howrah	22°35′N	088°20′E	–	11:33:06.3	282	211	15	–			–			–				12:30:22.0	198	131	2	286	0.878	0.849	
Hubli	15°21′N	075°10′E	–	11:38:08.5	299	220	24	–			–			–				12:39:57.3	20	303	10	283	0.857	0.822	
Hyderabad	17°23′N	078°29′E	531	11:36:34.6	294	217	22	–			–			–				12:37:39.7	19	305	8	284	0.936	0.924	
Indore	22°43′N	075°50′E	–	11:28:03.3	289	216	27	–			–			13:28:17.7	111	44	1	12:31:20.9	200	130	13	281	0.973	0.971	
Jabalpur	23°10′N	079°57′E	–	11:29:38.9	285	214	23	–			–			–				12:30:51.6	199	131	9	283	0.928	0.913	
Jadabpur	22°29′N	088°23′E	–	11:33:13.4	282	211	15	–			–			–				12:30:28.0	198	131	2	286	0.880	0.852	
Jagdalpur	19°04′N	082°02′E	–	11:35:35.4	290	215	19	12:35:03.1	118	46	12:35:51.7	279	208	–				12:35:27.5	19	307	5	284	1.015	1.000	00m49s
Jaipur	26°55′N	075°49′E	–	11:22:23.5	283	215	29	–			–			13:22:56.3	116	53	3	12:25:46.5	200	134	15	280	0.878	0.850	
Jalgaon	21°01′N	075°34′E	–	11:30:14.5	291	217	27	12:33:14.3	173	102	12:33:40.3	226	154	13:30:16.9	108	40	0	12:33:27.3	20	308	12	282	1.017	1.000	00m26s
Jalgaon	21°03′N	076°32′E	–	11:30:43.3	290	217	26	12:32:59.3	97	26	12:33:55.6	302	231	–				12:33:27.5	200	128	11	282	1.016	1.000	00m56s
Jambusar	22°03′N	072°48′E	–	11:27:03.9	291	218	30	12:31:37.8	173	102	12:32:06.1	226	155	13:29:54.3	109	41	2	12:31:52.0	20	309	15	281	1.018	1.000	00m28s
Jamnagar	22°28′N	070°04′E	–	11:24:27.0	293	220	33	–			–			13:29:58.9	108	40	5	12:30:47.3	20	309	18	280	0.985	0.986	
Jamshedpur	22°48′N	086°11′E	–	11:32:22.1	282	212	17	–			–			–				12:30:35.2	198	131	4	285	0.889	0.862	
Jeypore	18°51′N	082°35′E	–	11:36:01.3	290	215	19	12:35:13.9	127	55	12:35:59.9	270	199	–				12:35:36.9	19	307	5	285	1.014	1.000	00m46s
Jhansi	25°26′N	078°35′E	–	11:26:03.9	283	214	25	–			–			13:23:59.3	115	52	0	12:27:58.9	199	133	12	282	0.888	0.862	
Jodhpur	26°17′N	073°02′E	–	11:21:12.6	286	217	32	–			–			13:24:34.3	114	50	4	12:26:12.6	200	133	17	279	0.917	0.900	
Jullundur	31°19′N	075°34′E	–	11:16:25.7	278	215	30	–			–			13:16:27.0	121	62	5	12:19:25.0	199	138	17	278	0.787	0.736	
Kanpur	26°28′N	080°21′E	–	11:25:44.0	281	213	24	–			–			–				12:26:38.9	199	134	11	282	0.850	0.814	
Karanja	20°29′N	077°29′E	–	11:31:58.1	291	216	24	12:33:44.4	137	65	12:34:34.0	262	190	–				12:34:09.3	19	308	10	283	1.016	1.000	00m50s
Khamgaon	20°41′N	076°34′E	–	11:31:14.3	291	217	26	12:33:38.0	165	94	12:34:10.3	234	162	–				12:33:54.2	19	308	11	282	1.016	1.000	00m32s
Khavda	23°51′N	069°43′E	–	11:22:04.2	291	220	34	12:28:20.0	78	8	12:29:18.5	323	254	13:28:23.3	110	43	6	12:28:49.4	200	131	19	279	1.019	1.000	00m59s
Khetia	21°40′N	074°35′E	–	11:28:45.6	291	217	28	12:32:04.0	114	43	12:33:04.6	285	214	13:29:51.9	109	41	1	12:32:34.4	20	309	13	281	1.017	1.000	01m01s
Kolhapur	16°42′N	074°13′E	–	11:35:43.5	298	219	26	–			–			–				12:38:27.4	20	304	11	283	0.883	0.855	
Kondagaon	19°36′N	081°40′E	–	11:34:48.6	289	215	20	–			–			–				12:34:54.9	199	127	6	284	0.998	0.999	
Koraput	18°49′N	082°43′E	–	11:36:06.0	290	215	18	12:35:14.7	125	53	12:36:00.9	272	200	–				12:35:37.9	19	307	5	285	1.014	1.000	00m46s
Kota	25°11′N	075°50′E	–	11:24:43.0	285	216	28	–			–			13:25:15.4	114	49	2	12:28:07.9	200	132	14	281	0.917	0.899	
Lakhpat	23°49′N	068°47′E	–	11:21:19.7	292	220	35	12:28:05.9	142	72	12:29:06.7	259	189	13:28:33.7	109	42	7	12:28:36.4	21	311	20	278	1.019	1.000	01m01s
Limbdi	22°34′N	071°48′E	–	11:25:36.5	291	219	31	12:30:34.2	144	73	12:31:28.6	256	186	13:29:32.4	109	41	3	12:31:01.5	20	310	16	280	1.018	1.000	00m54s
Lucknow	26°51′N	080°55′E	122	11:25:33.7	280	213	24	–			–			–				12:26:08.4	199	134	10	282	0.837	0.798	
Ludhiana	30°54′N	075°51′E	–	11:17:11.5	278	215	30	–			–			13:17:01.7	121	61	5	12:20:05.1	199	138	17	278	0.794	0.743	
Madras	13°05′N	080°17′E	16	11:42:53.9	300	219	18	–			–			–				12:41:49.3	19	301	4	285	0.840	0.801	
Madurai	09°56′N	078°07′E	–	11:46:57.2	306	221	18	–			–			–				12:45:00.2	19	298	5	285	0.739	0.674	
Maliya	23°05′N	070°46′E	–	11:24:04.0	291	219	33	12:29:35.2	127	57	12:30:39.6	273	203	13:29:08.3	109	42	5	12:30:07.5	20	310	18	280	1.018	1.000	01m04s
Malkapur	20°53′N	076°12′E	–	11:30:46.3	291	217	26	12:33:16.2	148	77	12:34:01.6	251	180	–				12:33:39.0	20	308	12	282	1.016	1.000	00m45s
Mangalore	12°52′N	074°53′E	–	11:41:45.0	303	221	23	–			–			–				12:42:31.4	20	301	9	284	0.789	0.738	
Mangrul Pir	20°19′N	077°21′E	–	11:32:07.7	291	216	25	–			–			–				12:34:21.1	19	308	10	283	0.999	1.000	
Meerut	28°59′N	077°42′E	–	11:20:59.4	279	214	28	–			–			13:19:21.8	119	59	2	12:23:05.9	199	136	14	280	0.818	0.774	
Moradabad	28°50′N	078°47′E	–	11:21:53.1	279	214	26	–			–			13:19:12.0	119	59	2	12:23:23.8	199	136	13	280	0.812	0.766	
Morvi	22°49′N	070°50′E	–	11:24:31.0	292	219	33	12:30:15.9	176	105	12:30:43.6	224	154	13:29:26.4	109	41	4	12:30:29.8	20	310	18	280	1.018	1.000	00m28s
Mul	20°04′N	079°40′E	–	11:33:29.0	290	216	22	12:34:12.5	74	3	12:34:55.9	324	253	–				12:34:34.2	199	128	8	283	1.015	1.000	00m43s
Murtajapur	20°44′N	077°23′E	–	11:31:35.0	290	216	25	12:33:23.6	99	28	12:34:19.0	300	229	–				12:33:51.4	199	128	10	283	1.016	1.000	00m55s
Mysore	12°18′N	076°39′E	–	11:43:08.5	303	220	21	–			–			–				12:42:59.6	19	300	7	284	0.790	0.738	
Nadiad	22°42′N	072°52′E	–	11:26:10.3	291	218	30	12:30:49.2	43	333	12:31:14.6	357	286	13:29:08.5	109	42	3	12:31:02.0	200	130	16	280	1.018	1.000	00m25s
Nagpur	21°09′N	079°06′E	–	11:31:50.4	289	215	23	–			–			–				12:33:19.9	199	129	9	283	0.982	0.981	
Nandura	20°50′N	076°27′E	–	11:30:58.4	291	217	26	12:33:18.2	140	69	12:34:07.9	259	188					12:33:43.1	20	308	11	282	1.016	1.000	00m50s
Narasannapeta	18°25′N	084°03′E	–	11:36:55.6	290	215	17	12:35:27.2	124	52	12:36:11.9	273	201	–				12:35:49.6	18	307	3	285	1.014	1.000	00m45s
Nasik	19°59′N	073°48′E	–	11:30:41.7	294	218	28	–			–			13:31:52.0	106	37	1	12:34:35.1	20	307	13	282	0.960	0.954	
New Delhi	28°36′N	077°12′E	212	11:21:08.8	280	214	28	–			–			13:20:05.9	119	57	3	12:23:35.4	199	135	14	280	0.830	0.789	
Nowrangapur	19°14′N	082°33′E	–	11:35:32.4	289	215	19				–			–				12:35:12.2	199	127	5	285	1.014	1.000	
Padra	22°14′N	073°05′E	–	11:26:59.5	291	218	30	12:31:08.7	116	45	12:32:11.7	284	213	13:29:37.7	109	41	2	12:31:40.3	20	309	15	281	1.017	1.000	01m03s
Palkonda	18°36′N	083°45′E	–	11:36:38.1	290	215	17	12:35:17.9	105	33	12:36:04.7	292	220	–				12:35:41.3	199	127	4	285	1.014	1.000	00m47s
Pandharkawada	20°01′N	078°32′E	–	11:33:04.4	291	216	23	12:34:24.2	161	89	12:34:57.8	237	166	–				12:34:41.1	19	308	9	283	1.016	1.000	00m34s
Parlakimidi	18°46′N	084°05′E	–	11:36:30.9	289	214	17	–			–			–				12:35:27.0	198	127	4	285	1.000	1.000	
Parvatipuram	18°47′N	083°26′E	–	11:36:20.0	289	215	18	12:35:10.8	87	16	12:35:55.0	310	238	–				12:35:33.0	199	127	4	285	1.014	1.000	00m44s
Patna	25°36′N	085°07′E	–	11:28:52.0	279	211	19	–			–			–				12:27:25.4	198	133	6	284	0.833	0.792	

TABLE 32 - continued
LOCAL CIRCUMSTANCES FOR INDIA - III
TOTAL SOLAR ECLIPSE OF 1999 AUGUST 11

Location Name	Latitude	Longitude	Elev.	First Contact U.T.	P	V	Alt	Second Contact U.T.	P	V	Third Contact U.T.	P	V	Fourth Contact U.T.	P	V	Alt	Maximum Eclipse U.T.	P	V	Alt	Azm	Eclip. Mag.	Eclip. Obs.	Umbral Durat.
			m	h m s	°	°	°	h m s	°	°	h m s	°	°	h m s	°	°	°	h m s	°	°	°	°			
INDIA																									
Patur	20°27′N	076°56′E	–	11:31:44.6	291	217	25	–			–			–				12:34:11.5	19	308	11	283	0.999	1.000	
Petlad	22°28′N	072°48′E	–	11:26:27.7	291	218	30	12:30:48.4	97	26	12:31:50.7	303	233	13:29:25.7	109	42	3	12:31:19.6	200	129	15	281	1.018	1.000	01m02s
Pondicherry	11°56′N	079°53′E	–	11:44:23.9	302	219	18	–			–			–				12:42:57.2	19	300	4	285	0.807	0.759	
Pune (Poona)	18°32′N	073°52′E	–	11:32:50.7	296	219	27	–			–			13:33:15.2	104	33	0	12:36:19.8	20	306	13	282	0.925	0.910	
Raipur	21°14′N	081°38′E	–	11:32:46.2	287	214	21	–			–			–				12:33:03.7	199	129	7	284	0.960	0.953	
Rajahmundry	16°59′N	081°47′E	–	11:38:08.1	293	216	18	–			–			–				12:37:43.1	19	305	5	285	0.952	0.944	
Rajkot	22°18′N	070°47′E	–	11:25:15.1	292	219	32	–			–			13:30:02.9	108	40	4	12:31:10.4	20	309	17	280	0.988	0.989	
Rajpipla	21°47′N	073°34′E	–	11:27:57.4	291	218	29	12:32:01.1	165	94	12:32:36.5	234	163	13:30:00.8	109	41	2	12:32:18.9	20	309	14	281	1.017	1.000	00m35s
Ranchi	23°21′N	085°20′E	–	11:31:30.1	282	212	18	–			–			–				12:30:06.0	198	131	5	285	0.882	0.854	
Rapar	23°34′N	070°38′E	–	11:23:14.5	291	219	33	12:29:05.8	58	348	12:29:46.9	343	273	13:28:34.9	110	43	5	12:29:26.4	200	130	18	279	1.018	1.000	00m41s
Raurkela	22°13′N	084°53′E	–	11:32:40.2	284	213	18	–			–			–				12:31:29.4	198	130	5	285	0.912	0.892	
Raver	21°15′N	076°02′E	–	11:30:10.7	290	217	26	12:32:43.4	94	23	12:33:39.7	305	234	–				12:33:11.6	200	129	12	282	1.017	1.000	00m56s
Saharanpur	29°58′N	077°33′E	–	11:19:38.2	278	214	28	–			–			13:17:56.2	120	61	3	12:21:41.0	199	137	15	279	0.798	0.749	
Salem	11°39′N	078°10′E	–	11:44:27.6	303	220	19	–			–			–				12:43:27.4	19	300	5	285	0.785	0.732	
Salur	18°32′N	083°13′E	–	11:36:34.9	290	215	18	12:35:35.2	156	84	12:36:07.2	241	169	–				12:35:51.2	19	307	4	285	1.014	1.000	00m32s
Sangli	16°52′N	074°34′E	–	11:35:38.8	298	219	26	–			–			–				12:38:17.1	20	304	11	283	0.890	0.865	
Sankheda	22°10′N	073°35′E	–	11:27:25.2	291	218	29	12:31:19.2	96	25	12:32:19.8	304	233	13:29:34.6	109	41	2	12:31:49.6	200	129	15	281	1.017	1.000	01m01s
Sendhwa	21°41′N	075°06′E	–	11:29:02.8	290	217	28	12:32:09.9	79	8	12:33:01.6	321	250	13:29:41.8	109	41	1	12:32:35.8	200	129	13	282	1.017	1.000	00m52s
Shahada	21°28′N	074°18′E	–	11:28:52.2	291	218	28	12:32:37.3	180	108	12:32:58.3	220	149	13:30:10.0	108	40	1	12:32:47.8	20	309	14	281	1.017	1.000	00m21s
Shegaon	20°47′N	076°41′E	–	11:31:09.8	291	217	25	12:33:20.9	134	62	12:34:13.2	265	194	–				12:33:47.1	19	308	11	282	1.016	1.000	00m52s
Shirpur	21°21′N	074°53′E	–	11:29:23.1	291	217	28	12:32:37.1	151	80	12:33:22.3	248	177	13:30:07.6	108	40	1	12:32:59.8	20	309	13	282	1.017	1.000	00m45s
Sholapur	17°41′N	075°55′E	–	11:35:05.8	296	218	25	–			–			–				12:37:23.7	20	305	10	283	0.922	0.906	
Sojitra	22°33′N	072°43′E	–	11:26:17.0	291	218	30	12:30:42.3	89	18	12:31:42.1	311	241	13:29:21.1	109	42	3	12:31:12.3	200	130	16	280	1.018	1.000	01m00s
Srikakulam	18°18′N	083°54′E	–	11:37:01.9	290	215	17	12:35:46.1	166	94	12:36:11.2	231	159	–				12:35:58.7	19	306	4	285	1.014	1.000	00m25s
Srinagar	34°05′N	074°49′E	–	11:12:13.6	275	215	32	–			–			13:12:04.3	124	67	7	12:15:02.6	199	141	19	276	0.739	0.675	
Surat	21°10′N	072°50′E	–	11:28:22.4	293	218	30	–			–			13:30:52.4	107	39	2	12:33:00.1	20	308	15	281	0.980	0.979	
Surendranagar	22°42′N	071°41′E	–	11:25:19.6	291	219	32	12:30:18.5	129	59	12:31:20.6	271	200	13:29:24.5	109	41	4	12:30:49.6	20	310	17	280	1.018	1.000	01m02s
Taloda	21°34′N	074°13′E	–	11:28:40.6	291	218	28	12:32:19.3	158	87	12:33:00.1	241	170	13:30:04.8	108	41	1	12:32:39.8	20	309	14	281	1.017	1.000	00m41s
Tekkali	18°37′N	084°14′E	–	11:36:43.8	289	214	17	12:35:16.9	69	357	12:35:52.7	328	256	–				12:35:34.8	198	127	3	285	1.014	1.000	00m36s
Thana	19°12′N	072°58′E	–	11:31:21.1	295	219	29	–			–			13:32:51.1	105	34	1	12:35:27.0	20	306	14	282	0.934	0.920	
Tiruchchirappal…	10°49′N	078°41′E	–	11:45:45.7	304	221	18	–			–			–				12:44:08.7	19	299	5	285	0.767	0.709	
Trivandrum	08°29′N	076°55′E	–	11:48:57.5	309	223	19	–			–			–				12:46:23.9	19	297	5	285	0.689	0.613	
Tuticorin	08°47′N	078°08′E	–	11:48:39.9	308	222	18	–			–			–				12:45:58.9	19	297	4	285	0.707	0.636	
Ujjain	23°11′N	075°46′E	–	11:27:22.8	288	216	27	–			–			13:27:46.0	111	45	1	12:30:44.9	200	130	13	281	0.963	0.958	
Ulhasnagar	19°13′N	073°07′E	–	11:31:24.9	295	219	29	–			–			13:32:48.0	105	35	1	12:35:26.8	20	306	14	282	0.935	0.923	
Varanasi (Benar…	25°20′N	083°00′E	–	11:28:21.4	281	212	21	–			–			–				12:27:59.1	199	133	8	283	0.855	0.820	
Vijayawada	16°31′N	080°37′E	–	11:38:25.0	294	217	19	–			–			–				12:38:22.1	19	305	5	284	0.932	0.917	
Vishakhapatnam	17°42′N	083°18′E	–	11:37:37.2	291	215	17	–			–			–				12:36:43.1	19	306	4	285	0.982	0.981	
Wani	20°04′N	078°57′E	–	11:33:11.2	290	216	23	12:34:10.1	120	49	12:35:02.9	278	207	–				12:34:36.6	19	308	9	283	1.016	1.000	00m53s
Warangal	18°00′N	079°35′E	–	11:36:08.8	293	216	21	–			–			–				12:36:53.9	19	306	7	284	0.961	0.954	
Yaval	21°10′N	075°42′E	–	11:30:06.5	291	217	27	12:32:48.8	129	57	12:33:44.5	271	199	13:30:04.7	108	41	0	12:33:16.7	20	308	12	282	1.017	1.000	00m56s
Yavatmal	20°24′N	078°08′E	–	11:32:23.1	290	216	24	12:33:47.2	111	39	12:34:42.2	288	216	–				12:34:14.8	19	308	10	283	1.016	1.000	00m55s

TABLE 33
LOCAL CIRCUMSTANCES FOR SRI LANKA & MALDIVES

Location Name	Latitude	Longitude	Elev.	First Contact U.T.	P	V	Alt	Second Contact U.T.	P	V	Third Contact U.T.	P	V	Fourth Contact U.T.	P	V	Alt	Maximum Eclipse U.T.	P	V	Alt	Azm	Eclip. Mag.	Eclip. Obs.	Umbral Durat.
			m	h m s	°	°	°	h m s	°	°	h m s	°	°	h m s	°	°	°	h m s	°	°	°	°			
SRI LANKA																									
Colombo	06°56′N	079°51′E	7	11:51:33.5	310	223	15	–			–			–				12:47:09.0	19	296	2	285	0.670	0.590	
Dehiwala-Mount …	06°51′N	079°52′E	–	11:51:41.0	310	223	15	–			–			–				12:47:12.6	19	296	2	285	0.668	0.587	
Jaffna	09°40′N	080°00′E	–	11:47:35.7	305	221	16	–			–			–				12:44:55.5	19	298	3	285	0.747	0.684	
Kandy	07°18′N	080°38′E	–	11:51:02.4	308	222	14	–			–			–				12:46:41.9	19	296	1	285	0.686	0.610	
Kotte	06°54′N	079°54′E	–	11:51:36.6	310	223	15	–			–			–				12:47:09.9	19	296	2	285	0.670	0.589	
Moratuwa	06°46′N	079°53′E	–	11:51:48.5	310	223	15	–			–			–				12:47:16.2	19	296	2	285	0.666	0.585	
MALDIVES																									
Male	04°10′N	073°30′E	–	11:55:48.8	319	228	19	–			–			–				12:50:08.3	20	292	6	285	0.538	0.436	

TABLE 34
LOCAL CIRCUMSTANCES FOR ASIA - I
AFGHANISTAN, ARMENIA , AZERBAIJAN, BAHRAIN, BHUTAN , GEORGIA & KYRGYZSTAN
TOTAL SOLAR ECLIPSE OF 1999 AUGUST 11

Location Name	Latitude	Longitude	Elev.	First Contact U.T.				Second Contact U.T.			Third Contact U.T.			Fourth Contact U.T.				Maximum Eclipse U.T.					Eclip. Mag.	Eclip. Obs.	Umbral Durat.
			m	h m s	P °	V °	Alt °	h m s	P °	V °	h m s	P °	V °	h m s	P °	V °	Alt °	h m s	P °	V °	Alt °	Azm °			
AFGHANISTAN																									
Herat	34°20′N	062°12′E	–	10:58:04.1	283	226	45	–			–			13:12:55.6	119	61	17	12:09:14.9	201	142	31	268	0.853	0.820	
Kabul	34°31′N	069°12′E	1815	11:06:14.0	278	219	38	–			–			13:12:25.0	122	65	12	12:12:34.7	200	141	24	273	0.780	0.727	
Mazar-E Sharif	36°42′N	067°06′E	–	11:00:51.7	276	221	40	–			–			13:08:39.5	124	68	15	12:08:00.6	200	144	27	270	0.758	0.699	
Qandahar	31°32′N	065°30′E	1055	11:06:19.6	284	222	41	–			–			13:17:45.5	117	57	13	12:15:43.6	201	138	26	272	0.877	0.850	
Bangladesh																									
Barisal	22°42′N	090°22′E	–	11:33:19.6	280	211	13	–			–			–				12:29:42.1	198	131	1	287	0.862	0.829	
Chittagong	22°20′N	091°50′E	–	11:33:52.5	280	210	11	–			–			–				12:26 Set	–	–	0	287	0.849	0.812	
Comilla	23°27′N	091°12′E	–	11:32:40.0	279	210	12	–			–			–				12:28:38.0	198	132	0	287	0.840	0.800	
Dacca	23°43′N	090°25′E	–	11:32:16.6	279	210	13	–			–			–				12:28:33.5	198	132	1	286	0.839	0.799	
Dinajpur	25°38′N	088°38′E	–	11:29:53.7	277	210	16	–			–			–				12:26:46.3	198	134	3	286	0.807	0.760	
Gulshan	23°49′N	090°27′E	–	11:32:10.7	279	210	13	–			–			–				12:28:26.2	198	132	1	286	0.836	0.796	
Jamalpur	24°55′N	089°56′E	–	11:30:56.5	278	210	14	–			–			–				12:27:18.9	198	133	2	286	0.815	0.769	
Jessore	23°10′N	089°13′E	–	11:32:38.9	280	211	14	–			–			–				12:29:29.6	198	131	2	286	0.859	0.825	
Khulna	22°48′N	089°33′E	–	11:33:05.7	281	211	14	–			–			–				12:29:49.0	198	131	1	286	0.865	0.833	
Mirpur	23°47′N	090°21′E	–	11:32:11.8	279	210	13	–			–			–				12:28:30.1	198	132	1	286	0.837	0.797	
Naogaon	24°47′N	088°56′E	–	11:30:52.3	278	210	15	–			–			–				12:27:42.6	198	133	3	286	0.824	0.781	
Narayanganj	23°37′N	090°30′E	–	11:32:23.6	279	210	13	–			–			–				12:28:38.8	198	132	1	287	0.840	0.801	
Narsingdi	23°55′N	090°43′E	–	11:32:07.0	279	210	13	–			–			–				12:28:15.0	198	132	1	287	0.832	0.790	
Nasirabad	24°45′N	090°24′E	–	11:31:12.2	278	210	14	–			–			–				12:27:23.2	198	133	2	286	0.815	0.769	
Nawabganj	24°36′N	088°17′E	–	11:30:55.0	279	211	16	–			–			–				12:28:04.2	198	133	3	286	0.833	0.792	
Pabna	24°00′N	089°15′E	–	11:31:46.2	279	210	14	–			–			–				12:28:32.4	198	132	2	286	0.840	0.800	
Rajshahi	24°22′N	088°36′E	–	11:31:14.5	279	211	15	–			–			–				12:28:16.3	198	132	3	286	0.836	0.795	
Rangpur	25°45′N	089°15′E	–	11:29:55.2	277	210	15	–			–			–				12:26:29.9	198	134	3	286	0.801	0.751	
Saidpur	25°47′N	088°54′E	–	11:29:48.0	277	210	15	–			–			–				12:26:32.1	198	134	3	286	0.802	0.753	
Sirajganj	24°27′N	089°43′E	–	11:31:23.3	278	210	14	–			–			–				12:27:54.5	198	133	2	286	0.826	0.784	
Sylhet	24°54′N	091°52′E	–	11:31:17.2	277	209	12	–			–			–				12:26:48.6	198	133	1	287	0.803	0.753	
Tangail	24°15′N	089°55′E	–	11:31:38.1	279	210	14	–			–			–				12:28:05.2	198	132	2	286	0.830	0.788	
Tungi	23°53′N	090°24′E	–	11:32:06.1	279	210	13	–			–			–				12:28:22.5	198	132	1	286	0.835	0.794	
ARMENIA																									
Jerevan	40°11′N	044°30′E	–	10:20:51.5	286	257	60	–			–			12:55:29.7	116	64	35	11:42:03.1	201	155	48	245	0.925	0.912	
Kirovakan	40°48′N	044°30′E	–	10:20:06.3	285	257	60	–			–			12:54:18.1	117	66	35	11:40:59.7	201	156	48	244	0.912	0.895	
Leninakan	40°48′N	043°50′E	–	10:18:51.3	286	259	60	–			–			12:53:50.5	116	66	36	11:40:08.1	201	157	49	243	0.919	0.904	
AZERBAIJAN																									
Baku	40°23′N	049°51′E	–	10:30:18.5	283	245	56	–			–			12:58:12.1	119	67	30	11:48:02.7	201	153	43	251	0.862	0.832	
Gäncä	40°40′N	046°22′E	–	10:23:43.1	284	253	59	–			–			12:55:45.6	118	66	33	11:43:32.9	201	155	46	247	0.894	0.872	
Sumgait	40°36′N	049°38′E	–	10:29:39.2	283	246	56	–			–			12:57:40.1	119	68	30	11:47:25.8	201	153	43	251	0.860	0.829	
BAHRAIN																									
Al-Manamah	26°13′N	050°35′E	–	10:54:47.2	302	237	57	–			–			13:22:11.2	103	35	25	12:13:21.3	23	314	40	268	0.857	0.825	
BHUTAN																									
Thimbu	27°28′N	089°39′E	–	11:28:12.9	274	209	15	–			–			–				12:24:20.1	198	135	3	286	0.761	0.701	
GEORGIA																									
Batumi	41°38′N	041°38′E	–	10:13:43.7	286	264	61	–			–			12:50:35.4	116	67	38	11:35:47.4	201	160	50	239	0.925	0.912	
Kutaisi	42°15′N	042°40′E	–	10:15:00.7	284	261	60	–			–			12:50:10.2	117	69	37	11:36:09.3	201	160	49	239	0.900	0.880	
Rustavi	41°33′N	045°02′E	–	10:20:12.5	284	256	59	–			–			12:53:11.2	118	68	35	11:40:23.3	201	157	48	244	0.890	0.867	
Suchumi	43°01′N	041°02′E	–	10:11:08.3	284	265	60	–			–			12:47:25.0	117	70	38	11:32:41.3	201	162	51	236	0.901	0.881	
Tbilisi	41°43′N	044°49′E	–	10:19:36.9	284	256	59	–			–			12:52:42.9	118	68	35	11:39:49.9	201	157	48	243	0.889	0.865	
KYRGYZSTAN																									
Frunze	42°54′N	074°36′E	–	11:00:56.6	264	216	33	–			–			12:55:28.2	133	85	12	12:00:32.6	199	149	22	270	0.582	0.486	
Os	40°33′N	072°48′E	–	11:02:04.4	268	217	35	–			–			13:00:33.2	130	79	12	12:03:55.5	199	147	23	271	0.636	0.551	

TABLE 35
LOCAL CIRCUMSTANCES FOR ASIA - II
KAZAKHSTAN, MONGOLIA , MYANMAR, NEPAL , TAJIKISTAN , THAILAND, TURKMENISTAN & UZBEKISTAN
TOTAL SOLAR ECLIPSE OF 1999 AUGUST 11

Location Name	Latitude	Longitude	Elev.	First Contact U.T.	P	V	Alt	Second Contact U.T.	P	V	Third Contact U.T.	P	V	Fourth Contact U.T.	P	V	Alt	Maximum Eclipse U.T.	P	V	Alt	Azm	Eclip. Mag.	Eclip. Obs.	Umbral Durat.
			m	h m s	°	°	°	h m s	°	°	h m s	°	°	h m s	°	°	°	h m s	°	°	°	°			
KAZAKHSTAN																									
Akt'ubinsk	50°17'N	057°10'E	–	10:30:59.4	266	235	46	–			–			12:39:30.2	132	91	27	11:37:37.1	199	161	37	248	0.607	0.516	
Alma-Ata	43°15'N	076°57'E	775	11:02:43.1	263	214	31	–			–			12:54:12.1	135	86	11	12:00:38.8	199	149	21	272	0.558	0.458	
Celinograd	51°10'N	071°30'E	–	10:47:55.5	257	219	35	–			–			12:38:06.5	140	99	18	11:44:50.1	198	158	27	261	0.480	0.372	
Cimkent	42°18'N	069°36'E	–	10:56:26.6	268	220	38	–			–			12:57:30.2	131	81	16	11:59:37.9	199	149	26	267	0.633	0.547	
Dzambul	42°54'N	071°22'E	–	10:57:38.5	266	218	36	–			–			12:56:03.0	132	83	15	11:59:21.5	199	150	25	268	0.608	0.517	
Gurjev	47°07'N	051°56'E	–	10:26:10.7	273	243	51	–			–			12:45:02.5	127	83	30	11:38:29.0	200	160	41	245	0.709	0.640	
Karaganda	49°50'N	073°10'E	–	10:51:12.2	258	218	34	–			–			12:40:53.0	139	97	17	11:47:54.4	198	156	25	264	0.487	0.379	
Kokcetav	53°17'N	069°25'E	–	10:43:17.9	256	222	36	–			–			12:33:27.6	140	102	20	11:40:05.3	198	161	28	257	0.466	0.356	
Kustanaj	53°10'N	063°35'E	–	10:36:38.4	259	228	40	–			–			12:33:44.8	138	99	24	11:37:05.2	199	162	32	252	0.511	0.405	
Kzyl-Orda	44°48'N	065°28'E	–	10:48:23.4	268	224	41	–			–			12:52:29.8	131	84	20	11:53:05.7	200	153	30	262	0.627	0.540	
Pavlodar	52°18'N	076°57'E	–	10:51:58.2	253	215	31	–			–			12:34:40.7	143	103	16	11:44:51.5	198	158	23	265	0.427	0.314	
Petropavlovsk	54°54'N	069°06'E	–	10:41:11.5	254	222	36	–			–			12:29:44.3	142	105	21	11:37:02.8	198	163	28	256	0.447	0.336	
Semipalatinsk	50°28'N	080°13'E	–	10:56:57.2	253	212	29	–			–			12:37:58.5	142	101	13	11:49:00.4	198	156	21	269	0.429	0.317	
Sevcenko	43°35'N	051°05'E	–	10:28:33.2	278	244	54	–			–			12:52:15.0	123	75	30	11:43:45.3	201	156	42	248	0.785	0.734	
Temirtau	50°05'N	072°56'E	–	10:50:40.4	257	218	34	–			–			12:40:21.6	139	97	17	11:47:22.1	198	157	25	264	0.485	0.377	
Ural'sk	51°14'N	051°22'E	–	10:21:33.4	268	243	49	–			–			12:35:41.7	131	92	31	11:30:58.2	199	165	40	239	0.640	0.556	
Ust'-Kamenogorsk	49°58'N	082°38'E	–	10:59:26.9	252	211	27	–			–			12:38:21.3	143	102	11	11:50:23.1	198	156	19	272	0.420	0.307	
MONGOLIA																									
Ulaanbaatar	47°55'N	106°53'E	1307	11:11:53.3	243	200	9	–			–			–				11:52:09.4	196	155	3	290	0.319	0.207	
MYANMAR																									
Bassein	16°47'N	094°44'E	–	11:39:25.1	286	212	6	–			–			–				12:08	Set	–	–	0	286	0.511	0.405
Mandalay	22°00'N	096°05'E	77	11:34:20.4	279	209	7	–			–			–				12:10	Set	–	–	0	287	0.628	0.540
Monywa	22°05'N	095°08'E	–	11:34:16.4	279	210	8	–			–			–				12:13	Set	–	–	0	287	0.668	0.588
Moulmein	16°30'N	097°38'E	46	11:39:19.4	286	212	3	–			–			–				11:57	Set	–	–	0	286	0.321	0.208
Pegu	17°20'N	096°29'E	–	11:38:42.1	285	212	5	–			–			–				12:02	Set	–	–	0	286	0.413	0.299
Rangoon	16°47'N	096°10'E	–	11:39:16.0	286	212	5	–			–			–				12:02	Set	–	–	0	286	0.413	0.300
Sittwe (Akyab)	20°09'N	092°54'E	–	11:36:08.2	283	211	9	–			–			–				12:19	Set	–	–	0	287	0.756	0.695
Taunggyi	20°47'N	097°02'E	–	11:35:25.3	280	210	6	–			–			–				12:04	Set	–	–	0	287	0.502	0.395
NEPAL																									
Kathmandu	27°43'N	085°19'E	1348	11:26:33.9	276	211	20	–			–			–				12:24:44.7	198	135	7	284	0.785	0.732	
TAJIKISTAN																									
Dusanbe	38°35'N	068°48'E	–	11:00:18.5	273	220	39	–			–			13:04:57.9	127	73	14	12:05:38.3	200	145	26	269	0.707	0.636	
Leninabad	40°17'N	069°37'E	–	10:59:01.3	270	220	38	–			–			13:01:33.7	129	77	15	12:03:06.7	200	147	26	269	0.668	0.589	
THAILAND																									
Chiang Mai	18°47'N	098°59'E	–	11:37:01.0	282	210	3	–			–			–				11:53	Set	–	–	0	286	0.297	0.186
Krung Thep/Bankok	13°45'N	100°31'E	16	11:41:12.7	288	213	0	–			–			–				11:42	Set	–	–	0	286	0.010	0.001
TURKMENISTAN																									
Aschabad	37°57'N	058°23'E	–	10:47:28.8	280	231	49	–			–			13:05:46.4	121	66	22	12:00:19.3	201	147	35	262	0.820	0.778	
Cardzou	39°06'N	063°34'E	–	10:53:11.7	276	225	44	–			–			13:04:08.1	125	71	19	12:01:57.0	200	147	31	265	0.746	0.685	
Tasauz	41°50'N	059°58'E	–	10:44:37.7	275	230	47	–			–			12:58:20.0	126	75	23	11:54:42.3	200	151	34	260	0.730	0.665	
UZBEKISTAN																									
Andizan	40°45'N	072°22'E	–	11:01:22.8	268	217	35	–			–			13:00:14.2	130	79	13	12:03:25.8	199	147	24	270	0.637	0.551	
Angren	41°01'N	070°12'E	–	10:58:44.0	269	219	37	–			–			13:00:01.9	130	79	15	12:02:06.6	199	148	25	269	0.650	0.568	
Buchara	39°48'N	064°25'E	–	10:53:22.5	274	225	43	–			–			13:02:46.9	126	73	18	12:01:15.1	200	147	30	265	0.725	0.659	
Circik	41°29'N	069°35'E	–	10:57:27.2	269	220	38	–			–			12:59:10.0	130	79	15	12:01:02.2	199	148	26	268	0.648	0.564	
Dzizak	40°06'N	067°50'E	–	10:57:12.5	272	221	40	–			–			13:02:05.5	128	76	16	12:02:35.5	200	147	27	267	0.688	0.613	
Fergana	40°23'N	071°46'E	–	11:01:13.3	269	218	36	–			–			13:01:03.9	130	78	13	12:03:49.6	199	147	24	270	0.648	0.565	
Karsi	38°53'N	065°48'E	–	10:56:20.8	274	223	42	–			–			13:04:34.4	126	72	17	12:03:37.5	200	146	29	267	0.729	0.663	
Kokand	40°33'N	070°57'E	–	11:00:08.6	269	219	37	–			–			13:00:51.6	129	78	14	12:03:13.3	199	147	25	269	0.652	0.570	
Margilan	40°28'N	071°44'E	–	11:01:04.8	269	218	36	–			–			13:00:54.3	130	78	13	12:03:40.4	199	147	24	270	0.647	0.563	
Namangan	41°00'N	071°40'E	–	11:00:20.2	268	218	36	–			–			12:59:51.4	130	79	14	12:02:44.4	199	148	24	270	0.638	0.553	
Nukus	42°50'N	059°29'E	–	10:42:40.1	274	231	47	–			–			12:56:11.6	127	77	23	11:52:34.4	200	152	35	258	0.716	0.648	
Samarkand	39°40'N	066°48'E	–	10:56:32.3	273	222	41	–			–			13:03:00.5	127	74	16	12:02:48.8	200	147	28	267	0.705	0.634	
Taskent	41°20'N	069°18'E	–	10:57:19.4	269	220	38	–			–			12:59:29.9	130	79	15	12:01:10.0	200	148	26	268	0.653	0.570	
Urgenc	41°33'N	060°38'E	–	10:45:56.7	274	229	46	–			–			12:59:00.1	126	75	22	11:55:41.6	200	150	34	260	0.729	0.664	

TABLE 36
LOCAL CIRCUMSTANCES FOR CHINA
TOTAL SOLAR ECLIPSE OF 1999 AUGUST 11

Location Name	Latitude	Longitude	Elev.	First Contact U.T.	P	V	Alt	Second Contact U.T.	P	V	Third Contact U.T.	P	V	Fourth Contact U.T.	P	V	Alt	Maximum Eclipse U.T.	P	V	Alt	Azm	Eclip. Mag.	Eclip. Obs.	Umbral Durat.
			m	h m s	°	°	°	h m s	°	°	h m s	°	°	h m s	°	°	°	h m s	°	°	°	°			
CHINA																									
Aksu	41°10′N	080°20′E	–	11:08:00.8	263	212	28	–			–			12:57:21.5	134	84	8	12:04:50.0	198	147	17	275	0.566	0.468	
Baicheng	41°46′N	081°52′E	–	11:08:27.0	261	211	27	–			–			12:55:39.0	135	86	7	12:04:04.8	198	148	16	276	0.545	0.444	
Baiyin	36°47′N	104°07′E	–	11:21:23.5	256	203	7	–			–			–				11:59 Set	–	–	0	290	0.472	0.362	
Baoji	34°23′N	107°09′E	–	11:23:10.3	258	203	4	–			–			–				11:43 Set	–	–	0	289	0.305	0.194	
Baoshan	25°09′N	099°09′E	–	11:31:21.9	273	207	6	–			–			–				12:01 Set	–	–	0	287	0.505	0.398	
Baotou (Paotow)	40°40′N	109°59′E	–	11:18:02.8	250	201	4	–			–			–				11:44 Set	–	–	0	291	0.323	0.210	
Changzhi	36°11′N	113°08′E	–	11:20:52.9	254	202	0	–			–			–				11:23 Set	–	–	0	290	0.026	0.005	
Chengdu (Chengt…	30°39′N	104°04′E	–	11:26:24.7	264	204	4	–			–			–				11:49 Set	–	–	0	288	0.370	0.256	
Chongqing (Chun…	29°34′N	106°35′E	261	11:26:56.8	264	204	2	–			–			–				11:40 Set	–	–	0	288	0.228	0.127	
Chuxiong	25°02′N	101°30′E	–	11:31:16.3	272	207	4	–			–			–				11:52 Set	–	–	0	287	0.352	0.238	
Deyang	31°14′N	104°22′E	–	11:25:55.0	263	204	4	–			–			–				11:49 Set	–	–	0	288	0.369	0.255	
Dongchuan	26°10′N	103°01′E	–	11:30:09.0	270	206	3	–			–			–				11:47 Set	–	–	0	287	0.292	0.182	
Dukou	26°40′N	101°39′E	–	11:29:53.4	270	206	5	–			–			–				11:53 Set	–	–	0	288	0.396	0.281	
Enshi	30°17′N	109°19′E	–	11:25:53.2	262	204	0	–			–			–				11:28 Set	–	–	0	288	0.036	0.008	
Fuling	29°42′N	107°21′E	–	11:26:42.6	264	204	1	–			–			–				11:35 Set	–	–	0	288	0.140	0.062	
Gejiu (Kokiu)	23°22′N	103°06′E	–	11:32:25.0	274	207	2	–			–			–				11:43 Set	–	–	0	287	0.187	0.095	
Guangyuan	32°26′N	105°52′E	–	11:24:49.4	261	204	4	–			–			–				11:45 Set	–	–	0	289	0.320	0.207	
Guiyang (Kweiya…	26°35′N	106°43′E	–	11:29:11.0	268	206	1	–			–			–				11:33 Set	–	–	0	288	0.066	0.020	
Hami (Kumul)	42°48′N	093°27′E	–	11:13:40.8	254	205	17	–			–			12:48:27.3	140	94	1	12:02:31.3	197	149	8	283	0.455	0.344	
Hanzhong	33°08′N	107°02′E	–	11:24:08.9	260	203	3	–			–			–				11:42 Set	–	–	0	289	0.275	0.166	
Hohhot	40°51′N	111°40′E	–	11:17:44.6	249	200	3	–			–			–				11:37 Set	–	–	0	291	0.259	0.152	
Jingning	35°25′N	105°56′E	–	11:22:27.6	257	203	5	–			–			–				11:50 Set	–	–	0	289	0.390	0.276	
Jiuquan (Suzhou)	39°45′N	098°34′E	–	11:18:11.4	255	203	12	–			–			–				12:06:04.0	197	147	3	287	0.475	0.366	
Leshan	29°34′N	103°45′E	–	11:27:18.8	265	205	4	–			–			–				11:49 Set	–	–	0	288	0.357	0.243	
Linfen	36°05′N	111°32′E	–	11:21:16.3	254	202	1	–			–			–				11:29 Set	–	–	0	290	0.116	0.047	
Linhe	40°51′N	107°30′E	–	11:18:01.0	250	201	6	–			–			–				11:54 Set	–	–	0	291	0.398	0.283	
Luzhou	28°54′N	105°27′E	–	11:27:38.1	266	205	2	–			–			–				11:41 Set	–	–	0	288	0.231	0.129	
Mianyang	31°30′N	104°49′E	–	11:25:39.8	263	204	4	–			–			–				11:48 Set	–	–	0	288	0.352	0.238	
Neijiang	29°35′N	105°03′E	–	11:27:09.3	265	205	3	–			–			–				11:44 Set	–	–	0	288	0.279	0.170	
Pingliang	35°27′N	107°10′E	–	11:22:20.5	257	202	4	–			–			–				11:45 Set	–	–	0	289	0.333	0.219	
Qujing	25°32′N	103°41′E	–	11:30:34.0	271	206	2	–			–			–				11:44 Set	–	–	0	287	0.227	0.126	
Shihezi	44°18′N	086°02′E	–	11:08:20.1	256	208	24	–			–			12:49:00.1	139	93	6	12:00:21.2	198	150	14	278	0.477	0.368	
Shijiazhuang	38°03′N	114°28′E	–	11:19:18.0	251	201	0	–			–			–				11:21 Set	–	–	0	290	0.020	0.003	
Shiyan	30°22′N	104°27′E	–	11:26:36.1	264	204	4	–			–			–				11:47 Set	–	–	0	288	0.340	0.226	
Shuicheng	26°41′N	104°50′E	–	11:29:27.9	269	206	2	–			–			–				11:41 Set	–	–	0	288	0.193	0.099	
Suining	30°31′N	105°34′E	–	11:26:21.4	263	204	3	–			–			–				11:43 Set	–	–	0	288	0.278	0.169	
Taiyuan	37°55′N	112°30′E	–	11:19:47.0	252	201	1	–			–			–				11:28 Set	–	–	0	290	0.127	0.053	
Tianshui	34°30′N	105°58′E	–	11:23:11.2	258	203	4	–			–			–				11:48 Set	–	–	0	289	0.369	0.255	
Tongchuan	35°01′N	109°01′E	–	11:22:27.6	257	202	2	–			–			–				11:37 Set	–	–	0	289	0.223	0.122	
Wanxian	30°52′N	108°22′E	–	11:25:39.1	262	204	1	–			–			–				11:33 Set	–	–	0	288	0.118	0.048	
Weinan	34°29′N	109°29′E	–	11:22:47.2	257	202	2	–			–			–				11:34 Set	–	–	0	289	0.179	0.088	
Wuhai	39°39′N	106°41′E	–	11:19:00.7	252	201	6	–			–			–				11:55 Set	–	–	0	291	0.414	0.300	
Wulumuqi (Urumc…	43°48′N	087°35′E	906	11:09:49.9	256	207	22	–			–			12:49:20.6	139	93	5	12:01:13.9	198	150	13	279	0.475	0.365	
Wuzhong	37°57′N	106°10′E	–	11:20:24.6	254	202	6	–			–			–				11:53 Set	–	–	0	290	0.419	0.305	
Xiaguan	25°34′N	100°14′E	–	11:30:56.2	272	207	5	–			–			–				11:57 Set	–	–	0	287	0.450	0.338	
Xi'an (Sian)	34°15′N	108°52′E	–	11:23:03.3	258	203	2	–			–			–				11:36 Set	–	–	0	289	0.206	0.109	
Xianyang	34°22′N	108°42′E	–	11:22:59.5	258	203	2	–			–			–				11:37 Set	–	–	0	289	0.220	0.120	
Xining (Sining)	36°38′N	101°55′E	–	11:21:26.6	257	203	8	–			–			–				12:08 Set	–	–	0	290	0.511	0.405	
Xinxian	38°25′N	112°48′E	–	11:19:22.3	251	201	1	–			–			–				11:28 Set	–	–	0	290	0.128	0.054	

TABLE 37

SOLAR ECLIPSES OF SAROS SERIES 145

First Eclipse: 1639 Jan 04 — Duration of Series: 1370.3 yrs.
Last Eclipse: 3009 Apr 17 — Number of Eclipses: 77

Saros Summary: Partial: 34 Annular: 1 Total: 41 Hybrid: 1

Date	Eclipse Type	Gamma	Mag./ Width	Center Durat.	Date	Eclipse Type	Gamma	Mag./ Width	Center Durat.
1639 Jan 04	Pb	1.565	0.001		2360 Mar 18	T	-0.017	181	04m33s
1657 Jan 14	P	1.554	0.017		2378 Mar 29	T	-0.048	193	04m51s
1675 Jan 25	P	1.543	0.034		2396 Apr 09	T	-0.085	206	05m12s
1693 Feb 05	P	1.527	0.059		2414 Apr 20	T	-0.128	217	05m33s
1711 Feb 17	P	1.507	0.091		2432 Apr 30	T	-0.178	229	05m56s
1729 Feb 27	P	1.481	0.134		2450 May 12	T	-0.233	241	06m19s
1747 Mar 11	P	1.450	0.186		2468 May 22	T	-0.293	252	06m41s
1765 Mar 21	P	1.412	0.251		2486 Jun 02	T	-0.358	263	06m59s
1783 Apr 01	P	1.367	0.329		2504 Jun 14	T	-0.427	275	07m10s
1801 Apr 13	P	1.315	0.420		2522 Jun 25	T	-0.499	287	07m12s
1819 Apr 24	P	1.258	0.522		2540 Jul 05	T	-0.572	300	07m04s
1837 May 04	P	1.193	0.638		2558 Jul 16	T	-0.646	315	06m43s
1855 May 16	P	1.125	0.762		2576 Jul 27	T	-0.720	334	06m12s
1873 May 26	P	1.051	0.897		2594 Aug 07	T	-0.792	361	05m32s
1891 Jun 06	A	0.975	33	00m06s	2612 Aug 18	T	-0.862	406	04m45s
1909 Jun 17	H	0.896	51	00m24s	2630 Aug 30	T	-0.930	512	03m54s
1927 Jun 29	T	0.816	77	00m50s	2648 Sep 09	Ts	-0.992	-	02m49s
1945 Jul 09	T	0.736	92	01m15s	2666 Sep 20	P	-1.050	0.919	
1963 Jul 20	T	0.657	101	01m40s	2684 Oct 01	P	-1.103	0.817	
1981 Jul 31	T	0.579	108	02m02s	2702 Oct 13	P	-1.150	0.727	
1999 Aug 11	T	0.506	112	02m23s	2720 Oct 23	P	-1.191	0.648	
2017 Aug 21	T	0.437	115	02m40s	2738 Nov 04	P	-1.225	0.584	
2035 Sep 02	T	0.373	116	02m54s	2756 Nov 14	P	-1.255	0.528	
2053 Sep 12	T	0.314	116	03m04s	2774 Nov 25	P	-1.278	0.486	
2071 Sep 23	T	0.262	116	03m11s	2792 Dec 06	P	-1.297	0.451	
2089 Oct 04	T	0.217	115	03m14s	2810 Dec 17	P	-1.311	0.426	
2107 Oct 16	T	0.178	114	03m16s	2828 Dec 27	P	-1.322	0.405	
2125 Oct 26	T	0.146	112	03m15s	2847 Jan 08	P	-1.331	0.389	
2143 Nov 07	T	0.121	111	03m14s	2865 Jan 18	P	-1.339	0.374	
2161 Nov 17	T	0.101	110	03m13s	2883 Jan 30	P	-1.348	0.359	
2179 Nov 28	T	0.087	110	03m12s	2901 Feb 10	P	-1.359	0.339	
2197 Dec 09	T	0.077	111	03m13s	2919 Feb 21	P	-1.371	0.316	
2215 Dec 21	T	0.070	114	03m14s	2937 Mar 04	P	-1.389	0.285	
2233 Dec 31	T	0.065	117	03m18s	2955 Mar 15	P	-1.411	0.245	
2252 Jan 12	T	0.061	123	03m23s	2973 Mar 25	P	-1.439	0.194	
2270 Jan 22	T	0.056	129	03m29s	2991 Apr 06	P	-1.472	0.134	
2288 Feb 02	T	0.050	138	03m38s	3009 Apr 17	Pe	-1.514	0.059	
2306 Feb 14	T	0.040	147	03m49s					
2324 Feb 25	Tm	0.026	158	04m02s					
2342 Mar 08	T	0.008	169	04m16s					

Eclipse Type:
- P - Partial
- A - Annular
- T - Total
- H - Hybrid (Annular/Total)
- Pb - Partial Eclipse (Saros Series Begins)
- Pe - Partial Eclipse (Saros Series Ends)
- Ts - Total Eclipse (no southern limit)

Note: Mag./Width column gives either the eclipse magnitude (for partial eclipses) or the umbral path width in kilometers (for total and annular eclipses).

TABLE 38

CLIMATOLOGICAL STATISTICS FOR AUGUST ALONG THE ECLIPSE PATH OF THE TOTAL SOLAR ECLIPSE OF 1999 AUGUST 11

Station	Latitude	Longitude	T max °F	T min °F	Days with Rain	Days with TRW	Days with Scat. Cloud	POR	Hours of Sun-shine	% of Possible Sun-shine	Mean Cloud Cover (10ths)	Days with Fog
England												
Falmouth	50°08'N	5°02'W	67	56	7.5	0.7	1.0	4				
The Lizard	49°57'N	5°12'W	67	56	4.5	0.4	6.5	8				
Plymouth	50°22'N	4°07'W	66	55	7.3	1	5.8	11	6.4	43		
Bill of Portland	50°32'N	2°27'W	65	58	5.3	0.8	6.5	11				
Culdrose	50°05'N	5°15'W	67	56	4.5	0.4	6.5	8				
Bournemouth	50°47'N	1°50'W	68	55	6.6	2.0	6.7	11				
Guernsey	49°26'N	2°36'W	68	57	4.6	0.3	8.3	11				
France												
Cherbourg	49°39'N	1°28'W	68	58	7.4	2.0	n/a	n/a				
Caen	49°10'N	0°26'W	73	53	6.0	3.0	n/a	n/a				
Le Toquet	50°31'N	1°37'E	70	56	8.2	n/a	6.6	3				
Le Havre	42°39'N	0°05'E	73	56	7.0	2.2	10.0	8			8.0	2.4
Evreux	49°01'N	1°13'E	71	53	5.0	2.5	2.7	14				
Pontoise	49°06'N	2°02'E	74	55	5.9	3.0	5.4	11				
Beauvais	49°27'N	2°06'E	73	53	6.1	4.0	2.0	4				
Creil	49°15'N	2°31'E	73	54	6.8	4.0	3.5	2				
Paris - Le Bourget	48°58'N	2°26'E	75	56	6.5	4.0	7.2	9	6.6	45	6.9	1.5
Paris - Orly	48°43'N	2°24'E	75	56	6.7	3.2	5.4	11				
Laon	49°38'N	3°33'E	72	54	6.2	4.4	2.3	15				
Reims	49°19'N	4°03'E	76	54	6.2	4.0	5.1	9				
Suippes	49°09'N	4°38'E	75	55	7.6	2.0	3.7	5				
Vouziers	49°16'N	4°45'E	75	55	7.1	2.0	3.7	5				
Etain	49°14'N	5°40'E	71	53	5.4	3.4	3.3	12				
Metz	49°04'N	6°08'E	75	55	6.6	3.0	5.3	4				
Dizier	48°38'N	4°58'E	75	54	7.1	n/a	8.5	3				
Toul	48°47'N	5°59'E	72	54	8.2	4.7	3.2	16				
Nancy	48°42'N	6°14'E	73	54	6.6	5.0	2.2	13				
Phalsbourg	48°46'N	7°12'E	71	55	8.6	5.0	3.6	13				
Strasbourg	48°32'N	7°37'E	76	55	7.7	5.0	8.7	12	7.0	48	7.4	3.5
Chambly	49°01'N	5°52'E	69	53	6.2	2.1	2.9	11				
Gros-Tenquin	49°01'N	6°43'E	75	55	6.6	3.0	5.3	4				
Montmedy	49°27'N	5°25'E	71	53	8.3	3.4	3.3	12				
Rocroi	49°54'N	4°25'E	69	54	8.6	2.0	1.5	8				
Belgium												
Charleroi	50°27'N	4°27'E	69	54	7.7	2.0	1.5	8				
Florennes	50°14'N	4°39'E	69	54	9.5	2.0	1.5	8				
St. Hubert	50°02'N	5°24'E	67	51	8.5	4.0	1.8	8				
Luxembourg												
Luxembourg	49°37'N	6°12'E	70	54	8.3	4.5	3.3	8	6.5	44	5.9	4

Abbreviations:

Tmax - average daily maximum temperature (°F).
Tmin - average daily minimum temperature (°F).
TRW - thunderstorms.
POR - period of record (years).

TABLE 38 - continued

CLIMATOLOGICAL STATISTICS FOR AUGUST ALONG THE ECLIPSE PATH OF THE TOTAL SOLAR ECLIPSE OF 1999 AUGUST 11

Station	Latitude	Longitude	T max °F	T min °F	Days with Rain	Days with TRW	Days with Scat. Cloud	POR	Hours of Sun-shine	% of Possible Sun-shine	Mean Cloud Cover (10ths)	Days with Fog
Germany												
Trier	49°43'N	6°36'E	74	52	7.2	4.3	4.6	6				
Ramstein	49°26'N	7°36'E	72	57	7.3	3.8	2.5	16				
Sembach	49°30'N	7°52'E	70	53	8.1	4.2	2.4	15				
Stuttgart	48°41'N	9°12'E	74	55	6.8	4.0	4.1	15				
Karlsruhe	49°01'N	8°23'E	74	56	7.6	6.8	6.8	14	7.0	48	6.1	1.7
Ohringen	49°12'N	9°31'E	76	55	7.2	n/a	6.0	6				
Wurzburg	49°48'N	9°54'E	73	54	6.0	4.0	6.1	6				
Stotten	48°40'N	9°52'E	69	55	8.2	4.7	6.3	6				
Ulm	48°24'N	9°59'E	72	52	7.9	6.5	5.9	10				
Weissenburg	49°02'N	10°58'E	72	53	7.5	4.4	6.1	8				
Augsburg	48°23'N	10°51'E	72	54	8.4	5.5	6.9	11	7.2	49	5.1	1.7
Furstenfeld	48°12'N	11°16'E	73	54	8.6	5.4	4.6	12				
Munchen (Riem)	48°08'N	11°42'E	72	54	8.9	6.0	7.5	10				
Erding	48°19'N	11°56'E	74	54	8.0	4.5	4.9	12				
Passau	48°35'N	13°29'E	72	54	8.3	3.4	7.1	10				
Austria												
Linz	48°14'N	14°11'E	73	56	8.8	4.0	9.0	9				
Salzburg	47°47'N	13°00'E	73	55	9.9	4.0	8.0	11	7.0	48	5.5	1.6
Aigen	47°32'N	14°08'E	74	49	10.4	3.7	9.4	9				
Zeltweg	47°12'N	14°44'E	73	48	9.3	n/a	5.2	5				
Wien (Vienna)	48°15'N	16°22'E	73	58	7.0	6.0	6.8	5	8.1	56	4.9	0.8
Schwechat	48°07'N	16°34'E	79	57	2.4	1.9	11.0	5				
Hungary												
Szombathely	47°17'N	16°37'E	77	55	6.2	5.7	7.5	8	7.7	53	5.3	0.5
Keszthely	46°46'N	17°12'E	81	60	2.4	0.7	14.0	5				
Pecs	46°06'N	18°13'E	81	57	5.3	3.3	10.8	5	9.0	63	4.0	0.8
Budapest	47°31'N	19°02'E	81	61	4.7	4.4	8.2	8	8.7	60	4.3	0.1
Szolnok	47°11'N	20°13'E	84	58	2.8	1.8	10.9	8				
Szeged	46°15'N	20°06'E	82	58	4.1	4.6	12.0	8	9.2	64	3.9	0.3
Romania												
Bucharest	44°30'N	26°06'E	86	60	3.6	5.4	14.5	16			3.8	1.4
Cimpulung	45°17'N	25°02'E	75	50	4.9	5.4	10.8	5				
Deva	45°53'N	22°54'E	83	56	6.4	6.3	13.2	16				
Sibiu	45°48'N	24°09'E	78	55	7.1	6.4	11.5	10				
Omulurf	45°38'N	25°27'E	49	39	7.4	7.8	2.9	16				
Oradea	47°03'N	21°56'E	82	58	5.8	5.7	11.5	9				
Arad	46°11'N	21°19'E	82	58	4.8	6.4	11.5	16				
Timisoara	45°45'N	21°14'E	84	58	5.2	4.7	9.5	11			3.5	0.4
Turnu-Severin	44°38'N	22°38'E	86	61	4.5	4.6	14.6	16				
Constanta	44°11'N	28°40'E	80	63	2.8	2.8	18.0	16			2.9	1.8
Turnu-Magurele	43°45'N	24°52'E	86	61	3.3	2.6	17.3	10				
Bulgaria												
Varna	43°12'N	27°55'E	81	64	1.5	2.3	17.0	16	10.1	71	3.0	0.8
Burgas	42°29'N	27°29'E	82	64	2.6	1.8	19.4	12				
Sliven	42°41'N	26°16'N	84	63	2.5	2.5	15.2	15				
Rousse	43°52'N	25°58'E	86	64	3.4	5.3	17.5	15				

TABLE 38 - continued

CLIMATOLOGICAL STATISTICS FOR AUGUST ALONG THE ECLIPSE PATH OF THE TOTAL SOLAR ECLIPSE OF 1999 AUGUST 11

Station	Latitude	Longitude	T max °F	T min °F	Days with Rain	Days with TRW	Days with Scat. Cloud	POR	Hours of Sun-shine	% of Possible Sun-shine	Mean Cloud Cover (10ths)	Days with Fog
Turkey												
Zonguldak	42°27'N	31°40'E	78	64	5.1	4.4	16.2	7				
Sinop	42°01'N	35°05'E	78	68	3.6	3.6	17.6	7				
Sivas	39°49'N	36°54'E	82	51	0.5	1.2	19.1	6		86		0.2
Ankara	39°57'N	32°42'E	88	58	0.7	1.2	16.1	10		87		0.1
Samsun	41°17'N	36°20'E	80	65	2.3	3.0	13.2	10		66		0
Yesilkoy (Istanbul)	40°58'N	28°49'E	84	66	2.6	1.8	16.6	11		80		0.6
Kastamonu	41°22'N	33°46'E	80	54	3.4	3.2	13.3	6				
Merzifon	40°49'N	35°31'E	85	58	0.6	1.7	15.2	11				
Erhac	38°26'N	38°05'E	94	65	0.1	0.0	24.8	10				
Erzincan	39°42'N	39°31'E	89	59	0.7	3.0	16.8	11				
Erzurum	39°57'N	41°10'E	80	53	1.2	3.4	11.7	10		82		0
Van	38°28'N	43°20'E	82	56	0.3	1.6	24.2	7				
Malatya	38°21'N	38°15'E	94	65	0.1	0.0	24.8	10				
Diyarbakir	37°53'N	40°13'E	101	70	0.0	0.0	23.1	10		49 ?		0
Batman	37°55'N	41°07'E	101	70	0.0	0.0	23.1	10				
Siirt	37°53'N	41°52'E	98	73	0.2	0.6	27.6	6				
Syria												
Al Qamishli	37°01'N	41°11'E	105	75	0.0	0.0	29.3	11		91		0
Iraq												
Mosul	36°18'N	43°08'E	110	70	0.1	0.1	27.3	14				
Kirkuk	35°28'N	44°21'E	108	79	0.1	0.2	28.3	13				
Khanaqin	34°18'N	45°26'E	110	78	0.1	0.1	26.6	11				
Iran												
Rezaiyeh	37°32'N	45°05'E	90	63	0.2	0.8	26.2	5				
Kermanshah	34°19'N	47°07'E	99	55	0.1	0.0	25.7	10				
Hamadan	34°38'N	48°31'E	90	57	0.4	0.2	21.7	6				
Esfahan	32°37'N	51°42'E	96	63	0.2	0.2	24.4	10				
Dezful	32°26'N	48°24'E	113	80	0.0	0.0	26.4	14				
Yazd	31°54'N	54°24'E	100	71	0.0	0.0	25.3	7				
Kerman	30°15'N	56°57'E	95	62	0.1	0.4	20.6	7				
Shiraz	29°32'N	52°35'E	97	65	0.0	0.4	21.6	9				
Zahedan	29°27'N	60°54'E	96	63	0.0	0.1	24.2	8				
Pakistan												
Jiwani	25°04'N	61°48'E	88	79	0.2	0.0	12.8	10				
Karachi	24°54'N	67°09'E	88	79	2.8	1.8	4.9	5				
India												
Bhuj	23°16'N	69°40'E	88	76	4.8	1.0	2.7	10			5.9	0
Rajkot	22°18'N	70°47'E	89	75	7.8	1.0	1.6	8			6.5	0
Amahdabad	23°03'N	72°37'E	90	77	11.0	1.0	0.3	10	4.3	33	6.7	0
Surat	21°12'N	72°50'E	87	77	10.4	0.4	0.8	9				
Baroda	22°19'N	73°13'E	90	77	11.9	1.0	0.3	6				
Aurangabad	19°52'N	75°24'E	85	70	7.4	2.0	0.2	10				
Indore	22°43'N	75°48'E	83	71	11.0	3.0	0.2	10				
Nagpur	21°05'N	79°02'E	87	75	13.9	5.0	0.3	10				
Raipur	21°10'N	81°44'E	86	75	15.0	4.0	0.7	10			6.9	0
Jagdalpur	19°05'N	82°02'E	83	71	2.8	12.5	0.2	7			6.9	0
Visakhapatnum	17°43'N	83°13'E	89	79	7.8	5.0	1.5	10			6.5	0
Gopalpur	19°16'N	84°52'E	88	79	10.7	9.6	0.6	10				

TABLE 39

35 MM FIELD OF VIEW AND SIZE OF SUN'S IMAGE FOR VARIOUS PHOTOGRAPHIC FOCAL LENGTHS

Focal Length	Field of View	Size of Sun
28 mm	49° x 74°	0.2 mm
35 mm	39° x 59°	0.3 mm
50 mm	27° x 40°	0.5 mm
105 mm	13° x 19°	1.0 mm
200 mm	7° x 10°	1.8 mm
400 mm	3.4° x 5.1°	3.7 mm
500 mm	2.7° x 4.1°	4.6 mm
1000 mm	1.4° x 2.1°	9.2 mm
1500 mm	0.9° x 1.4°	13.8 mm
2000 mm	0.7° x 1.0°	18.4 mm
2500 mm	0.6° x 0.8°	22.9 mm

Image Size of Sun (mm) = Focal Length (mm) / 109

TABLE 40

SOLAR ECLIPSE EXPOSURE GUIDE

ISO		***f/Number***								
25		**1.4**	**2**	**2.8**	**4**	**5.6**	**8**	**11**	**16**	**22**
50		**2**	**2.8**	**4**	**5.6**	**8**	**11**	**16**	**22**	**32**
100		**2.8**	**4**	**5.6**	**8**	**11**	**16**	**22**	**32**	**44**
200		**4**	**5.6**	**8**	**11**	**16**	**22**	**32**	**44**	**64**
400		**5.6**	**8**	**11**	**16**	**22**	**32**	**44**	**64**	**88**
800		**8**	**11**	**16**	**22**	**32**	**44**	**64**	**88**	**128**
1600		**11**	**16**	**22**	**32**	**44**	**64**	**88**	**128**	**176**
Subject	***Q***	***Shutter Speed***								
Solar Eclipse										
Partial[1] - 4.0 ND	**11**	—	—	—	1/4000	1/2000	1/1000	1/500	1/250	1/125
Partial[1] - 5.0 ND	**8**	1/4000	1/2000	1/1000	1/500	1/250	1/125	1/60	1/30	1/15
Baily's Beads[2]	**11**	—	—	—	1/4000	1/2000	1/1000	1/500	1/250	1/125
Chromosphere	**10**	—	—	1/4000	1/2000	1/1000	1/500	1/250	1/125	1/60
Prominences	**9**	—	1/4000	1/2000	1/1000	1/500	1/250	1/125	1/60	1/30
Corona - 0.1 Rs	**7**	1/2000	1/1000	1/500	1/250	1/125	1/60	1/30	1/15	1/8
Corona - 0.2 Rs[3]	**5**	1/500	1/250	1/125	1/60	1/30	1/15	1/8	1/4	1/2
Corona - 0.5 Rs	**3**	1/125	1/60	1/30	1/15	1/8	1/4	1/2	1 sec	2 sec
Corona - 1.0 Rs	**1**	1/30	1/15	1/8	1/4	1/2	1 sec	2 sec	4 sec	8 sec
Corona - 2.0 Rs	**0**	1/15	1/8	1/4	1/2	1 sec	2 sec	4 sec	8 sec	15 sec
Corona - 4.0 Rs	**-1**	1/8	1/4	1/2	1 sec	2 sec	4 sec	8 sec	15 sec	30 sec
Corona - 8.0 Rs	**-3**	1/2	1 sec	2 sec	4 sec	8 sec	15 sec	30 sec	1 min	2 min

Exposure Formula: $t = f^2 / (I \times 2^Q)$

where: t = exposure time (sec)
f = *f*/number or focal ratio
I = ISO film speed
Q = brightness exponent

Abbreviations: ND = Neutral Density Filter.
Rs = Solar Radii.

Notes: [1] Exposures for partial phases are also good for annular eclipses.
[2] Baily's Beads are extremely bright and change rapidly.
[3] This exposure also recommended for the 'Diamond Ring' effect.

F. Espenak - 1997 Feb

Maps of the Umbral Path

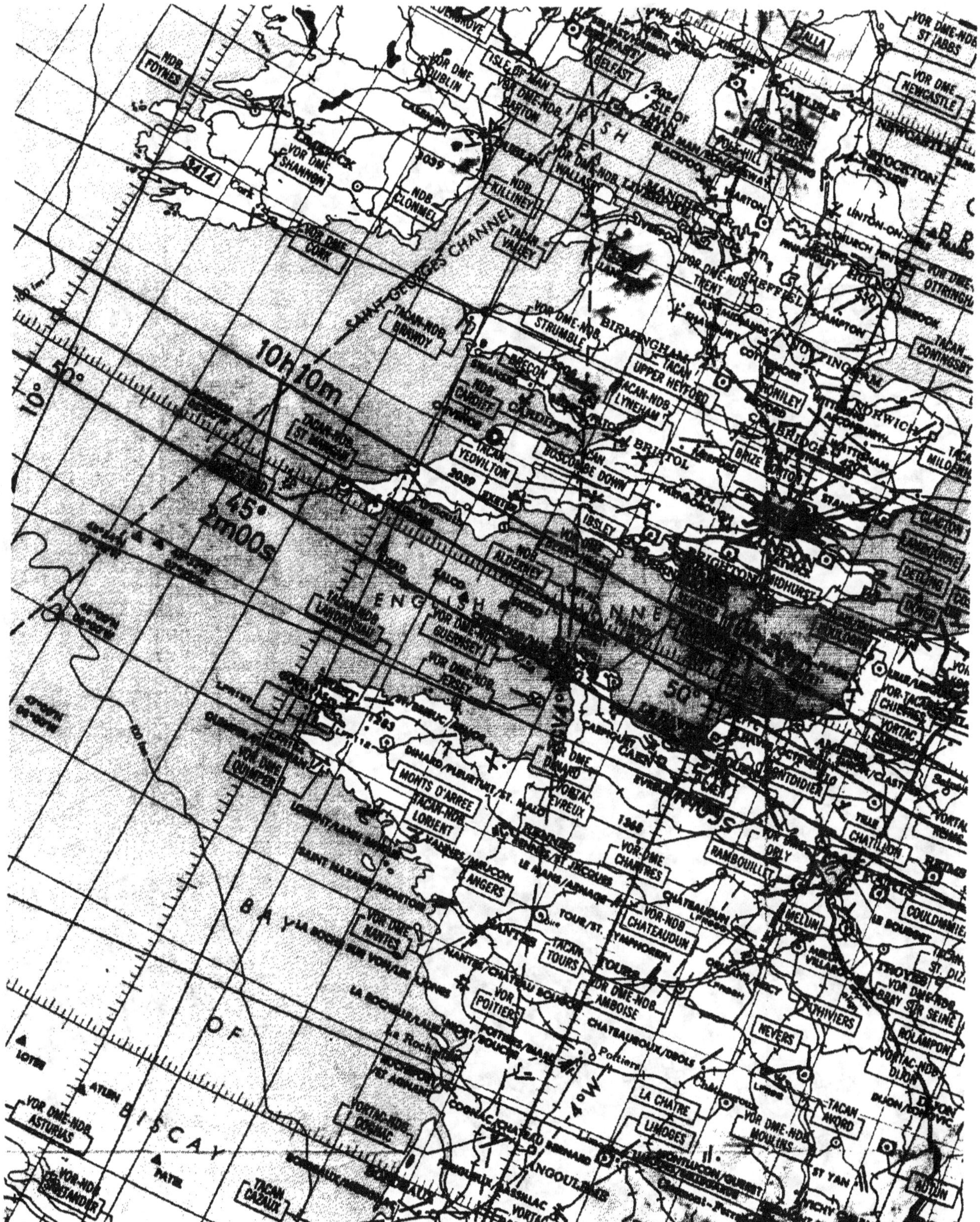
NDB FOYNES
VOR DME DUBLIN
ISLE OF MAN
BELFAST
VOR DME NEWCASTLE
VOR DME SHANNON
NDB CLONNEL
VOR DME CORK
NDB KILLINEY
ST GEORGES CHANNEL
TACAN VALLEY
TACAN-NDB BRAWDY
10h 10m
50°
10°
45°
2m00s
VOR DME-NDB STRUMBLE
BIRMINGHAM
TACAN UPPER HEYFORD
TACAN-NDB LYNEHAM
BRISTOL
BRECON
CARDIFF
TACAN YEOVILTON
BOSCOMBE DOWN
NORWICH
IBSLEY
ENGLISH CHANNEL
NDB ALDERNEY
VOR DME-NDB GUERNSEY
VOR DME-NDB JERSEY
MONTS D'ARREE
TACAN-NDB LORIENT
VOR DME QUIMPER
VOR DME CHARTRES
RAMBOUILLET
ORLY
CHATILLON
MELUN
ANGERS
VOR-NDB CHATEAUDUN
TACAN TOURS
VOR POITIERS
VOR DME-NDB AMBOISE
NEVERS
VOR DME NANTES
B A Y
O F
B I S C A Y
LA CHATRE
LIMOGES
VOR DME-NDB MOULINS
VORTAC-NDB COGNAC
TACAN AVORD
VORTAC-NDB DIJON
ROLAMPONT
AUTUN
VOR DME-NDB ASTURIAS
VOR-NDB SANTANDER
TACAN CAZAUX
ANGOULEME
4°W

OLDENBURG
ELBE
PAMPUS
HAMBURG
FRIEDLAND
OSNABRUCK
REKKEN
GUTERSLOH
EINDHOVEN
DORTMUND
BERLIN
LEINE
LICHTENAU
LEIPZIG
HERMSDORF
FLORENNES
MONTMEDY
TACAN
DINKELSBUHL
NÜRNBERG
KARLSRUHE
ALLERSBERG
STUTTGART
LUXEUIL
STRAUBING
STOCKERAU
TURIN
TREZZO
GENOVA
LJUBLJANA
KOSTAJNICA
OMARSKA
CRES

NDB
KALINOVKA
KHMELNITSKIY/RUZHICHNAYA
VINNITSA EAST
VOR DME
KOSICE
NDB
NEMIROV
CHORTKOV
IVANO FRANKOVSK
NDB
VERHNEYE/VYOTSKOYE
KOLOMYYA
UZHGOROD
NDB
JEL SAVA
JANOVCE
HUNGARY
NDB
MALAYA DOBRON
MUKACHEVO
NDB
MOGLIEV PODO
NDB
KHUST
CHERNOVTSY
MOLDAVSKAYA SSR
UKRAINSKAYA SSR
VOR DME-NDB
SAJOHIDVEG
MEZOKOVESD
SOVIET UNION
NDB
SUCEAVA
NDB
BAIA MARE
ROMANIA
MARKULESHTY
SUCEAVA/SALCEA
DEBRECEN
VOR DME
BUDAPEST-FERIHEGY
NDB
ORADEA
NDB
SATU MARE
NDB
SHIRYAYEV
RAKOCZIFALVA
VOR DME-NDB
BEKES
KECSKEMET
IASI NORTH
NDB
IASI
IASI
VOR DME-NDB
BUGAC
CLUJ
WARNING
NDB
CLUJ/SOMESENI
KISHINEV
KISHINEV SOUTHEAST
OROSHAZA
VOR DME-NDB
ARAD
NDB
TIRGU MURES
TIRGU MURES/VIDRASAU
LUIZI CALUGARA
ARAD
NDB
TIRGU MURES
NDB
SIBIU/TURNISOR
NDB
BRASOV
NDB
PETROVKA
SOMBOR
YUGOSLAVIA
ROMANIA
NDB
DEVA
SIBIU
NDB
GALATI
NDB
RENI
OSIJEK
VOR DME
NDB
TIMISOARA/GIARMATA
WARNING
VOR DME-NDB
STREJNIC
NDB
CARANSEBES
DERVERTA
VOR DME
BELGRADE
KOVIN
BELGRADE
GALATI
NDB
URZICENI
NDB
FETESTI
NDB
FANDAREI
VOR DME-NDB
FLORESTI
PLOIESTI
VOR DME-NDB
VALJEVO
NDB
CURTEA DE ARGES
TITU/BOTENI
VOR DME-NDB
TOPOLA
NDB
BRADISTE
CRAIOVA
BALACI
BUCHAREST
CONSTANTA
NDB
KOZLODUJ
VIDIN/SMURDAN
CARACAL NEW
NDB
RUSSE
LAMBRINOVO
NDB
ALEKSANDROVAC
OBRVA
NDB
GACKO
ROMANIA
BULGARIA
Danube
OSTROVO
NORTHWEST
RIBNO
DOLNA MITROPOLIYA
VOR-NDB
GOLYAMA
KAMENETS
BUKHOVTSI
IVANGRAD
NDB
GORNA
VOR DME
TARGOVISHTE
VOR DME
VARNA
NDB
BOZHOURISHTE
GORNA ORYAKHOVITSA
BALKAN MOUNTAINS
VOR DME
SOFIA
PRISTINA
NDB
LYULYAKOVO
TOPOLNITSA
BEZMER
BURGAS
VOR-NDB
STARA ZAGORA
PLOVDIV
RAVNETS
VOR DME
BURGAS
NDB
PRILEP
NDB
KRIVNOVO
NDB
OHRID
NDB
CROUCHOUTIGOVO
TURKEY
BULGARIA
VOR DME-NDB
RADOVETS
VORTAC
ISTANBUL
TIRANE
VOR DME-NDB
TEKIRDAG
VOR DME-NDB
KASTORIA
BUFFER ZONE
TURKEY
BUFFER ZONE
THESSALONIKI
ALEXANDROUPOLIS
SEA OF MARMARA
VOR DME-NDB
ALEX
KESAN
TACAN
BANDIRMA
NDB
TOPKAPI
TACAN-NDB
LARISA
VOR DME-VORTAC
THESSALONIKI
VOR DME TACAN-NDB
LIMNOS
NDB
BIGA
KERKIRA
TACAN
BALIKESIR
TACAN-NDB
PREVEZA
VOR DME-NDB
YIANNENA
VOR TACAN-NDB
NEA ANGHIALOS
AEGEAN SEA
NEA ANGHIALOS
LESVOS
NDB
BALIKESIR

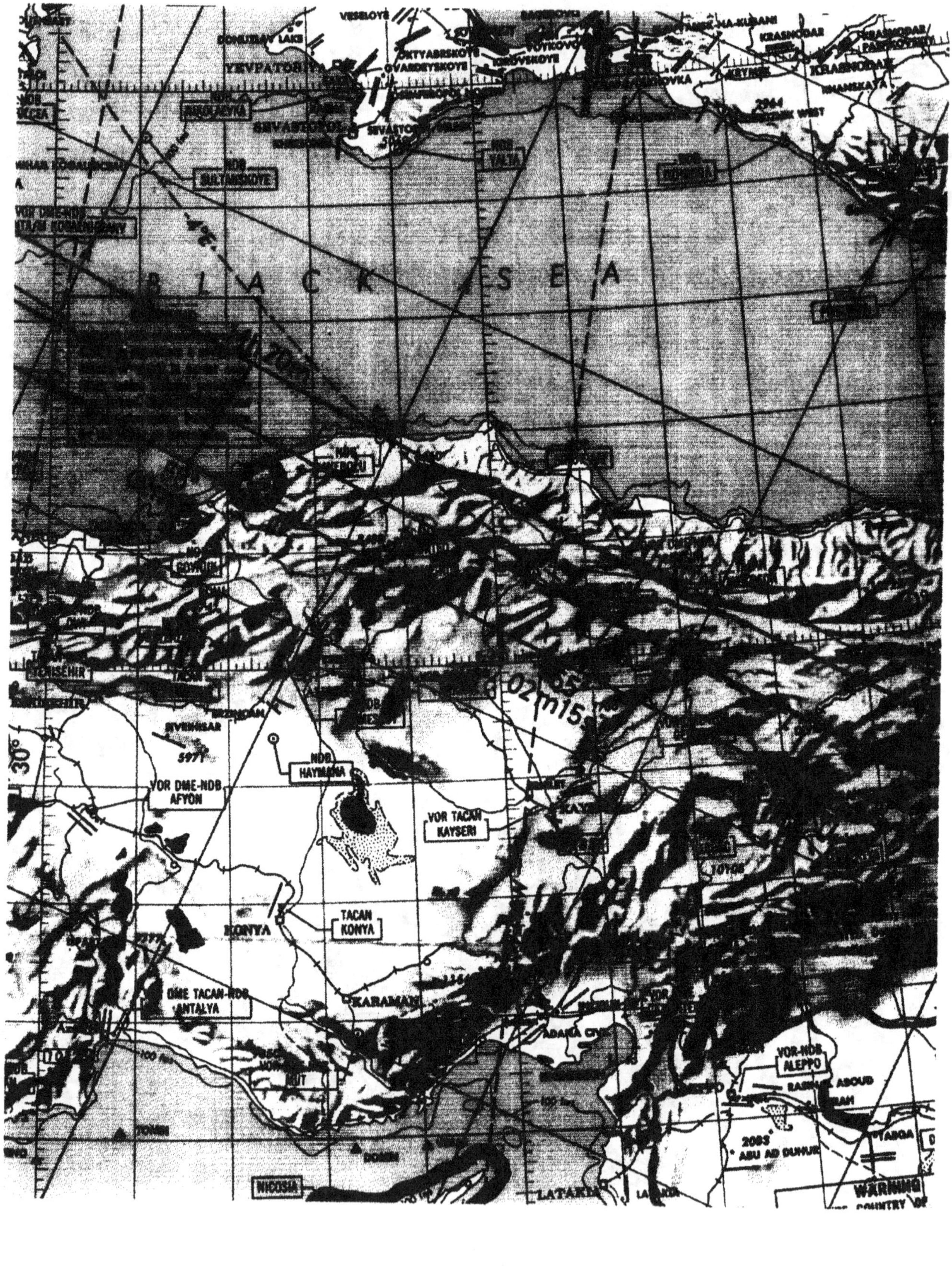
VESELOYE
OKTYABRSKOYE
GVARDEYSKOYE
KRASNODAR
YEVPATORIYA
SEVASTOPOL
YALTA
NDB
SULTANSKOYE
B L A C K S E A
SIVRIHISAR
5971
VOR DME-NDB
AFYON
NDB
HAYMANA
VOR TACAN
KAYSERI
KONYA
TACAN
KONYA
DME TACAN-NDB
ANTALYA
KARAMAN
VOR-NDB
ALEPPO
2083
ABU AD DUHUR
NICOSIA
LATAKIA
WARNING

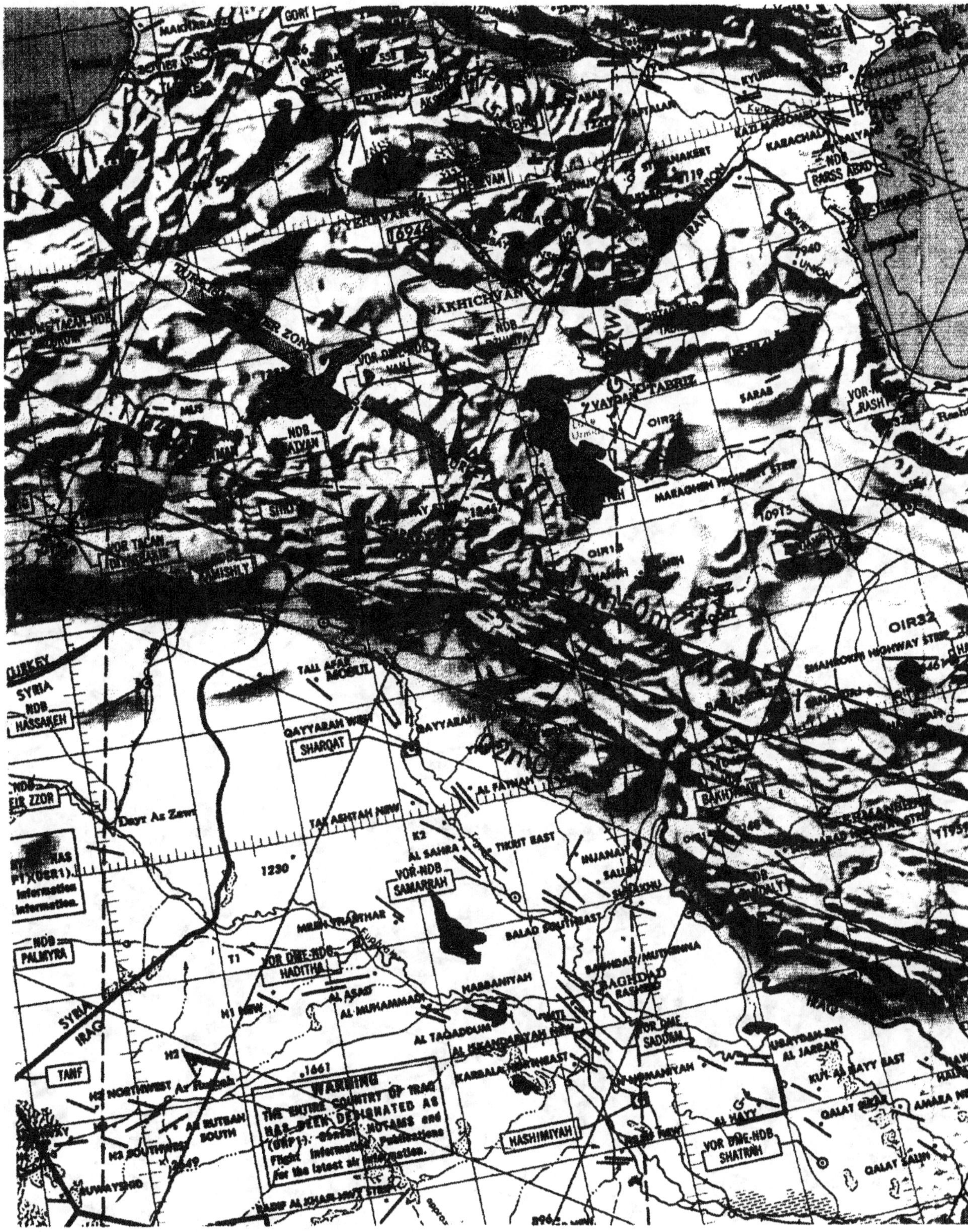
TURKEY
SYRIA
NDB HASSAKEH
TALL AFAR
MOSUL
QAYYARAH WEST
SHARQAT
AL FATHAH
Dayr Az Zawr
1230
K2
AL SAHRA
TIKRIT EAST
VOR-NDB SAMARRAH
BALAD SOUTHEAST
NDB PALMYRA
T1
VOR DME-NDB HADITHA
AL ASAD
H1 NEW
AL MUHAMMADI
HABBANIYAH
BAGHDAD
AL TAQADDUM
VOR DME SADDAM
SYRIA
IRAQ
TANF
H2
1661
WARNING
for the latest air information.
HASHIMIYAH
VOR DME-NDB SHATRAH
AL HAYY
UBAYDAH BIN
AL JARRAH
KUT AL HAYY EAST
GALAT SIKAR
AMARA NEW
QALAT SALEH
NAKHICHVAN
TABRIZ
SARAB
MARAGHEH HIGHWAY STRIP
OIR32
SHAHROKHI HIGHWAY STRIP
STEPANAKERT
PARSS ABAD

VOR TACAN MASHHAD
NDB NOSHAHR
NDB BABSAR
VORTAC TEHRAN
NDB DEHNAMAK
NDB VARAMIN
GREAT SALT DESERT
NDB SAVEH
NDB MALAYER
ANARAK
VORTAC YAZD
DARBAND
01m40s
VORTAC ESFAHAN
12h10
NDB AHWAZ
AHWAZ
BASRAH
30°
NDB FAILAKA
NDB AHMADI
WAFRA
KUWAIT INTL
OKP1
oil field
KUVER

GV48°W
VOR KANDAHAR
BOST
Farah
DILARAM
DEHE MAZAR
HELMAND
IRAN
AFGHANISTAN
DARBAND
2°E
VORTAC KERMAN
30°
Zahedan
VORTAC-NDB ZAHEDAN
PAKISTAN WEST ADIZ
12930
VASHT
01m 26s
29°
VOR-NDB PANJGUR
12h 20 m
STRAIT OF HORMUZ
OMAN
VOR DME SHARJAH
SHINAS
VORTAC CHAH BAHAR
GULF OF OMAN
PAKISTAN
SOUTH ADIZ
GV48°W

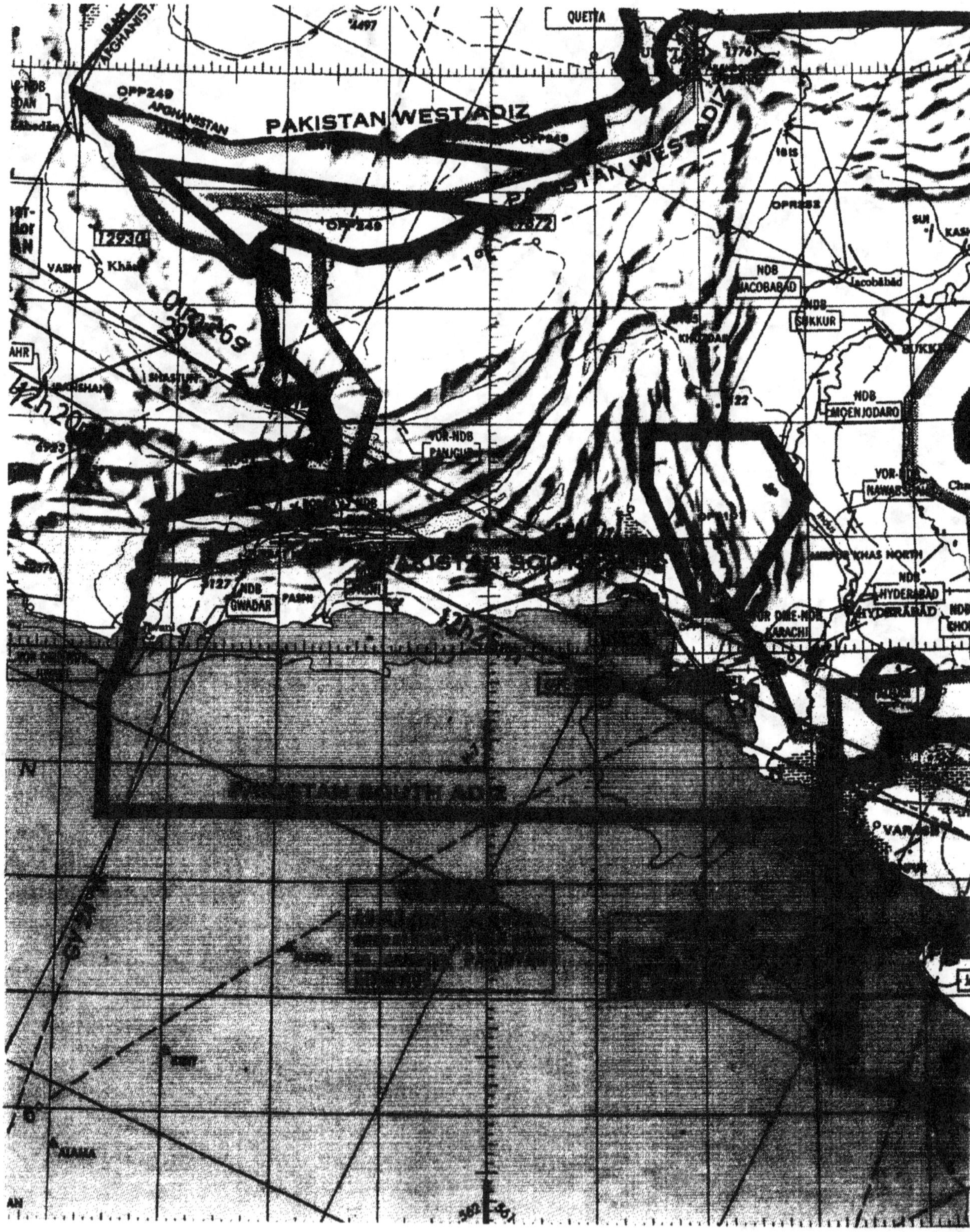

QUETTA
OPP249
AFGHANISTAN
PAKISTAN WEST ADIZ
OPP249
NDB
JACOBABAD
Jacobābād
NDB
SUKKUR
NDB
MOENJODARO
VOR-NDB
PANJGUR
NDB
GWADAR
PASNI
NDB
HYDERABAD
KARACHI
PAKISTAN SOUTH ADIZ

Phalodi
VIR121
Jaipur
VIR103
AGRA
Gwalior
JODHPUR
VOR-NDB
JODHPUR
WEST INDIA ADIZ
DELHI ADIZ
NDB
GWALIOR
VIR119
UTARLAI
VIR120
VIR136
NDB
KOTA
Kota
VIR112
NDB Nimach
UDAIPUR
VOR DME-NDB
MANDASOR
VOR DME-NDB
BHOPAL
Bhopal
MAHESANA
VOR DME-NDB
AHMEDABAD
RATLAM
AHMADABAD
DAHOD
INDORE
NDB
INDORE
NDB
RAJKOT
RAJKOT
VAR49A
NDB
PORBANDAR
BHAVNAGAR
VOR-NDB
BHAVNAGAR
JUNAGADH
WEST INDIA ADIZ
SURAT
NDB
SONGADH
VAR36
DAMAN
NDB
AKOLA
WASHIM
AURANGABAD
VOR-NDB
AURANGABAD
Nanded
5400
BOMBAY
AHMADNAGAR
BOMBAY ADIZ
PUNE
VAR42
NDB
BIDAR
HYDERABAD
SOLAPUR
NDB
VIKARABAD
BIJAPUR
NDB
GULBARGA
20°

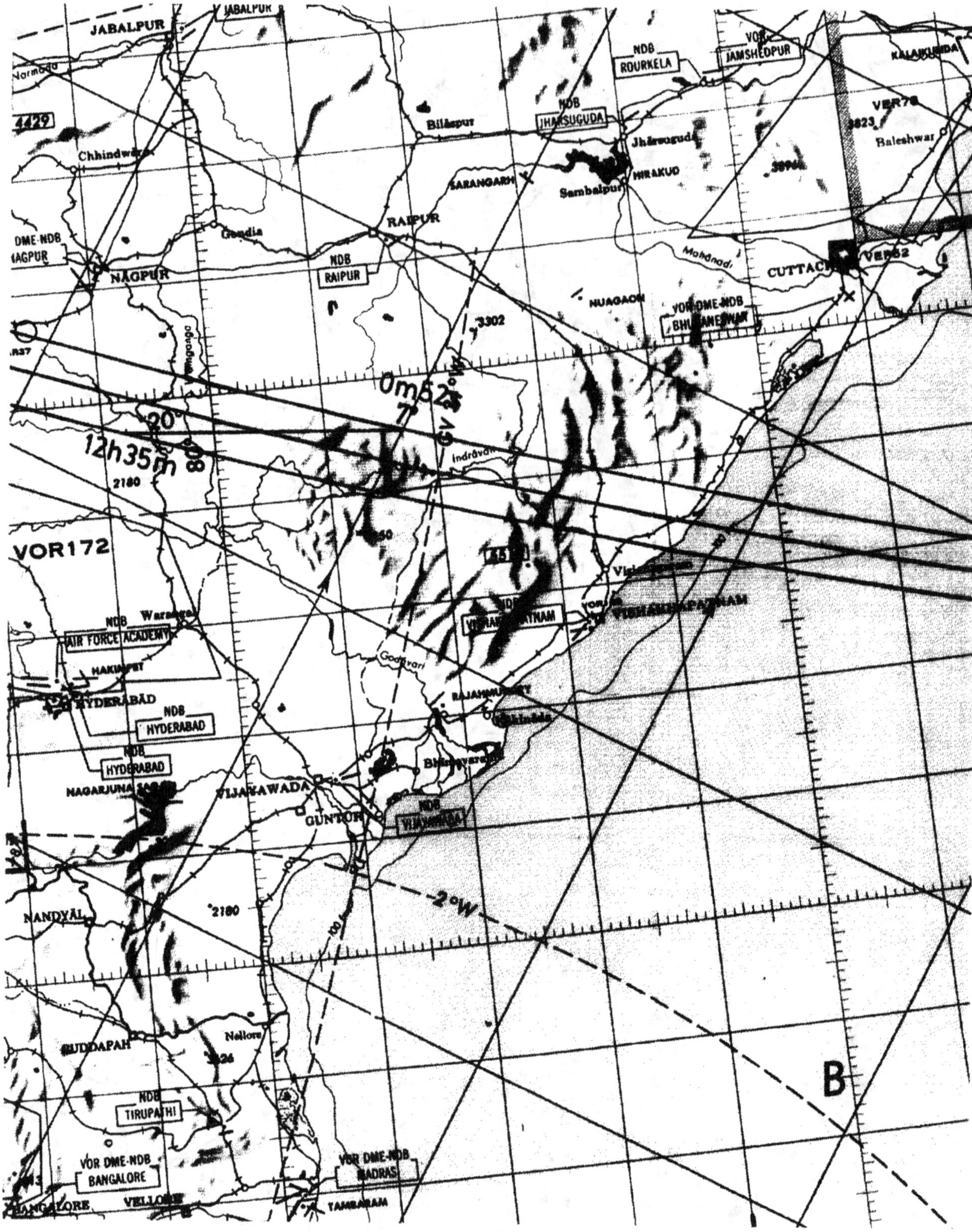
JABALPUR
NDB ROURKELA
VOR JAMSHEDPUR
NDB JHARSUGUDA
Bilaspur
Chhindwara
Jharsuguda
Sambalpur
HIRAKUD
SARANGARH
RAIPUR
Gondia
DME-NDB NAGPUR
NAGPUR
NDB RAIPUR
CUTTACK
Baleshwar
Mahanadi
NUAGAON
VOR DME-NDB BHUBANESWAR
3302
0m52
20°
12h35m
2180
VOR172
NDB
Warangal
AIR FORCE ACADEMY
HAKIMPET
HYDERABAD
NDB HYDERABAD
NDB HYDERABAD
NAGARJUNA SAGAR
VIJAYAWADA
GUNTUR
NDB VIJAYAWADA
Godavari
RAJAHMUNDRY
Kakinada
Bhimavaram
VISHAKHAPATNAM
NANDYAL
2180
2°W
CUDDAPAH
Nellore
NDB TIRUPATHI
VOR DME-NDB BANGALORE
VOR DME-NDB MADRAS
BANGALORE
VELLORE
TAMBARAM
B

www.ingramcontent.com/pod-product-compliance
Lightning Source LLC
Chambersburg PA
CBHW081727170526
45167CB00009B/3730